Non-Equilibrium Thermodynamics in Multiphase Flows

SOFT AND BIOLOGICAL MATTER

"Soft and Biological Matter" is a series of authoritative books covering established and emergent areas in the realm of soft matter science, including biological systems spanning from the molecular to the mesoscale. It aims to serve a broad interdisciplinary community of students and researchers in physics, chemistry, biophysics and materials science.

Pure research monographs in the series as well as those of more pedagogical nature, will emphasize topics in fundamental physics, synthesis and design, characterization and new prospective applications of soft and biological matter systems. The series will encompass experimental, theoretical and computational approaches.

Both authored and edited volumes will be considered.

Series Editors

Wenbing Hu, *Nanjing University, Nanjing, China*
Roland Netz, *Technical University Berlin, Berlin, Germany*
Roberto Piazza, *Politecnico di Milano, Milan, Italy*
Peter Schall, *University of Amsterdam, Amsterdam, the Netherlands*
Gerard C.L. Wong, *University of California, Los Angeles, USA*

Proposals should be sent to one of the Series Editors, or directly to the managing editor at Springer:

Dr. Maria Bellantone
Springer-SBM
van Godewwijckstraat 30
3311 GX Dordrecht
The Netherlands
maria.bellantone@springer.com

For further volumes:
www.springer.com/series/10783

Roberto Mauri

Non-Equilibrium Thermodynamics in Multiphase Flows

Roberto Mauri
Department of Chemical Engineering,
 Industrial Chemistry and Material Science
University of Pisa
Pisa, Italy

ISSN 2213-1736   ISSN 2213-1744 (electronic)
Soft and Biological Matter
ISBN 978-94-017-8358-3   ISBN 978-94-007-5461-4 (eBook)
DOI 10.1007/978-94-007-5461-4
Springer Dordrecht Heidelberg New York London

Printed on acid-free paper

Springer is part of Springer Science+Business Media (www.springer.com)

Alle mie bimbe, Maria, Lucia e Giulietta

Preface

As stated by Sybren de Groot and Peter Mazur[1] "Non-equilibrium thermodynamics provides us with a general framework for the macroscopic description of irreversible processes." Its basic step is to extend the meaning of all thermodynamic properties that we typically associate with equilibrium states, so that we can define and thus apply them locally to systems out of equilibrium. It should be stressed, however, that a general theory of non-equilibrium thermodynamics can be developed only when the driving forces and the resulting fluxes (e.g. the temperature gradient and the induced diffusive heat flux) can be assumed to be linearly related to each other. This assumption is satisfied when the driving forces are not very strong and when their time rates of change are not very fast, so that the entire system does not deviate substantially from its equilibrium state. Although the assumption of linear force-flux relations seems very limiting, though, in reality it applies very well to most cases. Think, for example, to the laws of Fick, Fourier and Newton, relating the diffusive fluxes of mass, heat and momentum to their respective driving forces, i.e. the gradients of chemical potential, temperature and velocity: these relations appear to apply very well even to systems far from equilibrium, i.e. when the driving forces are not small. In any case, we do not have to forget that the general case of far from equilibrium systems is heavily non-linear and can only be tackled using non-equilibrium statistical mechanics.

This book is divided into two parts. The first part presents the theory of non-equilibrium thermodynamics, reviewing its essential features and showing, when possible, some applications. After describing in Chap. 1 the local equilibrium assumption and the theory of fluctuations, in Chap. 2 we derive the celebrated Onsager's reciprocity relations and the fluctuation-dissipation theorem. Then, in Chaps. 3 and 4, we describe the most common ways to follow the evolution of stochastic systems, namely the Langevin and the Fokker-Planck equations. Chapters 5 and 6 are more advanced, as they describe alternative ways to study the trajectories of random systems, namely stochastic integration and path integrals.

---

[1]S.R. de Groot and P. Mazur, *Non-equilibrium Thermodynamics*, Dover (1962).

In the second part of this book, we show how the general theory of non-equilibrium thermodynamics can be applied to model multiphase flows and, in particular, to determine their constitutive relations. After deriving all balance equations in Chap. 7, in Chap. 8 we apply the results of non-equilibrium thermodynamics to derive the constitutive relations. Then, in Chap. 9, we describe the diffuse interface model of multiphase flows, as it is more fundamental than the classical, sharp interface theory and is therefore more suitable to be coupled to all non-equilibrium thermodynamics results. Finally, Chaps. 10 and 11 are devoted to determining the effective constitutive relations and the effective equations of complex fluids and composite materials. At the end, several appendices deal with some prerequisite topics in mathematics, statistical thermodynamics, analytical mechanics and transport phenomena.

This book grew out of graduate lectures on non-equilibrium thermodynamics which I have given at the City College of New York and at the University of Pisa to students of engineering and material science. No prior knowledge of statistical mechanics is required; the necessary prerequisites are only the equivalents of an introductory course on transport phenomena and one on thermodynamics.

Pisa   Roberto Mauri

Contents

Chapter 1
Introduction

Conventional thermodynamics is a static theory about systems that are in a state of stable equilibrium, establishing the relations among the variables that describe these equilibrium states. Therefore, the dynamic insight comes in two forms: (1) knowing the initial and final equilibrium states, one can authoritatively say whether such transformation can take place spontaneously or not; (2) if the change from an initial to a final state is so slow that the process can be assumed to be proceeding through a series of closely spaced equilibrium states, then such a process is called reversible and the entire path of time evolution of each of the state variables can be obtained from conventional thermodynamics. On the other hand, almost all the processes that we experience are irreversible and hence, most of the time, systems are not in a state of equilibrium as they evolve in time. A very simple and relevant example is that of thermal conduction, when a heat flux is induced by an imposed temperature difference, trying to re-establish the condition of thermal equilibrium, with uniform temperature. Obviously, the word "temperature" here does not indicate exactly a thermodynamic quantity, as it refers to a system that is out of equilibrium. Accordingly, we must extend the meaning of temperature, defining it locally, i.e. within a small volume and a short time interval, so that it can be defined in terms its mean value, neglecting all fluctuations.

In this first chapter, first we illustrate the principle of local equilibrium (Sect. 1.1), then we see how this assumption allows us to describe the fluctuations of any quantity around its thermodynamic, i.e. average, value, leading naturally to assuming a linear relation between the thermodynamic forces and their conjugated fluxes (Sect. 1.2). Finally, the physical meaning of entropy in out-of-equilibrium systems and a few applications to thermodynamics are illustrated in Sects. 1.3 and 1.4.

1.1 Local Equilibrium

Here we intend to see when a system can be considered, locally, at equilibrium, so that its thermodynamic quantities, such as temperature and pressure, can be still used

R. Mauri, *Non-Equilibrium Thermodynamics in Multiphase Flows*,
Soft and Biological Matter, DOI 10.1007/978-94-007-5461-4_1,
© Springer Science+Business Media Dordrecht 2013

and all thermodynamic equalities (e.g. the Gibbs-Duhem relation) are still valid. This is the condition of local equilibrium, which defines the minimum dimensions of any subsystem which then plays the role of a "material point". This size must be large enough to contain a large number of particles and neglect fluctuations, and yet small enough to neglect the effect of all macroscopic variations.

First of all, for a system composed of N particles at equilibrium, consider any extensive thermodynamic quantity A, fluctuating around its constant equilibrium value $\overline{A} = \langle A \rangle$, with time dependent fluctuation $\delta A = A - \langle A \rangle$. Clearly,

$$A = \sum_{i=1}^{N} a_i; \qquad \delta A = \sum_{i=1}^{N} \delta a_i, \qquad (1.1)$$

where a_i and $\delta a_i = a_i - a$ are the values of A and δA for the single i-th particle, with $a = \langle a \rangle = \sum a_i / N$. Then,

$$\langle A \rangle = \sum_{i=1}^{N} a_i = Na. \qquad (1.2)$$

In addition, let us assume that the fluctuations of the i-th particle are independent of those of the j-th particle, provided that $i \neq j$, i.e.,

$$\langle \delta a_i \delta a_j \rangle = (\delta a)^2 \delta_{ij}, \qquad (1.3)$$

where δa is the mean value of the single particle fluctuation. Consequently,[1]

$$(\delta A)^2 = \langle (\delta A)^2 \rangle = \sum_{i=1}^{N} \sum_{j=1}^{N} \langle \delta a_i \delta a_j \rangle = \sum_{i=1}^{N} \langle (\delta a)^2 \rangle = N(\delta a)^2, \qquad (1.4)$$

so that,

$$\frac{\delta A}{\overline{A}} = \frac{1}{\sqrt{N}} \frac{\delta a}{a}, \qquad (1.5)$$

showing that for large systems we can neglect fluctuations. In particular, classical thermodynamics corresponds to having $N \to \infty$.

At equilibrium, A is uniform and stationary; so, being at local equilibrium means that within a properly defined small volume (a) fluctuations are small, and (b) all macroscopic variations of A can be neglected. That means assuming that we can subdivide the system into small cells, which are large enough to contain very many particles, and yet are small enough to neglect the effects of all spatial and temporal macroscopic variations. Accordingly, these small cells play the role of the material points appearing in any continuum mechanics theory. So, if λ_{fl} is the size of the elementary cells, the following conditions must be satisfied.

---

[1]This relation is a particular case of the additive property of cumulants (A.30).

- λ_{fl} is large enough that the elementary cell contains a large number of particles, $N \gg 1$, so that $\delta A \ll \overline{A}$; this is the so-called continuum approximation;
- λ_{fl} is small enough so that the variations of A across that distance, $\lambda_{fl}|\overline{\nabla A}|$, due to the presence of a macroscopic gradient of A, is smaller (or at most equal) than the fluctuations δA.

This criterion must be supplemented with a condition stating that the characteristic fluctuation timescale, τ_{fl}, is much shorter than any macroscopic evolution time, i.e.:

- τ_{fl} is small enough so that the variations of A during that time, $\tau_{fl}\dot{\overline{A}}$, due to the presence of a macroscopic temporal variation of A, is smaller (or at most equal) than the fluctuations δA.

Therefore, the condition of local equilibrium is equivalent to assuming:

$$\frac{\lambda_{fl}}{\overline{\lambda}}; \frac{\tau_{fl}}{\overline{\tau}} \leq \frac{\delta A}{\overline{A}} \simeq \frac{1}{\sqrt{N}} \ll 1, \tag{1.6}$$

where $\overline{\tau} = \overline{A}/\dot{\overline{A}}$ and $\overline{\lambda} = \overline{A}/|\overline{\nabla A}|$ are typical evolution timescales and lengthscales, respectively. Basically, the condition of local equilibrium means that within a small volume of size λ_{fl} and for short time intervals τ_{fl} we can assume that A is uniform and constant, as we can neglect both its fluctuations and its spatial and temporal macroscopic variations.

For example, consider a gaseous system at ambient conditions, with number density $n \simeq \rho N_A/M_w \simeq 10^{20}$ molecules/cm^3, where $\rho \simeq 10^{-3}$ g/cm^3 is the mass density, $N_A = 6 \times 10^{23}$ molecules/mole is the Avogadro number and $M_w \simeq 10$ g/mole is the molecular mass. Assuming that all quantities must be evaluated with a 0.1 % precision, i.e. $\delta A/\overline{A} = 10^{-3}$, from (1.6) we see that the elementary volume must contain at least $N = 10^6$ particles, corresponding to a volume of 10^{-14} cm^3 and therefore $\lambda_{fl} \simeq 10^{-5}$ cm $= 0.1$ μm. This lengthscale is of the same magnitude as the mean free path and so it is reasonable to say that within the local equilibrium approximation we cannot describe phenomena involving smaller lengthscale. Applying (1.6) we see that the maximum temperature gradient that can be applied and yet satisfy the condition of local equilibrium is $\overline{\nabla T} \approx 10^{-3}\overline{T}/\lambda_{fl} \simeq 10^4$ K/cm, that is satisfied in any reasonable case. In addition, as the typical fluctuation velocity is (see Sect. 1.3) $v_{fl} \approx (kT/m)^{1/2} \approx 10^5$ cm/s, we find that $\tau_{fl} \approx \lambda_{fl}/v_{fl} \approx 10^{-10}$ s, showing that the fluctuation timescale is of the same magnitude as the mean collision time. So, we see that the typical local equilibrium condition on timescales is much less stringent than that about lengthscales. For liquid and solid systems, the condition of local equilibrium is satisfied even more easily, as ρ is 10^3 times larger and therefore λ_{fl} (and τ_{fl} as well) is 10 times smaller.

1.2 Fluctuations

Consider an isolated system, macroscopically at equilibrium, characterized by a variable A. Although on average $A = \overline{A}$, locally (i.e. within a subsystem at local equilibrium) we measure fluctuations $x = \delta A = A - \overline{A}$.

As it is explained in Appendix B, the probability that the subsystem experiences a displacement (i.e., a fluctuation) x from its equilibrium state, $x = 0$, is given by Einstein's fluctuation formula (B.17),

$$\Pi^{eq}(x) = C \exp\left[\frac{1}{k} S_{tot}(x)\right], \tag{1.7}$$

where C is a normalizing factor, k is the Boltzmann constant, while $S_{tot}(x)$ is the entropy of the whole isolated system (i.e. our subsystem plus its surrounding) when the subsystem is subjected to a fluctuation x.

Since $S_{tot}(x)$ reaches a maximum when $x = 0$, then $(\partial S_{tot}/\partial x)_0 = 0$ and $(\partial^2 S_{tot}/\partial x^2)_0 < 0$. Consequently, considering that $x/A \ll 1$, we obtain at leading order:

$$S_{tot}(x) = S_0 - \frac{k}{2} g x^2; \quad g > 0, \tag{1.8}$$

so that (1.7) reduces to the following Gaussian distribution,

$$\Pi^{eq}(x) = \sqrt{\frac{g}{2\pi}} \exp\left(-\frac{1}{2} g x^2\right), \tag{1.9}$$

where we have applied the normalizing condition,

$$\int_{-\infty}^{\infty} \Pi^{eq}(x)\, dx = 1, \quad \text{with} \quad \int_{-\infty}^{\infty} e^{-x^2}\, dx = \sqrt{\pi}. \tag{1.10}$$

Note that, as shown in Appendix A,

$$\langle x \rangle = \int_{-\infty}^{\infty} x \Pi^{eq}(x)\, dx = 0 \tag{1.11}$$

and

$$\langle x^2 \rangle = \int_{-\infty}^{\infty} x^2 \Pi^{eq}(x)\, dx = \frac{1}{g}, \tag{1.12}$$

so that Eq. (1.9) can be rewritten as:

$$\Pi^{eq}(x) = \frac{1}{\sqrt{2\pi \langle x^2 \rangle}} \exp\left\{-\frac{x^2}{2\langle x^2 \rangle}\right\}, \tag{1.13}$$

which coincides with Eq. (B.8).

When the state of the subsystem is described by a set of n independent variables A_i, where $i = 1, 2, \ldots, n$, with fluctuations x_i, these relations can be easily generalized as [1]:

$$S_{tot}(x) - S_0 = -\frac{k}{2} \sum_{i=1}^{n} \sum_{k=1}^{n} g_{ik} x_i x_k; \quad \|g_{ik}\| > 0, \tag{1.14}$$

so that we obtain the multi-variable Gaussian distribution,

$$\Pi^{eq}(x_i) = \sqrt{\|g_{ik}\|(2\pi)^{-n}} \exp\left\{-\frac{1}{2} \sum_{i=1}^{n} \sum_{k=1}^{n} g_{ik} x_i x_k\right\}, \tag{1.15}$$

where $\|g_{ik}\|$ denotes the norm, i.e. the determinant, of g_{ik}. Now define the thermodynamic driving force,

$$X_i = \frac{1}{k} \frac{\partial S_{tot}}{\partial x_i} = -\sum_{k=1}^{n} g_{ik} x_k, \tag{1.16}$$

tending to restore the $x_i = 0$ equilibrium state. Considering the inverse relation,

$$x_i = \frac{1}{k} \frac{\partial S_{tot}}{\partial X_i} = -\sum_{k=1}^{n} g_{ik}^{-1} X_k, \tag{1.17}$$

where $\sum_j g_{ij} g_{jk}^{-1} = \delta_{ik}$ and δ_{ik} denotes the Kronecker delta, the following equalities can be easily proved:

$$S_{tot} - S_0 = -\frac{1}{2} k \sum_{i=1}^{n} \sum_{k=1}^{n} x_i g_{ik} x_k = \frac{1}{2} k \sum_{i=1}^{n} x_i X_i = -\frac{1}{2} k \sum_{i=1}^{n} \sum_{k=1}^{n} X_i g_{ik}^{-1} X_k; \tag{1.18}$$

$$g_{ik} = -\frac{\partial X_i}{\partial x_k} = -\frac{1}{k} \frac{\partial^2 S_{tot}}{\partial x_i \partial x_k}; \tag{1.19}$$

$$g_{ik}^{-1} = -\frac{\partial x_i}{\partial X_k} = -\frac{1}{k} \frac{\partial^2 S_{tot}}{\partial X_i \partial X_k}. \tag{1.20}$$

Note the following equalities.

$$\langle x_i X_k \rangle = -\delta_{ik}, \tag{1.21}$$

$$\langle x_i x_k \rangle = g_{ik}^{-1}, \tag{1.22}$$

and

$$\langle X_i X_k \rangle = g_{ik}. \tag{1.23}$$

The first equality can be proved observing that:

$$\langle x_i X_k \rangle = \frac{1}{k} \int_{-\infty}^{\infty} x_i \frac{\partial S_{tot}}{\partial x_k} \Pi^{eq}(\mathbf{x}) \, d\mathbf{x}. \tag{1.24}$$

But $S_{tot} = S_0 + k \ln \Pi^{eq}$, so that

$$\langle x_i X_k \rangle = \int_{-\infty}^{\infty} x_i \frac{\partial \Pi^{eq}}{\partial x_k} \, d\mathbf{x} = \left[x_i \Pi^{eq} \right]_{-\infty}^{\infty} - \int_{-\infty}^{\infty} \Pi^{eq} \frac{\partial x_i}{\partial x_k} \, d\mathbf{x} = -\delta_{ik}, \tag{1.25}$$

where we have considered that Π^{eq} goes exponentially to zero as $|\mathbf{x}| \to \infty$ and the n variables are independent from each other, so that $\partial x_i / \partial x_k = \delta_{ik}$.

The second and third equalities can be easily proved from that (see Problem 1.2).

Finally, note that the speed with which the system returns to its equilibrium position depends on the entropy production which can be determined taking the time derivative of Eq. (1.18), obtaining:

$$\frac{d}{dt} \Delta S = k \sum_{i=1}^{n} \dot{x}_i X_i. \tag{1.26}$$

1.3 Physical Meaning of the Entropy Changes

As mentioned above, $S_{tot}(\mathbf{x})$ is the entropy of the whole, isolated, system, i.e. our subsystem at local equilibrium and its surroundings. Referring to the total entropy–total internal energy $S_{tot} - U_{tot}$ phase diagram of Fig. 1.1, the two curves are drawn at constant $\mathbf{x}$, one corresponding to the equilibrium points, with $\mathbf{x} = \mathbf{0}$, the other corresponding to non equilibrium points,[2] removed from equilibrium with a given fluctuation $\mathbf{x}_1$. The two curves are monotonically increasing, with the $\mathbf{x} = \mathbf{0}$ curve having a higher entropy, for a given energy, than the $\mathbf{x} = \mathbf{x}_1$ curve [2, Chap. 20]. Accordingly, in the figure we identify the three following points. Point a: it is the equilibrium point, with maximum entropy, $S_{tot}(a) = S_0$ at a fixed, given energy $U_{tot}(a) = U_0$; point b: it corresponds to a state where the subsystem has fluctuations $\mathbf{x}_1$, with entropy $S_{tot}(b) = S_1 < S_0$ and $U_{tot}(b) = U_0$ (the energy of our isolated system has to remain constant and therefore equal to U_0, while the entropy is less then its equilibrium value); point c: it is the equilibrium point of the system having entropy $S_{tot}(c) = S_1$, whose energy is $U_1 < U_0$.

Note that, if our system is isolated, the internal energy must be conserved and therefore we can only move from a to b; in other words, since $U_{tot}(c) < U_0$, our system needs an external source of work in order to be brought from a to c. In fact, $W_{min} = U_0 - U_1$ is the minimum work (as it is performed reversibly, i.e. at constant

---

[2] Note that, when $\mathbf{x} = \mathbf{x}_1$, the whole system is not at equilibrium, although, locally, the subsystem is at local equilibrium.

Fig. 1.1 Diagram S_{tot} vs U_{tot}

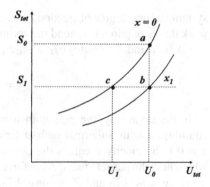

entropy) that an external source has to apply to bring the system out of equilibrium, from $\mathbf{x} = \mathbf{0}$ to $\mathbf{x} = \mathbf{x}_1$. Therefore we obtain:

$$-\Delta S_{tot} = (S_0 - S_1) = \left(\frac{\partial S}{\partial U}\right)_0 (U_0 - U_1), \qquad (1.27)$$

i.e.

$$\Delta S_{tot} = -\frac{W_{min}}{T_0}, \qquad (1.28)$$

where $T_0 = (\partial S_{tot}/\partial U_{tot})_0$ is the equilibrium temperature of the system.

From a different perspective, we may remember that the isolated system is composed of our subsystem, together with a heat reservoir at temperature T_0 and a work reservoir at pressure P_0. Therefore, $\Delta S_{tot} = \Delta S + \Delta S_T$, where ΔS is the entropy change of the subsystem, ΔS_T is the entropy change of the heat reservoir and we have considered that the work reservoir does not change its entropy. Now, $\Delta S_T = -Q_T/T_0$, where $Q_T = \Delta U + P_0 \Delta V$ is the heat entering the system (and thus leaving the heat reservoir), where ΔU and ΔV are the changes of the internal energy and the volume of the subsystem. Therefore, since the transformation is isothermal and isobaric, we can conclude that the total entropy variation is given by:

$$\Delta S_{tot} = -\frac{\Delta G}{T_0}, \qquad \Delta G = \Delta U - T_0 \Delta S + P_0 \Delta V, \qquad (1.29)$$

with $G = U - TS + PV$ denoting the Gibbs free energy of the subsystem. This shows that the minimum work equals the variation of the Gibbs free energy, so that g_{ik} can be determined from the relation

$$W_{min} = \Delta G = \frac{1}{2}kT_0 \sum_{i=1}^{n}\sum_{k=1}^{n} g_{ik}x_i x_k. \qquad (1.30)$$

Simple Examples   First, consider a particle immersed in a thermal bath at temperature T_0, located at position z and subjected to an elastic force $F' = -Az$, where A is the spring constant, which tends to restore its equilibrium configuration, $z = 0$. The system has one degree of freedom, with $x = z$, while $W_{min} = \frac{1}{2}Az^2$ is the minimum work that is required to extend the spring from $z = 0$ to z, and therefore we see that $g = A/kT_0$ and $X = -Az/kT_0$. Consequently, from Eq. (1.12) we obtain:

$$\langle z^2 \rangle = \frac{1}{g} = \frac{kT_0}{A}. \tag{1.31}$$

In the same way, the minimum work required to accelerate a free particle, at equilibrium with a thermal bath at temperature T_0, from its equilibrium velocity $v = 0$ to a velocity v equals the change in its kinetic energy, i.e. $W_{min} = \frac{1}{2}mv^2$, where m is the particle mass. As before, the system has one degree of freedom, with $x = v$, $g = m/kT_0$ and $X = -mv/kT_0$. Consequently, from Eq. (1.12) we obtain the Maxwell distribution, with:

$$\langle v^2 \rangle = \frac{1}{g} = \frac{kT_0}{m}. \tag{1.32}$$

These two examples are particular cases of the equipartition theorem, stating that, at equilibrium, each degree of freedom has the same mean energy, $\frac{1}{2}kT$ (see Sect. B.1).

1.4 Application to Thermodynamics

A subsystem, exchanging heat with a heat reservoir at constant temperature T_0 and work with a "work reservoir" at constant pressure P_0, will tend to reach a position of stable equilibrium, where its temperature and pressure coincide with those of the reservoirs. In general, as we saw in the previous section, the probability that the subsystem has a certain displacement $\mathbf{x}$ from equilibrium is:

$$\Pi^{eq} = K \exp\left(-\frac{\Delta G}{kT_0}\right), \tag{1.33}$$

where K is a normalization constant, while ΔG is the variation of the Gibbs free energy of the subsystem as it is brought out of equilibrium, from $\mathbf{x} = \mathbf{0}$ to $\mathbf{x}$. As we saw in (1.29),

$$\Delta G = \Delta U - T_0 \Delta S + P_0 \Delta V, \tag{1.34}$$

where the subscript "0" indicates the equilibrium state, which is determined by the external reservoirs.

Now, let us first consider a single-component system, which is defined by fixing the values of two independent variables, so that [2, Chap. 112]:

$$\Delta U = (\partial U/\partial S)_0 \Delta S + (\partial U/\partial V)_0 \Delta V$$
$$+ \frac{1}{2}\left[\frac{\partial^2 U}{\partial S^2}(\Delta S)^2 + 2\frac{\partial^2 U}{\partial S \partial V}(\Delta S \Delta V) + \frac{\partial^2 U}{\partial V^2}(\Delta V)^2\right] + \cdots. \tag{1.35}$$

Therefore, considering that $(\partial U/\partial S)_0 = T_0$ and $(\partial U/\partial V)_0 = -P_0$, we obtain:

$$\Delta G = \frac{1}{2}\left[\Delta S \Delta\left(\frac{\partial U}{\partial S}\right) + \Delta V \Delta\left(\frac{\partial U}{\partial V}\right)\right], \tag{1.36}$$

and from here,

$$\Delta G = \frac{1}{2}(\Delta S \Delta T - \Delta V \Delta P). \tag{1.37}$$

Choosing V and T as independent variables, consider the following equalities,

$$\Delta S = \left(\frac{\partial S}{\partial T}\right)_V \Delta T + \left(\frac{\partial S}{\partial V}\right)_T \Delta V = \frac{nc_V}{T}\Delta T + \left(\frac{\partial P}{\partial T}\right)_V \Delta V, \tag{1.38}$$

and

$$\Delta P = \left(\frac{\partial P}{\partial T}\right)_V \Delta T + \left(\frac{\partial P}{\partial V}\right)_T \Delta V = \left(\frac{\partial P}{\partial T}\right)_V \Delta T - \frac{1}{V\kappa_T}\Delta V, \tag{1.39}$$

where

$$nc_V = T\left(\frac{\partial S}{\partial T}\right)_V; \qquad \kappa_T = -\frac{1}{V}\left(\frac{\partial V}{\partial P}\right)_T \tag{1.40}$$

are the specific heat at constant volume and the isothermal compressibility, respectively, n are the number of moles within our subsystem and we have applied Maxwell's relation $(\frac{\partial S}{\partial V})_T = (\frac{\partial P}{\partial T})_V$. Finally we obtain:

$$\Delta G = \frac{1}{2}\left[\frac{nc_V}{T}(\Delta T)^2 + \frac{1}{V\kappa_T}(\Delta V)^2\right]. \tag{1.41}$$

This expression coincides with Eq. (1.30), provided that

$$x_1 = \Delta T; \qquad x_2 = \Delta V; \qquad g_{11} = \frac{nc_V}{kT^2}; \qquad g_{22} = \frac{1}{kTV\kappa_T}; \qquad g_{12} = 0. \tag{1.42}$$

Therefore we obtain:

$$\langle \Delta T \Delta V \rangle = 0; \qquad \langle (\Delta T)^2 \rangle = \frac{kT^2}{nc_V}; \qquad \langle (\Delta V)^2 \rangle = kTV\kappa_T, \tag{1.43}$$

showing that $c_V > 0$ and $\kappa_T > 0$. Defining

$$\delta T = \sqrt{\langle (\Delta T)^2 \rangle}; \qquad \delta V = \sqrt{\langle (\Delta V)^2 \rangle}, \tag{1.44}$$

these results can be rewritten as

$$\frac{\delta T}{T} = \sqrt{\frac{k}{nc_V}}; \qquad \frac{\delta V}{V} = \sqrt{\frac{kT\kappa_T}{V}}, \tag{1.45}$$

and for an ideal gas, where $c_V = \frac{3}{2}R$ and $\kappa_T = 1/P$, that yields:

$$\frac{\delta T}{T} = \sqrt{\frac{2}{3N}}; \qquad \frac{\delta V}{V} = \sqrt{\frac{1}{N}}, \qquad (1.46)$$

where $N = n N_A$ is the number of molecules in our subsystem.

Similar results are obtained when we choose other independent variables (see Problems 1.3 and 1.4). In particular, choosing S and P as independent variables, we obtain:

$$\langle \Delta S \Delta P \rangle = 0; \qquad \langle (\Delta S)^2 \rangle = k n c_P; \qquad \langle (\Delta P)^2 \rangle = \frac{kT}{V \kappa_S}, \qquad (1.47)$$

where

$$n c_P = T \left(\frac{\partial S}{\partial T} \right)_P; \qquad \kappa_S = -\frac{1}{V} \left(\frac{\partial V}{\partial P} \right)_S \qquad (1.48)$$

are the specific heat at constant pressure and the adiabatic compressibility, respectively.

Note that, as expected, we find that the fluctuations are of $\propto N^{-\frac{1}{2}}$.

Fluctuations in Binary Solutions   The previous analysis can be easily generalized to a two-component system ($1 =$ solvent; $2 =$ solute), as:

$$\Delta G = \frac{1}{2} \left[(\Delta S)(\Delta T) - (\Delta P)(\Delta V) + \sum_{i=1}^{2} (\Delta \mu_i)(\Delta n_i) \right] > 0, \qquad (1.49)$$

where n_i and μ_i are the number of moles and the molar chemical potential of component i. Therefore, we see that density fluctuations are not correlated to temperature and pressure fluctuations, i.e.,

$$\langle (\Delta T)(\Delta n_2) \rangle = \langle (\Delta P)(\Delta n_2) \rangle = 0. \qquad (1.50)$$

In addition, from $\Delta \mu_2 = (\partial \mu_2 / \partial n_2)_{T,P}(\Delta n_2)$, we obtain:

$$\langle (\Delta n_2)^2 \rangle = \frac{nkT}{(\partial \mu_2 / \partial x_2)_{T,P}}, \qquad (1.51)$$

where $x_2 = n_2/n$ is the molar fraction. In particular, for an ideal solution, $\mu_2 = \mu_2^0 + RT \ln x_2$, so that

$$\langle (\Delta n_2)^2 \rangle = \frac{nkT x_2}{RT} = \frac{n_2}{N_A} \quad \text{i.e.} \quad \frac{\delta N_2}{N_2} = \frac{1}{\sqrt{N_2}}, \qquad (1.52)$$

where $N_2 = n_2 N_A$ is the number of solute particles and $R = N_A k$ is the gas constant, with N_A denoting the Avogadro number.

1.5 Problems

Problem 1.1 Show that $\langle(\delta A)^4\rangle = 3N^2(\delta a)^4 = 3(\delta A)^4$.

Problem 1.2 Prove formulae (1.22) and (1.23).

Problem 1.3 Prove Eq. (1.47).

Problem 1.4 Determine $\langle\Delta T\,\Delta P\rangle$ and $\langle\Delta V\,\Delta P\rangle$.

References

1. de Groot, S.R., Mazur, P.: Non-Equilibrium Thermodynamics. Dover, New York (1984), Chap. VII.2
2. Landau, L.D., Lifshitz, E.M.: Statistical Physics, Part I. Pergamon, Elmsford (1980)

Chapter 2
Microscopic Reversibility

The *Principle of Microscopic Reversibility* was formulated by Richard Tolman [14] who stated that, at equilibrium, "any molecular process and the reverse of that process will be taking place on the average at the same rate". Applying this concept to macroscopic systems at local equilibrium leads to the rule of *detailed balances* (Sect. 2.2) and then, assuming linear relations between thermodynamic forces and fluxes, to the formulation of the celebrated reciprocity relations (Sect. 2.3) derived by Lars Onsager in 1931, and the fluctuation-dissipation theorem, (Sect. 2.4) proved by Herbert Callen and Theodore Welton in 1951. In this chapter, this vast subject matter is treated with a critical attitude, stressing all the hypotheses and their limitations.

2.1 Probability Distributions

Define:

- The simple probability $\Pi(\mathbf{x}, t)$ that the random variable $\mathbf{x}(t)$ has a certain value $\mathbf{x}$ at time t.[1]
- The joint probability $\Pi(\mathbf{x}_2, t_2; \mathbf{x}_1, t_1)$ that the random variable $\mathbf{x}(t)$ has a certain value $\mathbf{x}_2 = \mathbf{x}(t_2)$ at time t_2 and, also, that it has another value $\mathbf{x}_1 = \mathbf{x}(t_1)$ at time t_1.
- The conditional probability $\Pi(\mathbf{x}_2, t_2 | \mathbf{x}_1, t_1)$ that a random variable $\mathbf{x}$ has a certain value $\mathbf{x}_2 = \mathbf{x}(t_2)$ at time t_2, provided that at another (i.e. previous) time t_1 it has a value $\mathbf{x}_1 = \mathbf{x}(t_1)$.

By definition, when $t_2 > t_1$,

$$\Pi(\mathbf{x}_2, t_2; \mathbf{x}_1, t_1) = \Pi(\mathbf{x}_2, t_2 | \mathbf{x}_1, t_1) \, \Pi(\mathbf{x}_1, t_1). \qquad (2.1)$$

---

[1] Here and in the following, we use the same notation, $\mathbf{x}$, to indicate both the random variable and the value that it can assume. Whenever this might be confusing, different symbols will be used.

R. Mauri, *Non-Equilibrium Thermodynamics in Multiphase Flows*,
Soft and Biological Matter, DOI 10.1007/978-94-007-5461-4_2,
© Springer Science+Business Media Dordrecht 2013

In a stationary process all probability distributions are invariant under a time translation $t \to t + \tau$. Therefore for stationary processes the three distribution functions simplify as follows.

- $\Pi(\mathbf{x}, t) = \Pi(\mathbf{x})$ independent of t;
- $\Pi(\mathbf{x}_2, t_2; \mathbf{x}_1, t_1) = \Pi(\mathbf{x}_2, \tau; \mathbf{x}_1, 0)$;
- $\Pi(\mathbf{x}_2, t_2 | \mathbf{x}_1, t_1) = \Pi(\mathbf{x}_2, \tau | \mathbf{x}_1, 0)$,

where $\tau = t_1 - t_2$. In the following, when there is no ambiguity, the time 0 will be omitted.

If the process is also homogeneous, then all probability distributions are invariant under a space translation $\mathbf{x} \to \mathbf{x} + \mathbf{z}$. Therefore for stationary homogeneous processes the three distribution functions simplify as follows.

- $\Pi(\mathbf{x}) = \Pi$ independent of $\mathbf{x}$ and t;
- $\Pi(\mathbf{x}_1, \tau; \mathbf{x}_0, 0) = \Pi(\mathbf{z}, \tau)$;
- $\Pi(\mathbf{x}_1, \tau | \mathbf{x}_0, 0) = \Pi(\mathbf{z}, \tau)$,

where $\mathbf{z} = \mathbf{x}_1 - \mathbf{x}_0$. Note that for stationary and homogeneous processes the joint probability and the conditional probability are equal to each other; in fact, the ratio between them is given by the simple probability, which in this case is a constant.

Now, let us consider a stationary process. Its probability distribution functions are normalized as follows:

$$1 = \int \Pi(\mathbf{x}) \, d\mathbf{x} = \int \int \Pi(\mathbf{x}, \tau; \mathbf{x}_0, 0) \, d\mathbf{x} \, d\mathbf{x}_0. \tag{2.2}$$

Consequently,

$$1 = \int \Pi(\mathbf{x}, \tau | \mathbf{x}_0, 0) \, d\mathbf{x}, \tag{2.3}$$

showing that, since $\mathbf{x} = \mathbf{x}_0$ at $\tau = 0$, we have

$$\lim_{\tau \to 0} \Pi(\mathbf{x}, \tau | \mathbf{x}_0, 0) = \delta(\mathbf{x} - \mathbf{x}_0). \tag{2.4}$$

Based on these definitions, for any functions $f(\mathbf{x})$ and $g(\mathbf{x})$ we can define the averages,

$$\langle f(\mathbf{x}) \rangle = \int f(\mathbf{x}) \Pi(\mathbf{x}) \, d\mathbf{x}, \tag{2.5}$$

and

$$\langle f(\mathbf{x}_1) g(\mathbf{x}_0) \rangle = \int \int f(\mathbf{x}_1) g(\mathbf{x}_0) \Pi(\mathbf{x}_1, t_1; \mathbf{x}_0, t_0) \, d\mathbf{x}_1 \, d\mathbf{x}_0, \tag{2.6}$$

where $\mathbf{x}_1 = \mathbf{x}(t_1)$ and $\mathbf{x}_0 = \mathbf{x}(t_0)$. Obviously, while the first average is a constant, the second is a function of $\tau = t_1 - t_0$.

We can also define the conditional average of any functions $f(\mathbf{x})$ of the random variable $\mathbf{x}(t)$ as the mean value of $f(\mathbf{x})$ at time τ, assuming that $\mathbf{x}(0) = \mathbf{x}_0$, i.e.,

$$\langle f(\mathbf{x})\rangle_\tau^{\mathbf{x}_0} = \int f(\mathbf{x})\, \Pi[\mathbf{x}, \tau \,|\, \mathbf{x}_0, 0]\, d\mathbf{x}. \tag{2.7}$$

This conditional average depends on τ and on $\mathbf{x}_0$.

Now, substituting (2.1) into (2.6), we obtain the following equality,

$$\langle f(\mathbf{x}) g(\mathbf{x}_0)\rangle = \int\int f(\mathbf{x}) g(\mathbf{x}_0) \Pi(\mathbf{x}, \tau \,|\, \mathbf{x}_0, 0) \Pi(\mathbf{x}_0)\, d\mathbf{x}\, d\mathbf{x}_0, \tag{2.8}$$

that is:

$$\langle f(\mathbf{x}) g(\mathbf{x}_0)\rangle = \langle\langle f(\mathbf{x})\rangle_\tau^{\mathbf{x}_0} g(\mathbf{x}_0)\rangle. \tag{2.9}$$

2.2 Microscopic Reversibility

For a classical N-body system with conservative forces, microscopic reversibility is a consequence of the invariance of the equations of motion under time reversal and simply means that for every microscopic motion reversing all particle velocities also yield a solution. More precisely, the equations of motion of an N particle system are invariant under the transformation

$$\tau \to -\tau; \quad \mathbf{r} \to \mathbf{r}; \quad \mathbf{v} \to -\mathbf{v}, \tag{2.10}$$

where $\mathbf{r} \equiv \mathbf{r}^N$ and $\mathbf{v} \equiv \mathbf{v}^N$ are the positions and the velocities of the N particles. This leads to the so-called *principle of detailed balance*, stating that in a stationary situation each possible transition $(\hat{\mathbf{r}}, \hat{\mathbf{v}}) \to (\mathbf{r}, \mathbf{v})$ balances with the time reversed transition $(\mathbf{r}, -\mathbf{v}) \to (\hat{\mathbf{r}}, -\hat{\mathbf{v}})$, so that,

$$\Pi[\mathbf{r}(\tau), \mathbf{v}(\tau); \hat{\mathbf{r}}(0), \hat{\mathbf{v}}(0)] = \Pi[\hat{\mathbf{r}}(\tau), -\hat{\mathbf{v}}(\tau); \mathbf{r}(0), -\mathbf{v}(0)]. \tag{2.11}$$

As we saw in Sect. 1.1, this same condition can be applied when we deal with thermodynamic, coarse grained variables at local equilibrium, i.e. when $\tau \leq \tau_{fluct}$, where τ_{fluct} is the typical fluctuation time. In fact, for such very short times, forward motion is indistinguishable from backward motion, as they both are indistinguishable from fluctuations.

First, let us consider variables $x_i(t)$ that are invariant under time reversal, e.g. they are even functions of the particle velocities. In this case, another, perhaps more intuitive, way to write the principle of detailed balance is to assume that the conditional mean values of a variable at times τ and $-\tau$ are equal to each other, which means,

$$\langle \mathbf{x}\rangle_\tau^{\mathbf{x}_0} = \langle \mathbf{x}\rangle_{-\tau}^{\mathbf{x}_0} \tag{2.12}$$

or, equivalently,

$$\Pi(\mathbf{x}, \tau | \mathbf{x}_0, 0) = \Pi(\mathbf{x}, -\tau | \mathbf{x}_0, 0). \qquad (2.13)$$

Multiplying this last equation by $\Pi(\mathbf{x}_0)$, it can be rewritten as

$$\Pi(x_i, \tau; x_{0k}, 0) = \Pi(x_i, -\tau; x_{0k}, 0) = \Pi(x_i, 0; x_{0k}, \tau), \qquad (2.14)$$

where we have applied the stationarity condition. As expected, this equation is identical to (2.11), with $\mathbf{x} = \mathbf{r}$.

Now, define the correlation function for a stationary process as:

$$\langle x_i x_k \rangle(\tau) = \langle x_i(\tau) x_k(0) \rangle = \int \int x_i x_{0k} \Pi(\mathbf{x}, \tau; \mathbf{x}_0, 0) \, d\mathbf{x} \, d\mathbf{x}_0, \qquad (2.15)$$

for $\tau > 0$. Applying (2.14), we see that from microscopic reversibility we obtain:

$$\langle x_i x_k \rangle(\tau) = \langle x_i x_k \rangle(-\tau), \qquad (2.16)$$

that is,

$$\langle x_i(\tau) x_k(0) \rangle = \langle x_i(0) x_k(\tau) \rangle. \qquad (2.17)$$

From this expression, applying Eq. (2.9), we see that another formulation of microscopic reversibility is:

$$\langle x_{0k} \langle x_i \rangle_\tau^{\mathbf{x}_0} \rangle = \langle x_{0i} \langle x_k \rangle_\tau^{\mathbf{x}_0} \rangle. \qquad (2.18)$$

Now, consider the general case where x_i is an arbitrary variable which, under time reversal, transforms into the reversed variable according to the rule,

$$x_i \rightarrow \epsilon_i x_i, \qquad (2.19)$$

where $\epsilon_i = +1$, when the variable is even under time reversal and $\epsilon_i = -1$, when it is odd.[2] At this point, Eq. (2.18) can be generalized as:

$$\langle x_{0k} \langle x_i \rangle_\tau^{\mathbf{x}_0} \rangle = \epsilon_i \epsilon_k \langle x_{0i} \langle x_k \rangle_\tau^{\mathbf{x}_0} \rangle. \qquad (2.20)$$

In the following, we will denote by x those variables having $\epsilon = +1$, i.e. those remaining invariant under time reversal, and by y those variables having $\epsilon = -1$, i.e. those changing sign under time reversal, (e.g. velocity or angular momentum).[3]

---

[2]Note that, since $\Pi(x_i) = \Pi(\epsilon_i x_i)$, then $\langle x_i \rangle = \epsilon_i \langle x_i \rangle$, implying that all odd variables have zero stationary mean.

[3]In most of the literature, x- and y-variables are generally referred to as α- and β-variables.

2.3 Onsager's Reciprocity Relations

Assume the following linear phenomenological relations: (i.e. neglecting fluctuations)

$$\dot{x}_i = \sum_{j=1}^{n} L_{ij} X_j, \tag{2.21}$$

where the dot denotes time derivative, $\dot{x}$ are referred to as thermodynamic fluxes, while X are the generalized forces defined in (1.16). That means that this equation holds when we apply it to its conditional averages,

$$\langle \dot{x}_i \rangle_{\tau}^{x_0} = \sum_{j=1}^{n} L_{ij} \langle X_j \rangle_{\tau}^{x_0}. \tag{2.22}$$

The coefficients L_{ij} are generally referred to as Onsager's, or phenomenological, coefficients. Now take the time derivative of Eq. (2.18), considering that x_0 is constant:

$$\sum_{j=1}^{n} L_{ij} \langle x_{0k} \langle X_j \rangle_{\tau}^{x_0} \rangle = \sum_{j=1}^{n} L_{kj} \langle x_{0i} \langle X_j \rangle_{\tau}^{x_0} \rangle. \tag{2.23}$$

Considering that $\langle X_j \rangle_{\tau=0}^{x_0} = X_{0j}$ and $\langle x_{0i} X_{0j} \rangle = -\delta_{ij}$, we obtain:

$$L_{ik} = L_{ki}. \tag{2.24}$$

These are the celebrated reciprocity relations, derived by Lars Onsager [12, 13] in 1931.

In the presence of a magnetic field $\mathbf{B}$ or when the system rotates with angular velocity $\mathbf{\Omega}$, the operation of time reversal implies, besides the transformation (2.10), the reversal of $\mathbf{B}$ and $\mathbf{\Omega}$ as well. Therefore, the Onsager reciprocity relations become:

$$L_{ik}(\mathbf{B}, \mathbf{\Omega}) = L_{ki}(-\mathbf{B}, -\mathbf{\Omega}). \tag{2.25}$$

In the following, we will assume that $\mathbf{B} = 0$ and $\mathbf{\Omega} = 0$; however, we should keep in mind that in the presence of magnetic fields or overall rotations, the Onsager relations can be applied only when $\mathbf{B}$ and $\mathbf{\Omega}$ are reversed.

A clever way to express the Onsager coefficients L_{ij} can be obtained by multiplying Eq. (2.22) by x_{0k} and averaging:

$$\langle x_{0k} \langle \dot{x}_i \rangle_{\tau}^{x_0} \rangle = \sum_{j=1}^{n} L_{ij} \langle x_{0k} \langle X_j \rangle_{\tau}^{x_0} \rangle. \tag{2.26}$$

Now take $\tau = 0$ and apply Eq. (2.9) to obtain:

$$L_{ik} = -\langle \dot{x}_i x_k \rangle_0^{sym}, \tag{2.27}$$

where the superscript "sym" indicates the symmetric part of a tensor, i.e. $A_{ij}^{sym} = \frac{1}{2}(A_{ij} + A_{ji})$. This is one of the many forms of the fluctuation-dissipation theorem, which states that the linear response of a given system to an external perturbation is expressed in terms of fluctuation properties of the system in thermal equilibrium. Although it was formulated by Nyquist in 1928 to determine the voltage fluctuations in electrical impedances [11], the fluctuation-dissipation theorem was first proven in its general form by Callen and Welton [3] in 1951.

In (2.27), $\dot{\mathbf{x}}$ is the velocity of the random variable as it relaxes to equilibrium. Therefore, considering that $\mathbf{x}$ tends to $\mathbf{0}$ for long times, we see that $\mathbf{x}(0) = -\int_0^\infty \dot{\mathbf{x}}(t)\,dt$ and therefore the fluctuation-dissipation theorem can also be formulated through the following Green-Kubo relation:[4]

$$L_{ik} = \int_0^\infty \left\langle \dot{x}_i(0)\dot{x}_k(t)\right\rangle^{sym} dt, \tag{2.28}$$

showing that the Onsager coefficients can be expressed as the time integral of the correlation function between the velocities of the random variables at two different times.

Now, consider the opposite process, where the random variable evolves out of its equilibrium position $\mathbf{x} = \mathbf{0}$. Therefore, applying again Eq. (2.27), but with negative times, we obtain:

$$L_{ik} = \frac{1}{2}\frac{d}{dt}\langle x_i x_k\rangle_0^{sym}, \tag{2.29}$$

showing that the Onsager coefficients can be expressed as the temporal growth of the mean square displacements of the system variables from their equilibrium values.

These results are easily extended to the case where we have both x and y-variables, i.e. even and odd variables under time reversal. In this case, the phenomenological equations (2.21) can be generalized as:

$$\dot{x}_i = \sum_{j=1}^n L_{ij}^{(xx)} X_j + \sum_{j=1}^n L_{ij}^{(xy)} Y_j, \tag{2.30}$$

$$\dot{y}_i = \sum_{j=1}^n L_{ij}^{(yx)} X_j + \sum_{j=1}^n L_{ij}^{(yy)} Y_j, \tag{2.31}$$

where

$$X_i = \frac{1}{k}\frac{\partial \Delta S}{\partial x_i}; \qquad Y_i = \frac{1}{k}\frac{\partial \Delta S}{\partial y_i} \tag{2.32}$$

are the thermodynamic forces associated with the x and y-variables, respectively. With the help of these quantities, the reciprocity relations (2.24) were generalized

---

[4]The Green-Kubo relation is also called the fluctuation-dissipation theorem of the second kind. See [7, 8].

by Casimir in 1945 as [4]:

$$L_{ij}^{(xx)} = L_{ji}^{(xx)}; \qquad L_{ij}^{(xy)} = -L_{ji}^{(yx)}; \qquad L_{ij}^{(yy)} = L_{ji}^{(yy)}. \tag{2.33}$$

Substituting (2.30), (2.31) and (2.33) into the generalized form (1.26) of the entropy production term,

$$\frac{1}{k}\frac{dS}{dt} = \sum_{i=1}^{n} \dot{x}_i X_i + \sum_{i=1}^{n} \dot{y}_i Y_i, \tag{2.34}$$

we obtain:

$$\frac{1}{k}\frac{d}{dt}\Delta S = \sum_{i=1}^{n}\sum_{j=1}^{n} L_{ij}^{(xx)} X_i X_j + \sum_{i=1}^{n}\sum_{j=1}^{n} L_{ij}^{(yy)} Y_i Y_j. \tag{2.35}$$

This shows that neither the antisymmetric parts of the Onsager coefficients $\mathbf{L}^{(xx)}$ and $\mathbf{L}^{(yy)}$, nor the coupling terms between x and y-variables, $\mathbf{L}^{(xy)}$ and $\mathbf{L}^{(yx)}$, give any contribution to the entropy production rate.

It should be stressed that, when we apply the Onsager-Casimir reciprocity relations, we must make sure that the n variables (and therefore their time derivatives, or fluxes, as well) are independent from each other, and similarly for the thermodynamic forces.[5]

Comment 2.1 In the course of deriving the reciprocity relations, we have assumed that the same equations (2.21) govern both the macroscopic evolution of the system and the relaxation of its spontaneous deviations from equilibrium. This condition is often referred to as Onsager's postulate and is the basis of the Langevin equation (see Chap. 3).[6] The fluctuation-dissipation theorem, Eqs. (2.29) and (2.41), can be seen as a natural consequence of this postulate.

Comment 2.2 The simplest way to see the meaning of the fluctuation-dissipation theorem is to consider the free diffusion of Brownian particles (see Sect. 3.1). First, consider a homogeneous system, follow a single particle as it moves randomly around[7] and define a coefficient of self-diffusion as (one half of) the time derivative of its mean square displacement. Then, take the system out of equilibrium, and define the gradient diffusivity as the ratio between the material flux resulting from an imposed concentration gradient and the concentration gradient itself. As shown by Einstein in his Ph.D. thesis on Brownian motion [6], when the problem is linear (i.e. when particle-particle interactions are neglected), these two diffusivities are

---

[5]As shown in [10], when fluxes and forces are not independent, but still linearly related to one another, there is a certain arbitrariness in the choice of the independent variables, so that at the end the phenomenological coefficients can be chosen to satisfy the Onsager relations.

[6]Onsager stated that "the average regression of fluctuations will obey the same laws as the corresponding macroscopic irreversible process". See discussions in [9, 15].

[7]This process is sometimes called Knudsen *effusion*.

equal to each other, thus establishing perhaps the simplest example of fluctuation-dissipation theorem. Although we take this result for granted, it is far from obvious, as it states the equality between two very different quantities: on one hand, the fluctuations of a system when it is macroscopically at equilibrium; on the other hand, its dissipative properties as it approaches equilibrium.

2.4 Fluctuation-Dissipation Theorem

As we saw in the previous section, the *fluctuation-dissipation theorem* (FDT) connects the linear response relaxation of a system to its statistical fluctuation properties at equilibrium and it relies on Onsager's postulate that the response of a system in thermodynamic equilibrium to a small applied force is the same as its response to a spontaneous fluctuation.

First, let us derive the FDT in a very simple and intuitive way, following the original formulation by Callen and Greene [1, 2]. Assume that a constant thermodynamic force $X_0 = -F_0/kT$ is applied to the system for an infinite time $t < 0$ and then it is suddenly turned off at $t = 0$. Therefore, at $t = 0$ the system will have a non-zero position of stable equilibrium, x_0, such that

$$F_0 = \nabla_x W_{min} = kT g \cdot x_0, \qquad (2.36)$$

where $W_{min} = \frac{1}{2}kT x_0 \cdot g \cdot x_0 = F_0 \cdot x_0$ is the minimum work that the constant force, F_0, has to exert to displace the system to position x_0.

Now, in the absence of any external force, i.e. when $t > 0$, the mean value of the x-variable relaxes in time following Eq. (2.22), with,

$$\langle \dot{x} \rangle^{X_0}(t) = -M \cdot \langle x \rangle^{X_0}(t); \quad \text{i.e.,} \quad \langle x \rangle^{X_0}(t) = \exp(-Mt) \cdot x_0, \qquad (2.37)$$

where $M = L \cdot g$ is a constant phenomenological relaxation coefficient. Therefore, substituting (2.36) into (2.37) we obtain:

$$\langle x \rangle^{X_0}(t) = \chi(t) \cdot \frac{F_0}{kT}, \qquad (2.38)$$

where

$$\chi(t) = \exp(-Mt) \cdot g^{-1} \qquad (2.39)$$

is a time dependent relaxation coefficient.

On the other hand, the function χ is related to the correlation function at equilibrium, $\langle xx \rangle$. In fact, from the definition (2.15), substituting (2.37) we see that:

$$\langle xx \rangle(t) = \left\langle \langle x \rangle_t^{X_0} x_0 \right\rangle = \exp(-Mt) \cdot \langle x_0 x_0 \rangle = \exp(-Mt) \cdot g^{-1}. \qquad (2.40)$$

Comparing the last two equations, we conclude:

$$\langle \mathbf{x}\mathbf{x} \rangle (t) = \chi(t).$$ (2.41)

This relation represents the fluctuation-dissipation theorem.[8]

Note that, when $t = 0$, the relation (2.41) is identically satisfied, since $\langle \mathbf{x}\mathbf{x} \rangle (0) = \mathbf{g}^{-1}$, while $\chi(0) = \mathbf{g}^{-1}$.

The fluctuation-dissipation theorem can also be determined assuming a general, time-dependent driving force, $\mathbf{F}(t)$. In this case, due to the linearity of the process, we can write:

$$\langle \mathbf{x}(t) \rangle = \frac{1}{kT} \int_{-\infty}^{\infty} \kappa(t - t') \cdot \mathbf{F}(t') \, dt',$$ (2.42)

where $\kappa(t)$ is the generalized susceptibility, with $\kappa(t) = 0$ for $t < 0$. Denoting by $\widehat{\mathbf{x}}(\omega)$, $\widehat{\kappa}(\omega)$ and $\widehat{\mathbf{X}}(\omega)$, the Fourier transforms (C.1) of $\langle \mathbf{x}(t) \rangle$, $\kappa(t)$ and $\mathbf{X}(t)$, respectively, we have:

$$\widehat{\mathbf{x}}(\omega) = \frac{1}{kT} \widehat{\kappa}(\omega) \cdot \widehat{\mathbf{F}}(\omega).$$ (2.43)

In general, $\widehat{\kappa}(\omega)$ is a complex function, with $\widehat{\kappa} = \widehat{\kappa}^{(r)} + i\widehat{\kappa}^{(i)}$, where the superscripts (r) and (i) indicate the real and imaginary part. Since $\kappa(t)$ is real, we have:

$$\widehat{\kappa}(-\omega) = \widehat{\kappa}^*(\omega),$$ (2.44)

where the asterisk indicates complex conjugate, showing that $\widehat{\kappa}^{(r)}$ is an even function, while $\widehat{\kappa}^{(i)}$ is an odd function, i.e.,

$$\widehat{\kappa}^{(r)}(-\omega) = \widehat{\kappa}^{(r)}(\omega); \qquad \widehat{\kappa}^{(i)}(-\omega) = -\widehat{\kappa}^{(i)}(\omega).$$ (2.45)

Analogous relations exists regarding the correlation function $\langle \mathbf{x}\mathbf{x} \rangle (t)$. In fact, considering the microscopic reversibility (2.16) and the reality condition, we obtain:

$$\langle \widehat{\mathbf{x}\mathbf{x}} \rangle (\omega) = \langle \widehat{\mathbf{x}\mathbf{x}} \rangle^+(\omega) = \langle \widehat{\mathbf{x}\mathbf{x}} \rangle^*(-\omega),$$ (2.46)

i.e. the Fourier transform of the correlation function is a real and symmetric matrix.

As shown in Appendix C, using the causality principle, i.e. imposing that $\kappa(t) = 0$ for $t < 0$, we see that the generalized susceptibility is subjected to the Kramers-Kronig relation (C.17), so that $\kappa(t)$ can be related to $\chi(t)$ as [cf. Eq. (C.30)][9]

$$\chi(\omega) = \frac{2}{i\omega} \kappa(\omega).$$ (2.47)

---

[8]The same result can be obtained assuming that the constant thermodynamic force $\mathbf{X}_0$ is suddenly turned on at $t = 0$, so that for long times the system will have a non-zero position of stable equilibrium, $\mathbf{x}_F = -\mathbf{g}^{-1} \cdot \mathbf{X}_0$. In that case, redefining the random variable $\mathbf{x}$ as $\mathbf{x}_F - \mathbf{x}$, we find again Eq. (2.41).

[9]This is a somewhat simplified analysis. For more details, see [5].

Substituting this result into Eq. (2.41) and considering (2.46), we see that the fluctuation-dissipation theorem can be written in the following equivalent form:

$$\langle \widehat{\mathbf{xx}} \rangle^{(r)} (\omega) = \frac{2}{\omega} \left[\widehat{\boldsymbol{\kappa}}^{(i)} (\omega) \right]^{(s)}, \qquad (2.48)$$

where the superscripts (s) denotes the symmetric part of the tensor.

Note that, since

$$\langle \mathbf{xx} \rangle_0 = \int_{-\infty}^{\infty} \langle \widehat{\mathbf{xx}} \rangle (\omega) \, \frac{d\omega}{2\pi} = \frac{1}{\pi} \int_{-\infty}^{\infty} \frac{\widehat{\boldsymbol{\kappa}}^{(i)} (\omega)}{\omega} \, d\omega, \qquad (2.49)$$

using the dispersion equation (C.17) with $u = 0$, we obtain the obvious relation,

$$\langle \mathbf{xx} \rangle_0 = \widehat{\boldsymbol{\kappa}}^{(r)} (0) = \widehat{\boldsymbol{\kappa}} (0) = \mathbf{g}^{-1}, \qquad (2.50)$$

where we have used the fact that $\widehat{\boldsymbol{\kappa}}(0)$ is an even function.

This result can be easily extended to cross correlation functions between x- and y-type variables, considering that $\langle \widehat{\mathbf{xy}} \rangle (\omega)$ is an imaginary and antisymmetric matrix, i.e.,

$$\langle \widehat{\mathbf{xy}} \rangle (\omega) = - \langle \widehat{\mathbf{xy}} \rangle^{+} (\omega) = - \langle \widehat{\mathbf{xy}} \rangle^{*} (-\omega). \qquad (2.51)$$

At the end, the fluctuation-dissipation relation becomes,

$$\langle \widehat{\mathbf{xy}} \rangle^{(i)} (\omega) = \frac{2}{\omega} \left[\widehat{\boldsymbol{\kappa}}^{(r)} (\omega) \right]^{(a)}, \qquad (2.52)$$

where the superscripts (a) denotes the antisymmetric part of the tensor.

To better understand the meaning of the fluctuation-dissipation relation, consider the single variable case,[10]

$$\langle \dot{x} \rangle = -M \left(\langle x \rangle - \frac{F}{gkT} \right). \qquad (2.53)$$

Now, Fourier transforming this equation, we obtain (2.43) with,

$$\widehat{\kappa}(\omega) = \frac{M}{g(M - i\omega)} = \frac{1}{g - i\omega L^{-1}} = \frac{g + i\omega L^{-1}}{g^2 + \omega^2 L^{-2}}. \qquad (2.54)$$

On the other hand, the correlation function (2.40) gives:

$$\langle xx \rangle (t) = \frac{1}{g} e^{-Mt}, \qquad (2.55)$$

---

[10]Here, when the applied force F is constant, the equilibrium state will move from $\langle x \rangle = 0$ to $\langle x \rangle = F/(gkT)$.

whose Fourier transform yields:

$$\langle \widehat{xx} \rangle (\omega) = \frac{2M}{g(M^2 + \omega^2)} = \frac{2L^{-1}}{g^2 + \omega^2 L^{-2}} \tag{2.56}$$

thus showing that the FDT (2.48) is identically satisfied.

Identical results are obtained in the multi-variable case, where we have:

$$\langle \dot{\mathbf{x}} \rangle = \mathbf{L} \cdot \left(\langle \mathbf{X} \rangle + \frac{1}{kT} \mathbf{F} \right), \tag{2.57}$$

where $\mathbf{L} = \mathbf{M} \cdot \mathbf{g}^{-1}$ is the Onsager phenomenological coefficient, while $\mathbf{X} = -\mathbf{g} \cdot \mathbf{x}$, obtaining:

$$\widehat{\boldsymbol{\kappa}} = \left(\mathbf{g} - i\omega \mathbf{L}^{-1} \right)^{-1}. \tag{2.58}$$

Note that the symmetry of $\widehat{\boldsymbol{\kappa}}$ is a direct consequence of the Onsager reciprocity relation $\mathbf{L} = \mathbf{L}^+$.

Sometimes, it is convenient to consider the fluctuations of $\mathbf{x}$ as being caused by a random fictitious force $\mathbf{f}$, so that the instantaneous value of $\mathbf{x}$ (not its mean value, which is identically zero) is linearly related to $\mathbf{f}$ through the same generalized susceptibility κ that governs the relaxation of the system far from equilibrium,[11] i.e.,

$$\mathbf{x}(t) = \frac{1}{kT} \int_{-\infty}^{\infty} \boldsymbol{\kappa}(t - t') \cdot \mathbf{f}(t') \, dt'. \tag{2.59}$$

In this case, considering that $\widehat{\boldsymbol{\kappa}}(-\omega) = \widehat{\boldsymbol{\kappa}}^*(\omega)$, we have:

$$\langle \widehat{\mathbf{xx}} \rangle (\omega) = (kT)^{-2} \widehat{\boldsymbol{\kappa}}^*(\omega) \cdot \langle \widehat{\mathbf{ff}} \rangle (\omega) \cdot \widehat{\boldsymbol{\kappa}}(\omega); \tag{2.60}$$

then, we obtain:

$$\langle \widehat{\mathbf{ff}} \rangle (\omega) = (kT)^2 \frac{2}{\omega} \widehat{\boldsymbol{\kappa}} \cdot \widehat{\boldsymbol{\kappa}}^{(i)} \cdot \widehat{\boldsymbol{\kappa}}(\omega) = \frac{2}{\omega} (kT)^2 \mathrm{Im} \left\{ \left[\widehat{\boldsymbol{\kappa}}^* \right]^{-1} \right\}. \tag{2.61}$$

Therefore, when the generalized susceptibility can be expressed as Eq. (2.58), we obtain:

$$\langle \widehat{\mathbf{ff}} \rangle (\omega) = 2(kT)^2 \mathbf{L}^{-1} = 2kT \boldsymbol{\zeta}, \tag{2.62}$$

where $\boldsymbol{\zeta} = kT\mathbf{L}^{-1}$. In fact, in this case Eq. (2.57) becomes the Langevin equation (see next chapter),

$$\dot{\mathbf{x}} = -\mathbf{M} \cdot \mathbf{x} + \widetilde{\mathbf{J}}, \tag{2.63}$$

---

[11] This is clearly equivalent to the Onsager regression hypothesis. Note that here and in the following $\mathbf{x}$ denotes the fluctuation $(\mathbf{x} - \langle \mathbf{x} \rangle)$.

where $\widetilde{\mathbf{J}} = \frac{1}{kT}\mathbf{L} \cdot \mathbf{f}$ is the fluctuating flux, satisfying the following relation:

$$\langle \widetilde{\mathbf{J}}(0)\,\widetilde{\mathbf{J}}(t)\rangle = 2\mathbf{L}\,\delta(t), \tag{2.64}$$

where $\delta(t)$ is the Dirac delta. This shows that there is no correlation between the particle position $\mathbf{x}$ and the random force $\mathbf{f}$ (see Problem 2.2). In fact, it is this lack of correlation that is at the foundation of the Onsager regression hypothesis, therefore justifying the Langevin equation, as discussed in the next chapter.

2.5 Problems

Problem 2.1 Consider a small particle of arbitrary shape moving in an otherwise quiescent Newtonian fluid. In creeping flow conditions, determine the symmetry relations satisfied by the resistance matrix connecting velocity and angular velocity with the force and the torque that are applied to the particle.

Problem 2.2 Consider a driven 1D oscillator of mass m at frequency ω_0, with damping force $\zeta \dot{x}$, with x denoting the displacement from its equilibrium position, $x = 0$. Determine the spectrum of the random force.

References

1. Callen, H.B., Greene, R.F.: Phys. Rev. **86**, 702 (1952)
2. Callen, H.B., Greene, R.F.: Phys. Rev. **88**, 1387 (1952)
3. Callen, H.B., Welton, T.A.: Phys. Rev. **83**, 34 (1951)
4. Casimir, H.B.G.: Rev. Mod. Phys. **17**, 343 (1945)
5. de Groot, S.R., Mazur, P.: Non-Equilibrium Thermodynamics. Dover, New York (1962), Chap. VIII.4
6. Einstein, A.: Ann. Phys. (Berlin) **17**, 549 (1905)
7. Green, M.S.: J. Chem. Phys. **22**, 398 (1954)
8. Kubo, R.: J. Phys. Soc. Jpn. **12**, 570 (1957)
9. Marconi, U.M.B., et al.: Phys. Rep. **461**, 111 (2008)
10. Meixner, J.: Rheol. Acta **12**, 465 (1973)
11. Nyquist, H.: Phys. Rev. **32**, 110 (1928)
12. Onsager, L.: Phys. Rev. **37**, 405 (1931)
13. Onsager, L.: Phys. Rev. **38**, 2265 (1931)
14. Tolman, R.C.: The Principles of Statistical Mechanics, p. 163. Dover, New York (1938), Chap. 50
15. van Kampen, N.G.: Stochastic Processes in Physics and Chemistry. North-Holland, Amsterdam (1981), Chap. VIII.8

Chapter 3
Langevin Equation

The *Langevin equation* was proposed in 1908 by Paul Langevin [2] to describe Brownian motion, that is the apparently random movement of a particle immersed in a fluid, due to its collisions with the much smaller fluid molecules. As the Reynolds number of this movement is very low, the drag force is proportional to the particle velocity; this, so called, Stokes law represents a particular case of the linear phenomenological relations that are assumed to hold in irreversible thermodynamics. In this chapter, after a brief description of Brownian motion (Sect. 3.1), first we review the original Langevin approach in 1D (Sect. 3.2), then we generalize it to study the evolution of a set of random variables with linear phenomenological forces (Sect. 3.3). The most general case, with non-linear phenomenological forces, represents a non-trivial generalization of the Langevin equation and is studied in Chap. 5 within the framework of the theory of stochastic differential equations.

3.1 Brownian Motion

The random motion of suspended particles that are sufficiently large to be observed is termed Brownian motion, after the Scottish botanist Robert Brown, who in 1828 described this phenomenon from microscopic observations he had made of pollen grains suspended in water.

Let us begin by considering the continuous, one-dimensional translational Brownian motion as represented by a one-dimensional random walk problem. It is well known (see Appendix A) that the probability $\Pi_N(n)$ that, in a total of N steps, we make n steps to the right (and $N - n$ steps to the left) is given by the binomial distribution,

$$\Pi_N(n) = \frac{N!}{n!(N-n)!} p^n q^{N-n}, \tag{3.1}$$

where p and $q = 1 - p$ are the probabilities that each individual step is to the right or to the left, respectively. As shown in Appendix A, the sum of all probabilities is

R. Mauri, *Non-Equilibrium Thermodynamics in Multiphase Flows*,
Soft and Biological Matter, DOI 10.1007/978-94-007-5461-4_3,
© Springer Science+Business Media Dordrecht 2013

normalized to one, i.e. $\sum_n \Pi_N(n) = 1$, its mean value is $\langle n \rangle = \bar{n} = Np$ and its dispersion is $\langle (n - \bar{n})^2 \rangle = (\delta n)^2 = Npq$. Therefore, since $\delta n / \bar{n} \propto 1/\sqrt{N}$, we see that, as $N \gg 1$, the binomial distribution becomes sharper and sharper around its equilibrium value $\bar{n} = Np$. We can reach the same conclusion observing that, from (3.1), the peak value of Ω_N increases extremely fast with N, and therefore, as the area under the Π curve is normalized and equal to 1, its width must decrease inversely.

From the above considerations, it is not surprising that, when $N \gg 1$, the binomial distribution tends to the following Gaussian distribution (see Appendix A for a formal proof),

$$\Pi_N(n) = (2\pi Npq)^{-1/2} \exp\left[-\frac{(n - Np)^2}{2Npq}\right].$$
(3.2)

In particular, when, as in our case, $p = q = 1/2$, we can also express this result in terms of the actual displacement variable $x = m\ell$, where $m = 2(n - N/2)$ and ℓ is the length of each step. To do that, rewrite Eq. (3.2) as

$$\Pi_N(m) = \left(\frac{2}{\pi N}\right)^{1/2} \exp\left[-\frac{m^2}{2N}\right].$$
(3.3)

Now, if we regard $x = m\ell$ as a continuous variable, noting that m can only assume integer values, the range dx contains $dx/2\ell$ possible values of m, all of them occurring with the same probability $\Omega_N(m)$. Hence, the probability of finding the particle anywhere between x and $x + dx$ is simply obtained by summing $\Pi_N(m)$ over all values of m lying within dx. Therefore, defining a continuous probability density, $\Pi_N(x)$, we can write:

$$\Pi_N(x)dx = \Pi_N(m)\frac{dx}{2\ell},$$
(3.4)

obtaining,

$$\Pi_N(x) = \frac{1}{\sqrt{2\pi N}\ell} \exp\left[-\frac{x^2}{2N\ell^2}\right].$$
(3.5)

In general, for any values of p and q, we obtain:

$$\Pi_N(x) = \frac{1}{\sqrt{2\pi}\sigma} \exp\left[-\frac{(x - \bar{x})^2}{2\sigma^2}\right]$$
(3.6)

where

$$\bar{x} = N(p - q)\ell; \qquad \sigma = 2\sqrt{Npq\ell^2}.$$
(3.7)

As we have seen previously,

$$\bar{x} = \int_{-\infty}^{\infty} x\Pi(x)\,dx; \qquad \sigma^2 = \int_{-\infty}^{\infty} (x - \bar{x})^2 \Pi(x)\,dx.$$
(3.8)

Now, consider that the number of steps is proportional to the time,

$$N = t/\tau, \tag{3.9}$$

where τ is the mean interval of time between two successive steps (or collisions). Accordingly, treating time as a continuous variable, Eq. (3.5) can be rewritten as:

$$\Pi(x,t) = \frac{1}{2\sqrt{\pi D^s t}} \exp\left[-\frac{x^2}{4D^s t}\right], \tag{3.10}$$

where

$$D^s = \frac{\ell^2}{2\tau}, \tag{3.11}$$

is the self-diffusion coefficient, showing that

$$\langle x^2 \rangle = 2D^s t. \tag{3.12}$$

In the diffusion process of a solute, we expect that the concentration c at any x and t satisfies the diffusion equation,[1]

$$\frac{\partial c}{\partial t} = D\frac{\partial^2 c}{\partial x^2}, \tag{3.13}$$

where D is the gradient diffusion coefficient, i.e. the ratio between material flux and concentration gradient. Assuming that N_0 particles of solute are initially concentrated at the origin, this equation admits the following solution [see Eqs. (4.74)–(4.80) in Sect. 4.6],

$$c(x,t) = \frac{N_0}{2\sqrt{\pi D^g t}} \exp\left[-\frac{x^2}{4Dt}\right]. \tag{3.14}$$

Comparing Eq. (3.14) with (3.10) we see that they are identical to each other provided that $c = N_0\Pi$ and $D^s = D^g$, i.e. self-diffusivity equals gradient diffusivity. This is the fundamental point of the fluctuation-dissipation theorem. Note that the two diffusivity coefficients have two very different meanings. On one hand, D^s describes the mean square displacement of a tracer particles diffusing in an otherwise homogeneous system, while, on the other hand, D considers a non-homogeneous system and measures the ratio between mass flux and concentration gradient. Accordingly, it is not at all obvious that these two coefficients must be equal to each other.

Considering that in 3D the motions along the tree directions, x, y and z, are uncorrelated with each other, with $r^2 = x^2 + y^2 + z^2$, we find that the 3D probability

---

[1] Here we assume that c denotes the number of particles per unit volume (or length, in this 1D case), but it could be mass or moles, instead.

density function becomes,

$$\Pi(\mathbf{r}, t) = \frac{1}{8(\pi Dt)^{3/2}} \exp\left[-\frac{r^2}{4Dt}\right], \tag{3.15}$$

so that,

$$\langle r^2 \rangle = \langle x^2 \rangle + \langle y^2 \rangle + \langle z^2 \rangle = 6Dt. \tag{3.16}$$

Now, let us evaluate the diffusion coefficient D following Einstein's thermodynamic argument [1]. Consider a suspension of particles subjected to a potential field $\psi(\mathbf{r})$, such as gravity. In general, the particle flux is $\mathbf{J} = c\mathbf{v} - D\nabla c$, where c is the concentration, $\mathbf{v}$ the particle velocity, and D the gradient diffusivity. At equilibrium, $\mathbf{J} = 0$, i.e., the diffusive flux, $-D\nabla c$, balances the convective flux, $c\mathbf{v}$, with $\mathbf{v} = \mathbf{F}/\zeta$, where $\mathbf{F} = -\nabla\psi$ is the applied force and ζ is the drag coefficient. Therefore, we obtain:

$$D\nabla c = \mathbf{v}c = \frac{\mathbf{F}}{\zeta}c \quad \Rightarrow \quad D\nabla \ln c = -\frac{\nabla\psi}{\zeta}. \tag{3.17}$$

Note that here we use the steady drag law, $\mathbf{F} = \zeta\mathbf{v}$, where ζ is the drag coefficient,[2] even if the particle is changing its velocity very rapidly; this is justified because velocities are so small that fluid and particle inertia can be neglected. Now, consider that at equilibrium,

$$c(\mathbf{r}) = K \exp\left[-\frac{\psi(\mathbf{r})}{kT}\right] \quad \Rightarrow \quad \nabla \ln c = -\frac{\nabla\psi}{kT}, \tag{3.18}$$

so that, comparing (3.17) with (3.18) we obtain the Stokes-Einstein relation:

$$D = \frac{kT}{\zeta}. \tag{3.19}$$

This is another manifestation of the fluctuation-dissipation theorem.

3.2 One-Dimensional Langevin Equation

An alternative approach to study Brownian motion is to employ the Langevin equation. Here, we assume that the force acting on a single particle is a combination of a mean force, i.e. the frictional drag associated with the fluid response to particle motion, and a much faster, so called, Brownian force, $f(t)$, characterizing the very

---

[2]For a spherical particle of radius a, the Stokes law states that $\zeta = 6\pi\eta a$, where η is the fluid viscosity (see Sect. F.2.2).

rapid fluctuations, associated with molecular motion time scale ($\approx 10^{-10}$ s for gases; $\approx 10^{-13}$ s for liquids),[3]

$$m\frac{d^2x}{dt^2} = -\zeta\frac{dx}{dt} + f(t). \tag{3.20}$$

This equation cannot be used directly, since the mean values of velocity and acceleration are zero. Instead, multiplying it by x and rearranging, we obtain:

$$\frac{m}{2}\frac{d^2x^2}{dt^2} - m\left(\frac{dx}{dt}\right)^2 = xf - \frac{\zeta}{2}\frac{dx^2}{dt}, \tag{3.21}$$

and therefore, averaging all terms of this expression, we obtain:

$$\frac{m}{2}\frac{d^2\langle x^2\rangle}{dt^2} + \frac{\zeta}{2}\frac{d\langle x^2\rangle}{dt} = kT, \tag{3.22}$$

where we have considered that $\langle(dx/dt)^2\rangle = kT/m$ from equipartition, and that f fluctuates much more rapidly than x around its zero mean value. Now, solving this equation when $t \gg m/\zeta$, so that the first term on the LHS can be neglected, we obtain,

$$\langle x^2\rangle = 2Dt, \tag{3.23}$$

where $D = kT/\zeta$, in agreement with Eq. (3.19).

Now let us see how to proceed more rigorously. The starting equation is:

$$\dot{p} = -\frac{\zeta}{m}p + f(t), \tag{3.24}$$

where $p = m\,dx/dt$ is the particle momentum. The Brownian force has the following statistical properties:

$$\langle f(t)\rangle_t^{v_0} = 0, \tag{3.25}$$

and

$$\langle f(t)f(t+\tau)\rangle_\tau^{v_0} = 2Q\delta(\tau), \tag{3.26}$$

where Q is a constant, indicating that $f(t)$ fluctuates around its zero mean much more rapidly than p. In addition, Eq. (3.25) implies the following linear phenomenological relation,[4]

$$\langle\dot{p}\rangle_\tau^{v_0} = -\frac{\zeta}{m}\langle p\rangle_\tau^{v_0}. \tag{3.27}$$

---

[3] Here and in the following we denote by x both the position of the Brownian particle and the local coordinate. When this is confusing, the particle trajectory will be indicated by $X(t)$.

[4] Here we see that macroscopic regressions and microscopic fluctuations are governed by the same linear force, indicating that Onsager's postulate is automatically satisfied.

Therefore, since in Eq. (1.32) we have seen that for a free particle undergoing random walk the fluctuating variable is $x = p$ and the generalized force is $X = -p/mkT$, then Eq. (3.27) can be rewritten in terms of the Onsager coefficient L,

$$\langle \dot{x} \rangle_0^\tau = L \langle X \rangle_0^\tau; \quad L = kT\zeta. \tag{3.28}$$

Solving the Langevin equation with initial condition $p(0) = p_0$, we obtain:

$$p(t) = p_0 e^{-\zeta t/m} + \int_0^t e^{-\zeta(t-t')/m} f(t') dt'. \tag{3.29}$$

Taking the average (it is automatically a conditioned average as we have fixed that $p(0) = p_0$) we obtain:

$$\left\langle p(t) \right\rangle_t^{p_0} = p_0 e^{-\zeta t/m}. \tag{3.30}$$

Note that

$$\lim_{t \gg m/\zeta} \left\langle p(t) \right\rangle_t^{p_0} = 0. \tag{3.31}$$

Now define

$$\Delta p = p - \left\langle p(t) \right\rangle_t^{p_0} = \int_0^t e^{-\zeta(t-t')/m} f(t') dt'. \tag{3.32}$$

Squaring this expression and taking the average we obtain:

$$\left\langle (\Delta p)^2 \right\rangle_\tau^0 = \int_0^t \int_0^t e^{-\zeta(t-t')/m - \zeta(t-t'')/m} \left\langle f(t') f(t'') \right\rangle dt' dt'', \tag{3.33}$$

that is,

$$\left\langle (\Delta p)^2 \right\rangle_\tau^0 = \frac{Qm}{\zeta} \left(1 - e^{-2\zeta t/m} \right). \tag{3.34}$$

When $t \gg m/\zeta$, we have $\Delta p = p$ and $\langle p^2 \rangle = kTm$, so that:

$$\lim_{t \gg m/\zeta} \left\langle (\Delta p)^2 \right\rangle_\tau^0 = \frac{Qm}{\zeta} = kTm, \tag{3.35}$$

and we obtain:

$$Q = kT\zeta = L. \tag{3.36}$$

This is another form of the fluctuation-dissipation theorem, equating (apart from a constant proportionality term, depending on the heat bath) the intensity of the fluctuating force, Q, to the drag coefficient, and therefore to the Onsager coefficient, L.

3.3 Generalized Langevin Equation

Generalizing Eq. (3.20) we have:[5]

$$\dot{x}_i = -\sum_{k=1}^{n} M_{ik} x_k + \tilde{J}_i(t),$$ (3.37)

where M_{ik} is a phenomenological mobility coefficient and $\tilde{J}_i(t)$ is a generalized Brownian force, i.e. a random noise. If we identify $\dot{x}_i$ with a flux, then $\tilde{J}_i$ represents its fluctuating component, with

$$\langle \tilde{J}_i(t) \rangle = 0$$ (3.38)

and

$$\langle \tilde{J}_i(t) \tilde{J}_k(t+\tau) \rangle = 2Q_{ik}\delta(\tau),$$ (3.39)

where $Q_{ik} = \langle \tilde{J}_{i,0} \tilde{J}_{k,0} \rangle$. Compared to the 1D case, with $x_1 = v$, we have:

$$M = \frac{\zeta}{m}; \qquad Q = \frac{\langle f^2 \rangle}{2m^2} = \frac{kT\zeta}{m^2}.$$ (3.40)

Note that M_{ik} is proportional to the Onsager coefficient L_{ik}. In fact, defining from (2.34) the force thermodynamically conjugated to the flux $\dot{x}_i$, i.e. $kX_i = \partial \dot{S}/\partial \dot{x}_i$, from Eq. (2.22),

$$\langle \dot{x}_i \rangle_\tau^{\mathbf{x}_0} = \sum_{j=1}^{n} L_{ij} \langle X_j \rangle_\tau^{\mathbf{x}_0}; \quad \text{with } X_j = -\sum_{k=1}^{n} g_{jk} x_k,$$ (3.41)

we obtain (remind that $\mathbf{L}$ is a symmetric matrix):

$$M_{ik} = \sum_{j=1}^{n} L_{ij} g_{jk} \quad \text{i.e.} \quad L_{ik} = \sum_{j=1}^{n} M_{ij} g_{jk}^{-1},$$ (3.42)

and therefore the Langevin equation (3.37) can also be written in the following form,

$$\dot{x}_i = \sum_{k=1}^{n} L_{ik} X_k + \tilde{J}_i(t).$$ (3.43)

Proceeding as in the previous chapter, from (3.37) we obtain:

$$\langle x_i \rangle_t^{\mathbf{x}_0} = \sum_{k=1}^{n} e^{-M_{ik}t} x_{k,0}.$$ (3.44)

---

[5]Here we apply again Onsager's postulate, assuming that the same linear relation (3.37) describe both microscopic fluctuations and macroscopic regressions, where in the latter the fluctuating term can be neglected [see (3.41)].

Then define

$$\Delta x_i = x_i - \langle x_i \rangle_t^{\mathbf{x}_0}; \qquad \langle \Delta x_i \rangle_t^{\mathbf{x}_0} = 0, \tag{3.45}$$

obtaining:

$$\langle (\Delta x_i \Delta x_k) \rangle_t^{\mathbf{x}_0} = G_{ik}^{-1} - \sum_{j=1}^{n} \sum_{\ell=1}^{n} e^{-M_{ij}t} G_{j\ell}^{-1} e^{-M_{k\ell}t}, \tag{3.46}$$

where G_{ik} is a symmetric matrix (i.e. $G_{ik} = G_{ki}$ or $\mathbf{G} = \mathbf{G}^+$) defined through:

$$2Q_{ik} = \sum_{j=1}^{n} [M_{ij} G_{jk}^{-1} + G_{ij}^{-1} M_{kj}], \quad \text{i.e.} \quad \mathbf{Q} = [\mathbf{M} \cdot \mathbf{G}^{-1}]^{sym}, \tag{3.47}$$

with $\mathbf{A}^{sym} = \frac{1}{2}(\mathbf{A} + \mathbf{A}^+)$.

Now, when $t \gg \|\mathbf{M}\|^{-1}$, we have $\Delta x_i = x_i$, with $\langle x_i x_k \rangle = g_{ik}^{-1}$. Therefore, since $\langle (\Delta x_i \Delta x_k) \rangle_t^{\mathbf{x}_0} \to G_{ik}^{-1}$, we obtain:

$$\mathbf{G} = \mathbf{g}, \tag{3.48}$$

so that

$$\mathbf{Q} = [\mathbf{M} \cdot \mathbf{g}^{-1}]^{sym} = \mathbf{L}. \tag{3.49}$$

This is the fluctuation dissipation theorem, equating two quantities, namely $\mathbf{Q}$ and $\mathbf{L}$, that describe fluctuations and dissipation, respectively. Consequently Eq. (3.39) can be rewritten as:

$$\langle \tilde{J}_i(t) \tilde{J}_k(t + \tau) \rangle = 2L_{ik} \delta(\tau). \tag{3.50}$$

3.4 Problems

Problem 3.1 Derive the Stokes-Einstein relation using the simplified Langevin equation,

$$\zeta \dot{x} = f(t), \tag{3.51}$$

where the inertial term is neglected.

Problem 3.2 Show that the Stokes-Einstein relation can also be obtained multiplying Eq. (3.30) by v_0, averaging out over all possible values of v_0, and finally integrating twice over time.

Problem 3.3 Defining $\delta p = p - p_0$, from the general solution of the Langevin equation, Eq. (3.29), for $t \ll m/\zeta$ obtain the following result:

$$\langle \delta p \rangle_t^{p_0} = p_0 [e^{-\zeta t/m} - 1] = -\frac{\zeta}{m} p_0 t + O(t^2) \tag{3.52}$$

and

$$\langle (\delta p)^2 \rangle_t^{p_0} = \frac{Qm}{\zeta}\left(1 - e^{-2\zeta t/m}\right) = 2kT\zeta t + O\left(t^2\right). \tag{3.53}$$

Note that (3.53) is a particular case of the relation (2.29), considering that here $L = kT\zeta$.

Problem 3.4 Consider the Brownian motion of a particle with mass m immersed in a fluid and attracted to the origin through a linear force,

$$m\ddot{z} + \zeta\dot{z} + Az = f(t), \tag{3.54}$$

where $f(t)$ is a random noise. Show that $\langle f^2 \rangle$ is the same that we have found in the absence of any external force, i.e., $\langle f(t)f(t+\tau)\rangle_\tau^0 = 2kT\zeta\delta(\tau)$.

References

1. Einstein, A.: Brownian Movement. Dover, New York (1956), Chap. 5
2. Langevin, P.: C. R. Acad. Sci. (Paris) **146**, 530 (1908)

Chapter 4
Fokker-Planck Equation

The time evolution of the probability density function of a set of random variables is described by the *Fokker-Planck equation*, named after Adriaan Fokker and Max Planck. Originally, it was developed to describe the motion of Brownian particles and later was generalized to follow the evolution of a set of random variables with linear phenomenological constitutive relations. In this chapter, the Fokker-Planck equation is derived in the framework of Markov processes (Sects. 4.1 and 4.2), showing its most general solution (Sect. 4.3) and how the fluctuation-dissipation theorem follows from it (Sect. 4.4). Then, in Sect. 4.5, a counter example is illustrated, where we show that when the applied forces are non-conservative the fluctuation-dissipation theorem cannot be applied. Finally, in Sect. 4.6, we study the simplest case of Brownian motion, namely pure diffusion, also referred to as the Wiener process, named in honor of Norbert Wiener, stressing how the associated mathematical inconsistencies can be completely resolved only by applying the theory of stochastic differential equations (see Chap. 5).

4.1 Markov Processes

As we saw in Sect. 2.1, a stochastic process is described completely by the probability $\Pi[\ldots, \mathbf{x}(t_2), \mathbf{x}(t_1), \mathbf{x}(t_0)]$ that a time-dependent set of random variables, $\mathbf{X}(t)$, assumes values $\mathbf{x}_0$, $\mathbf{x}_1$, $\mathbf{x}_2$, etc. at times t_0, t_1, t_2, etc. Then, in terms of these, so-called, joint probability density functions we can define the conditional probability densities,

$$\Pi\big[\mathbf{x}(t_n), \mathbf{x}(t_{n-1}), \ldots, \mathbf{x}(t_{i+1}), \mathbf{x}(t_i), \mathbf{x}(t_{i-1}), \ldots, \mathbf{x}(t_1), \mathbf{x}(t_0)\big]$$
$$= \Pi\big[\mathbf{x}(t_n), \ldots, \mathbf{x}(t_{i+1})|\mathbf{x}(t_i), \ldots, \mathbf{x}(t_0)\big] \Pi\big[\mathbf{x}(t_i), \ldots, \mathbf{x}(t_0)\big], \quad (4.1)$$

where $t_n > t_{n-1} > \cdots > t_{i+1} > t_i > t_{i-1} > \cdots > t_1 > t_0$.[1]

---

[1] This condition is not required for the validity of Eq. (4.1), but it is essential in the definition of a Markov process.

R. Mauri, *Non-Equilibrium Thermodynamics in Multiphase Flows*,
Soft and Biological Matter, DOI 10.1007/978-94-007-5461-4_4,
© Springer Science+Business Media Dordrecht 2013

Now, let us define a Markov process. Loosely speaking, it consists of a process where the future is determined by the knowledge of only the present. Accordingly, it is required that the conditional probability is determined entirely by the knowledge of the most recent condition, i.e.,

$$\Pi\big[\mathbf{x}(t_n)|\mathbf{x}(t_{n-1}), \mathbf{x}(t_{n-2})\ldots\mathbf{x}(t_0)\big] = \Pi\big[\mathbf{x}(t_n)|\mathbf{x}(t_{n-1})\big]. \tag{4.2}$$

This condition means that we can express any joint probability distribution in terms of the product of conditional probabilities. In fact, using the definition (4.1),

$$\Pi\big[\mathbf{x}(t_2), \mathbf{x}(t_1)|\mathbf{x}(t_0) = \Pi\big[\mathbf{x}(t_2)|\mathbf{x}(t_1), \mathbf{x}(t_0)\big]\Pi\big[\mathbf{x}(t_1)|\mathbf{x}(t_0)\big]\big], \tag{4.3}$$

and applying (4.2), we obtain:

$$\Pi\big[\mathbf{x}(t_2), \mathbf{x}(t_1)|\mathbf{x}(t_0)\big] = \Pi\big[\mathbf{x}(t_2)|\mathbf{x}(t_1)\big]\Pi\big[\mathbf{x}(t_1)|\mathbf{x}(t_0)\big]. \tag{4.4}$$

Now consider the identity:

$$\Pi\big[\mathbf{x}(t_2)|\mathbf{x}(t_0)\big] = \int \Pi\big[\mathbf{x}(t_2), \mathbf{x}(t_1)|\mathbf{x}(t_0)\big]\,d\mathbf{x}_1, \tag{4.5}$$

where the integral is taken over all the possible values of $\mathbf{x}_1 = \mathbf{x}(t_1)$. When the process is Markovian, substituting (4.4) we obtain the so-called Chapman-Kolmogorov equation,

$$\Pi\big[\mathbf{x}(t_2)|\mathbf{x}(t_0)\big] = \int \Pi\big[\mathbf{x}(t_2)|\mathbf{x}(t_1)\big]\Pi\big[\mathbf{x}(t_1)|\mathbf{x}(t_0)\big]\,d\mathbf{x}_1. \tag{4.6}$$

4.2 Derivation of the Fokker-Planck Equation

Denoting $\delta\mathbf{x} = \mathbf{x} - \mathbf{x}_0$, define:

$$\langle\dot{x}_{i,0}\rangle = \lim_{\tau\to 0} \frac{\langle\delta x_i\rangle_\tau^0}{\tau} = V_i. \tag{4.7}$$

and

$$\lim_{\tau\to 0} \frac{\langle\delta x_i \delta x_k\rangle_\tau^0}{\tau} = 2Q_{ik} \tag{4.8}$$

and neglect higher-order moments, assuming that $(\langle\delta\mathbf{x}\rangle)^n$ goes to zero more rapidly than τ, i.e.,

$$\lim_{\tau\to 0} \frac{\langle(\delta\mathbf{x})^n\rangle_\tau^t}{\tau} = 0. \tag{4.9}$$

Here, $\mathbf{V}(\mathbf{x})$ and $\mathbf{Q}(\mathbf{x})$ are assumed to be well behaved functions of the random variable $\mathbf{x}$, expressing its phenomenological velocity and diffusivity [in fact, definition (4.8) is a Stokes-Einstein relation].

For any given function $R(\mathbf{x})$ that tends to 0 exponentially as $|\mathbf{x}| \to \infty$, consider the following integral:[2]

$$I = \int \dot{\Pi}[\mathbf{x}(t)|\mathbf{x}_0] R(\mathbf{x}) \, d\mathbf{x} = \lim_{\tau \to 0} \frac{1}{\tau} \int \{\Pi[\mathbf{x}(t+\tau)|\mathbf{x}_0] - \Pi[\mathbf{x}(t)|\mathbf{x}_0]\} R(\mathbf{x}) \, d\mathbf{x}, \tag{4.10}$$

where,

$$\Pi[\mathbf{x}(t+\tau)|\mathbf{x}_0] = \int \Pi[\mathbf{x}(t+\tau)|\mathbf{x}'(t)] \Pi[\mathbf{x}'(t)|\mathbf{x}_0] \, d\mathbf{x}'$$

$$= \int \Pi[\mathbf{x}(\tau)|\mathbf{x}'] \Pi[\mathbf{x}'(t)|\mathbf{x}_0] \, d\mathbf{x}'. \tag{4.11}$$

Now, change notation in the integrand of Eq. (4.10) as follows: $\mathbf{x}' \to \mathbf{x}_0$; $\mathbf{x}_0 \to \mathbf{x}_1$ for the first term and $\mathbf{x} \to \mathbf{x}_0$; $\mathbf{x}_0 \to \mathbf{x}_1$ for the second term:

$$I = \lim_{\tau \to 0} \frac{1}{\tau} \left\{ \int \int \Pi[\mathbf{x}(\tau)|\mathbf{x}_0] \Pi[\mathbf{x}_0(t)|\mathbf{x}_1] R(\mathbf{x}) \, d\mathbf{x}_0 \, d\mathbf{x} \right.$$

$$\left. - \int \Pi[\mathbf{x}_0(t)|\mathbf{x}_1] R(\mathbf{x}_0) \, d\mathbf{x}_0 \right\}. \tag{4.12}$$

So, expanding:

$$R(\mathbf{x}) = R(\mathbf{x}_0) + \delta\mathbf{x} \cdot (\nabla R)_0 + \frac{1}{2}(\delta\mathbf{x}\delta\mathbf{x}):(\nabla\nabla R)_0 + \cdots, \tag{4.13}$$

we obtain:

$$I = \int d\mathbf{x}_0 \Pi[\mathbf{x}_0(t)|\mathbf{x}_1] [\mathbf{V} \cdot (\nabla R)_0 + \mathbf{Q}:(\nabla\nabla R)_0] \tag{4.14}$$

where

$$\mathbf{V} = \lim_{\tau \to 0} \frac{1}{\tau} \int d\mathbf{x} \Pi[\mathbf{x}(\tau)|\mathbf{x}_0] \frac{\delta\mathbf{x}}{\tau}; \qquad \mathbf{Q} = \int d\mathbf{x} \Pi[\mathbf{x}(\tau)|\mathbf{x}_0] \frac{(\delta\mathbf{x}\delta\mathbf{x})}{2\tau}. \tag{4.15}$$

Subtracting this expression from (4.10) we obtain:

$$0 = \int d\mathbf{x} \{\dot{\Pi}[\mathbf{x}(t)|\mathbf{x}_0] R(\mathbf{x}) - \Pi[\mathbf{x}(t)|\mathbf{x}_0][\mathbf{V} \cdot \nabla R + \mathbf{Q}:\nabla\nabla R]\}. \tag{4.16}$$

Now consider the following equality:

$$\nabla_i \nabla_k (\Pi Q_{ik} R) = R \nabla_i \nabla_k (\Pi Q_{ik}) + \Pi Q_{ik} \nabla_i \nabla_k R + 2 \nabla_i R \nabla_k (\Pi Q_{ik}), \tag{4.17}$$

that is,

$$\nabla_i \nabla_k (\Pi Q_{ik} R) = 2 \nabla_i (\Pi Q_{ik} \nabla_k R) + R \nabla_i \nabla_k (\Pi Q_{ik}) - \Pi Q_{ik} \nabla_i \nabla_k R. \tag{4.18}$$

Substituting it into (4.16) we have:

$$0 = \int d\mathbf{x} R \{\dot{\Pi} + \nabla_i (V_i \Pi) - \nabla_i \nabla_k (Q_{ik} \Pi)\} + I_2 \tag{4.19}$$

---

[2]This derivation of the Fokker-Planck equation can be found in [1], reprinted in [6]. A rigorous and more general treatment can be found in Sect. 5.2.

where

$$I_2 = \int d\mathbf{x} \nabla_i \{ -V_i \Pi R + \nabla_k (\Pi Q_{ik} R) - 2\Pi Q_{ik} \nabla_k R \} = 0, \qquad (4.20)$$

since $R(x)$ tends exponentially to zero at infinity. Finally, considering that $R(x)$ is arbitrary, we obtain the Fokker-Planck equation (FPE):[3]

$$\dot{\Pi} = \nabla_i \left[-V_i \Pi + \nabla_k (Q_{ik} \Pi) \right] \qquad (4.21)$$

that is,

$$\dot{\Pi} + \nabla \cdot \mathbf{J} = 0, \qquad (4.22)$$

where $\mathbf{J}$ is the probability flux,

$$\mathbf{J} = \mathbf{V} \Pi - \nabla \cdot (\mathbf{Q} \Pi). \qquad (4.23)$$

The FPE must be solved with appropriate boundary conditions (in general, either homogeneous or periodic) and initial condition,

$$\Pi (\mathbf{x}(t)|\mathbf{x}_0) = \delta(\mathbf{x}_0), \qquad (4.24)$$

where $\delta(\mathbf{x})$ is Dirac delta function.

In particular, in the vicinity of the equilibrium state, as in our case, the diffusivity $\mathbf{Q}$ can be assumed to be constant, while the mean velocity is a linear function of the distance $\mathbf{x}$ from equilibrium,

$$\mathbf{V} = -\mathbf{M} \cdot \mathbf{x}. \qquad (4.25)$$

Then, the evolution of the random variable $\mathbf{x}$ defines the so-called *Ornstein-Uhlenbeck* process. For example, in the 1D motion of a free Brownian particle, when we compare the assumptions (4.7) and (4.8) with the results (3.40) of the 1D Langevin equation, we obtain (see Problems 1 and 2):

$$M = \frac{\zeta}{m}; \qquad Q = \frac{\langle f^2 \rangle}{2m^2} = \frac{kT\zeta}{m^2}. \qquad (4.26)$$

The solution of the FPE is the propagator, or Green function, of a general convection-diffusion equation,

$$\frac{\partial p}{\partial t} = -\nabla \cdot \mathbf{J}^p; \qquad \mathbf{J}^p = \mathbf{V}p - \nabla \cdot (\mathbf{Q}p), \qquad (4.27)$$

with $p = p(\mathbf{x}, t)$, subjected to appropriate homogeneous or periodic boundary conditions (see below) and with initial condition,

$$p(\mathbf{x}, 0) = p_0(\mathbf{x}). \qquad (4.28)$$

In fact, due to the linearity of the problem, we have:

$$p(\mathbf{x}, t) = \int_V \Pi(\mathbf{x}, t|\mathbf{x}_0) p_0(\mathbf{x}) \, d\mathbf{x}_0, \qquad (4.29)$$

where $\Pi(\mathbf{x}(t)|\mathbf{x}_0)$ satisfies the FPE (4.22), with initial conditions (4.24).

---

[3]The FPE is also known as the *Kolmogorov*, or as the *Smoluchowski equation*, named after Andrey Nikolaevich Kolmogorov and Marian Ritter von Smolan Smoluchowski, respectively.

Now, assume that for long times the FPE tends to a stationary solution, $\Pi^{eq}(\mathbf{x})$, with

$$\mathbf{J}^{eq} = \mathbf{V}\Pi^{eq} - \nabla \cdot \left(\mathbf{Q}\Pi^{eq}\right) = 0. \tag{4.30}$$

Then, defining a function $\Pi^{\dagger}(\mathbf{x}, t)$ by

$$\Pi(\mathbf{x}, t) = \tau \Pi^{eq}(\mathbf{x}) \, \Pi^{\dagger}(\mathbf{x}, t), \tag{4.31}$$

where τ is a characteristic volume, we can see by direct substitution that $q(\mathbf{x}, t)$ satisfies the following equation:

$$\frac{\partial \Pi^{\dagger}}{\partial t} = \mathbf{V} \cdot \nabla \Pi^{\dagger} + \mathbf{Q} \cdot \nabla\nabla \Pi^{\dagger}. \tag{4.32}$$

This is called the *backward Fokker-Planck equation*, or *backward Kolmogorov equation*, as it can be obtained moving backward in time, using the procedure that has been illustrated above (see also Sect. 5.2 for a more rigorous derivation).

A particularly simple case arises when $\mathbf{D} = D\mathbf{I}$, with D constant, and the generalized velocity field is potential, $\mathbf{V} = -\nabla\psi$; then we find: $c(\mathbf{x}) = K \exp\left(-\psi(\mathbf{x})/D\right)$, where K is a normalization constant.

Particular care should be taken in defining the boundary conditions. In fact, suppose that the solutions of the forward FPE, $\Pi(\mathbf{x}, t)$ and $\Pi^{eq}(x)$, satisfy no-flux, or reflecting, conditions at the boundary σ_τ of the domain volume τ, i.e.

$$\mathbf{n} \cdot \mathbf{J} = \mathbf{n} \cdot \mathbf{J}^{eq} = 0 \quad \text{on } \sigma_\tau, \tag{4.33}$$

where $\mathbf{n}$ is the unit vector perpendicular to σ_τ. Then, from the definition (4.31) we see that the solution $\Pi^{\dagger}(\mathbf{x}, t)$ of the backward FPE satisfies the following boundary condition:

$$\mathbf{n} \cdot \mathbf{D} \cdot \nabla\Pi^{\dagger} = 0 \quad \text{on } \sigma_\tau. \tag{4.34}$$

For absorbing or periodic boundary conditions, that is when $\Pi = \Pi^{eq} = 0$ on σ_τ or when Π and Π^{eq} are periodic, then the same boundary conditions apply as well to $\Pi^{\dagger}$, that is $\Pi^{\dagger} = 0$ on σ_τ or $\Pi^{\dagger}$ periodic, respectively.

4.3 General Solution of the FPE

Let us write the FPE (4.22) as:

$$\dot{\Pi} + \mathcal{L}[\Pi] = \delta(\mathbf{x} - \mathbf{x}_0)\delta(t), \quad \mathbf{x} \in \tau, \tag{4.35}$$

where $\Pi = \Pi(\mathbf{x}, t|\mathbf{x}_0)$ is the conditional probability distribution, τ denotes the domain volume, while the operator $\mathcal{L}$ is defined for any function $p(\mathbf{x})$ as,

$$\mathcal{L}[p] = \nabla \cdot \mathbf{J}^p; \quad \mathbf{J}^p = \mathbf{V}p - \nabla \cdot (\mathbf{D}p). \tag{4.36}$$

Assume that this problem must be solved with the no-flux, or reflecting, boundary conditions (4.33),

$$\mathbf{n} \cdot \mathbf{J}^p = \mathbf{n} \cdot \left[\mathbf{V}p - \nabla \cdot (\mathbf{D}p)\right] = 0 \quad \text{on } \sigma_\tau, \tag{4.37}$$

where σ_τ is the surface delimiting τ and $\mathbf{n}$ is a unit vector perpendicular to σ_τ; other cases, with absorbing, or periodic, boundary conditions, are simpler and will be mentioned later.

This is a classical Sturm-Liouville problem, and its solution reads,

$$\Pi(\mathbf{x}, t|\mathbf{x}_0) = \frac{1}{\tau} \sum_j A_j \phi_j(\mathbf{x}) e^{-\lambda_j t}, \tag{4.38}$$

where $\phi_j(\mathbf{x})$ and λ_j are the eigenfunctions and eigenvalues of the operator $\mathfrak{L}$,

$$\mathfrak{L}[\phi_j] = \lambda_j \phi_j \quad \text{in } \tau, \quad \text{with } \mathbf{n} \cdot \left[\mathbf{V}\phi_j - \nabla \cdot (\mathbf{D}\phi_j) \right] = 0 \quad \text{on } \sigma_\tau. \tag{4.39}$$

The coefficients A_j are constants to be determined through the initial condition as,

$$A_j = \int_\tau \Pi(\mathbf{x}, 0|\mathbf{x}_0) \phi_j^\dagger(\mathbf{x}) \, d\mathbf{x}, \tag{4.40}$$

where $\phi_j^\dagger$ are the eigenfunctions of the adjoint operator $\mathfrak{L}^\dagger$,

$$\mathfrak{L}^\dagger[\phi_j^\dagger] = \lambda_j^\dagger \phi_j^\dagger \quad \text{in } \tau, \tag{4.41}$$

defined such that, for $p = p(\mathbf{x}, t)$ and $q = q(\mathbf{x}, t)$,

$$\int_\tau \mathfrak{L}[p] q \, d\mathbf{x} = \int_\tau p \, \mathfrak{L}^\dagger[q] \, d\mathbf{x}. \tag{4.42}$$

Substituting (4.31) into (4.42), we see that $q(\mathbf{x}, t)$ satisfies the backward FPE (4.32) with boundary conditions (4.34), that is:

$$\mathfrak{L}^\dagger[q] = -\mathbf{V} \cdot \nabla q - \mathbf{D} : \nabla\nabla q; \quad \text{in } \tau; \quad \mathbf{n} \cdot \mathbf{D} \cdot \nabla q = 0 \quad \text{on } \sigma_\tau. \tag{4.43}$$

Accordingly, we see that:

$$\lambda_j^\dagger = \lambda_j; \quad \phi_j(\mathbf{x}) = \tau \Pi^{eq}(\mathbf{x}) \phi_j^\dagger(\mathbf{x}). \tag{4.44}$$

Due to the mutual adjointness of $\mathfrak{L}$ and $\mathfrak{L}^\dagger$, we know that $\lambda_j^\dagger$ equals the complex conjugate of λ_j, and therefore we conclude that λ_j are real. In addition, the sets $\{\phi_j\}$ and $\{\phi_j^\dagger\}$ are orthogonal to each other, i.e. $\int_\tau \phi_j \phi_k^\dagger \, d\mathbf{x} = 0$ when $j \neq k$. Since both ϕ_j and $\phi_j^\dagger$ are defined within an arbitrary constant factor, they can be normalized so that:

$$\frac{1}{\tau} \int_\tau \phi_j^\dagger \, d\mathbf{x} = \delta_{j0}, \tag{4.45}$$

and

$$\frac{1}{\tau} \int_\tau \phi_j \phi_k^\dagger \, d\mathbf{x} = \delta_{jk}. \tag{4.46}$$

Accordingly, we find that $\phi_0^\dagger(\mathbf{x}) = 1$, $\phi_0(\mathbf{x}) = \tau \Pi^{eq}(\mathbf{x})$ and $\lambda_0 = 0$. Consequently, Eq. (4.45) with $k = 0$ yields

$$\frac{1}{\tau} \int_\tau \phi_j \, d\mathbf{x} = \delta_{j0}. \tag{4.47}$$

Using this normalization, imposing that $\Pi(\mathbf{x}, 0|\mathbf{x}_0) = \delta(\mathbf{x} - \mathbf{x}_0)$, from (4.40) we find that $A_j = \phi_j^\dagger(\mathbf{x}_0)$. Therefore, the solution (4.38) can be written as:

$$\Pi(\mathbf{x}, t|\mathbf{x}_0) = \frac{1}{\tau} \sum_j \phi_j(\mathbf{x})\phi_j^\dagger(\mathbf{x}_0)e^{-\lambda_j t}. \tag{4.48}$$

Now, denoting by $\Pi^\dagger(\mathbf{x}, t|\mathbf{x}_0)$ the Green function of the backward Fokker-Planck equation, we can repeat the previous analysis, replacing ϕ_j with $\phi_j^\dagger$, therefore obtaining the following reciprocity relation:

$$\Pi(\mathbf{x}, t|\mathbf{x}_0, 0) = \Pi^\dagger(\mathbf{x}_0, 0|\mathbf{x}, t). \tag{4.49}$$

Thus, as $\Pi(\mathbf{x}, t|\mathbf{x}_0, 0)$ describes the effect at location $\mathbf{x}$ and time t of a point source initially located at $\mathbf{x}_0$, $\Pi^\dagger(\mathbf{x}_0, 0|\mathbf{x}, t)$ describes the inverse effect. Accordingly, the equality (4.49) is identical to the microscopic reversibility condition (2.13).

Finally, note that in deriving the general solution (4.38) we have used the completeness of the set $\{\phi_j\}$ (and consequently $\{\phi_j^\dagger\}$ as well), as that is always true for Sturm-Liouville problems; in particular, that means that we can apply (4.38) even when $t = 0$, obtaining:

$$\lim_{t \to 0} \Pi(\mathbf{x}, t|\mathbf{x}_0) = \frac{1}{\tau} \sum_j \phi_j(\mathbf{x})\phi_j^\dagger(\mathbf{x}_0) = \delta(\mathbf{x} - \mathbf{x}_0). \tag{4.50}$$

4.4 Fluctuation-Dissipation Theorem

As $t \to \infty$, the system tends to equilibrium, with

$$J_i^\Pi = 0; \quad \Pi = \Pi^{eq} = C \exp\left[-\frac{1}{2}\sum_{i=1}^{n}\sum_{k=1}^{n} g_{ik}x_i x_k\right]. \tag{4.51}$$

Substituting these expressions in (4.23), with $\mathbf{Q}$ constant and $\mathbf{V} = -\mathbf{M} \cdot \mathbf{x}$, as we are in the vicinity of the equilibrium state, we obtain:

$$0 = \sum_{k=1}^{n}\left(M_{ik}x_k\Pi^{eq} + Q_{ik}\nabla_k\Pi^{eq}\right), \quad \text{with } \nabla_k\Pi^{eq} = -\sum_{j=1}^{n} g_{kj}x_j\Pi^{eq}. \tag{4.52}$$

Therefore we conclude:

$$M_{ij} = \sum_{k=1}^{n} Q_{ik}g_{kj}, \tag{4.53}$$

that is:

$$Q_{ik} = \sum_{j=1}^{n}\left(M_{ij}g_{jk}^{-1}\right)^{sym} = L_{ik}, \quad \mathbf{Q} = \left(\mathbf{M} \cdot \mathbf{g}^{-1}\right)^{sym} = \mathbf{L}, \tag{4.54}$$

where L_{ik} are the Onsager coefficients. This is the form (3.49) and (2.64) of the fluctuation-dissipation theorem, equating two quantities, namely $\mathbf{Q}$ and $\mathbf{L}$, that describe fluctuations and dissipation, respectively and is also equivalent to the Kubo relation (2.29).

The Fokker-Planck equation can be solved exactly as,

$$\Pi\big[\mathbf{x}(t)|\mathbf{x}_0\big] = C \exp\left[-\frac{1}{2}\sum_{i=1}^{n}\sum_{k=1}^{n} V_{ik}\Delta x_i \Delta x_k \right], \tag{4.55}$$

where

$$\Delta x_i = x_i - \langle x_i \rangle_t^0; \quad \langle x_i \rangle_t^0 = x_{k,0}\exp(-M_{ik}t) \tag{4.56}$$

and

$$V_{ik}^{-1} = g_{ik}^{-1} - \sum_{j=1}^{n}\sum_{\ell=1}^{n} e^{-M_{ij}t}\, g_{j\ell}^{-1}\, e^{-M_{k\ell}t}, \tag{4.57}$$

where $\mathbf{g} = (\mathbf{Q}^{-1}\cdot\mathbf{M})^{sym}$ from the fluctuation-dissipation theorem.

Note that these results coincide with those obtained using the Langevin equation, namely $\langle x_i \rangle_t^0$ is the phenomenological value of Eq. (3.44)], while V_{ik}^{-1} is the variance $\langle (\Delta x_i \Delta x_k)\rangle_t^0$ of Eq. (3.46).

As a summary, the Fokker-Planck equation can be written once $\mathbf{M}$ and $\mathbf{Q}$ are determined, by following the following steps.

- write the phenomenological equation, $\langle \dot{\mathbf{x}} \rangle_t^0 = -\mathbf{M}\cdot\langle \mathbf{x} \rangle_t^0$.
- Compute: $\mathbf{Q} = (\mathbf{M}\cdot\mathbf{g}^{-1})^{sym}$.

Note that, for small time, i.e. $t \ll \|\mathbf{M}\|^{-1}$, $\langle \mathbf{x} \rangle_t^0 = (\mathbf{I}-\mathbf{M}t)\cdot\mathbf{x}_0$ and Eq. (4.55) simplifies as:

$$\Pi\big[\mathbf{x}(t)|\mathbf{x}_0\big] = C \exp\left[-\frac{1}{2t}\sum_{i=1}^{n}\sum_{k=1}^{n} Q_{ik}^{-1}\Delta x_i \Delta x_k \right], \tag{4.58}$$

that is the probability function is a Gaussian distribution with mean $\langle \mathbf{x} \rangle_t^0$ and variance tensor $\mathbf{Q}t$. That means that the system moves with a systematic drift, whose characteristic time is $\|\mathbf{M}\|^{-1}$, on which a Gaussian fluctuation is superimposed, with covariant matrix $\mathbf{Q}t$, that is, we can write:

$$\mathbf{x}(t) = \mathbf{x}_0 - \mathbf{M}\cdot\mathbf{x}_0 t + \mathbf{q}\, t^{1/2}, \tag{4.59}$$

where

$$\langle \mathbf{q}(t) \rangle = \mathbf{0}; \quad \langle \mathbf{q}(t)\mathbf{q}(t) \rangle = \mathbf{Q}. \tag{4.60}$$

Therefore, we see that

- trajectories are continuous, since $\mathbf{x}(t) \to \mathbf{x}_0$ as $t \to 0$;
- trajectories are nowhere differentiable in time, because of the $t^{1/2}$ dependence occurring in (4.59).

This is a fundamental flaw that has been resolved thanks to stochastic differentiation, as we will see in Chap. 5.

4.5 "Violation" of the FD Theorem

Until now, we have studied the evolution of a system that fluctuates around its equilibrium position and is subjected to conservative forces. As stated by the equipartition theorem, which is one of the formulations of the second law of thermodynamics, each degree of freedom contributes the potential associated with the conservative force by a fixed amount, equal to $\frac{1}{2}kT$. In addition, due to microscopic reversibility, the dissipating recovering force is related to diffusivity through the fluctuation dissipation theorem.

When the system is subjected to non-conservative forces, however, there is no potential associated with such forces, so that the motion of the system is irreversible, even at a fundamental level. Therefore, in that case, neither the equipartition principle nor the fluctuation dissipation theorem are valid any more, as the principle of microscopic reversibility cannot be applied.[4] To understand this point, consider a Brownian particle diffusing with diffusion coefficient $D = kT$, where the drag coefficient has been normalized, i.e. $\zeta = 1$. Assuming that the Brownian particle is subjected to a linear attracting force, $\mathbf{F} = -\mathbf{\Gamma} \cdot \mathbf{x}$, which tends to restore the equilibrium state $\mathbf{x} = \mathbf{0}$, the probability density satisfies the Fokker-Planck equation [3, 4],

$$\frac{\partial \Pi}{\partial t} + \nabla \cdot \mathbf{J} = 0; \quad \mathbf{J} = \mathbf{F}\Pi - kT \nabla \Pi, \tag{4.61}$$

with appropriate initial and boundary conditions. A typical example is a particle immersed in a linear shear flow and attached to the origin with a spring, so that $\mathbf{\Gamma}$ is the sum of a velocity gradient tensor $\mathbf{\Gamma}^{(1)}$, with $\mathrm{Tr}\{\mathbf{\Gamma}\} = 0$, and a spring constant $\mathbf{\Gamma}^{(2)} = K\mathbf{I}$. In this case, the force is the sum of a conservative and a non-conservative components, i.e. $\mathbf{F} = \mathbf{F}^{(c)} + \mathbf{F}^{(nc)}$, with $\mathbf{F}^{(c)} = -\mathbf{\Gamma}^{(s)} \cdot \mathbf{x}$ and $\mathbf{F}^{(nc)} = -\mathbf{\Gamma}^{(a)} \cdot \mathbf{x}$, where $\mathbf{\Gamma}^{(s)}$ and $\mathbf{\Gamma}^{(a)}$ are the symmetric and antisymmetric part of the $\mathbf{\Gamma}$ matrix, respectively. Naturally, if $\mathbf{\Gamma} = \mathbf{\Gamma}^{(s)}$, the applied force is conservative, i.e. $\mathbf{F} = -\nabla\phi$, with $\phi = \frac{1}{2} \sum_{ij} x_i \Gamma_{ij}^{(s)} x_j$ and all the results that we have obtained so far can be applied, with $\mathbf{\Gamma}^{(s)} = kT\mathbf{g}$. In particular, in this case the fluctuation-dissipation theorem (4.54), with $Q_{ij} = kT\delta_{ij}$, is identically satisfied, as well as the equipartition theorem, $\frac{1}{2} x_i \Gamma_{ij}^{(s)} x_j = \frac{1}{2}kT\delta_{ij}$.

As we saw in Sect. 3, an alternative approach to study Brownian motion is the Langevin equation (3.37),

$$\dot{\mathbf{x}} - \mathbf{F} = \mathbf{f}, \tag{4.62}$$

where $\dot{\mathbf{x}}$ represents the particle velocity, while $\mathbf{f}$ is the Brownian force. Now, since $\mathbf{J} = \Pi \dot{\mathbf{x}}$, from (4.61) we see that $\mathbf{f} = -kT \nabla \log P$, so that, integrating by parts we obtain:

$$\langle f_i x_j \rangle = kT\delta_{ij}, \tag{4.63}$$

---

[4]Since the FD theorem simply cannot be applied in this case, the word "violation" has been put between quotation marks in the title of this section.

where the bracket denotes the average of any function A of the random variable $\mathbf{x}$, i.e. $\langle A(\mathbf{x}) \rangle = \int A(\mathbf{x}) \Pi(\mathbf{x}) \, d\mathbf{x}$. Consequently, multiplying Eq. (4.62) by $\mathbf{x}$ and considering (4.63), we find:

$$(\Gamma_{ik} \sigma_{kj})^{sym} = kT \, \delta_{ij}, \tag{4.64}$$

with σ_{ij} indicating the correlation tensor, $\sigma_{ij} = \langle x_i x_j \rangle$. Therefore:

$$\Gamma_{ik} \sigma_{kj} = kT_{ij} = k(T\delta_{ij} + T_{ij}^{(a)}), \tag{4.65}$$

where $T_{ij}^{(a)}$ is an antisymmetric tensor. Obviously, the same result can be obtained solving the Fokker-Planck equation (4.61), obtaining at steady state the Gaussian distribution [2, 5],

$$P(\mathbf{x}) = C \exp\left(-\frac{\psi(\mathbf{x})}{kT}\right), \tag{4.66}$$

where C is a normalization factor, while

$$\psi = \frac{kT}{2} x_i \sigma_{ij}^{-1} x_j, \tag{4.67}$$

and σ_{ij} is related to Γ_{ij} through Eq. (4.65).

As we saw in Sect. 2.4, the fluctuation-dissipation theorem [see Eq. (2.41) with $\chi_{ij} = \Gamma_{ij}^{-1}$] is valid only when the antisymmetric temperature tensor is identically zero and the same is true for the equipartition theorem.

Now, consider as an example of application the 2D following form of the $\boldsymbol{\Gamma}$ matrix:

$$\boldsymbol{\Gamma} = \begin{pmatrix} a_1 & -b_2 \\ b_1 & a_2 \end{pmatrix}, \tag{4.68}$$

where a_1 and a_2 are both positive constant, i.e. there must be a recovering force pushing the particle back towards the origin. Then, we find the following solution of Eq. (4.64):

$$\sigma = \frac{kT}{a_+(a_1 a_2 + b_1 b_2)} \begin{pmatrix} a_2 a_+ + b_2 b_+ & (a_1 b_2 - a_2 b_1) \\ (a_1 b_2 - a_2 b_1) & a_1 a_+ + b_1 b_+ \end{pmatrix} \tag{4.69}$$

where $a_+ = a_1 + a_2$ and $b_+ = b_1 + b_2$. In addition, since the susceptibility χ_{ij} is defined such that $|x_i| = \chi_{ij} F_j$, in our case we obtain,

$$\chi = \boldsymbol{\Gamma}^{-1} = \frac{1}{(a_1 a_2 + b_1 b_2)} \begin{pmatrix} a_2 & -b_1 \\ b_2 & a_1 \end{pmatrix}. \tag{4.70}$$

At this point, we can determine the tensorial non equilibrium temperature (4.64), obtaining:

$$T_{ij} = T\left[\begin{pmatrix} 1 & 0 \\ 0 & 1 \end{pmatrix} + K\begin{pmatrix} 0 & -1 \\ 1 & 0 \end{pmatrix}\right]; \quad K = \frac{b_1 + b_2}{a_1 + a_2}. \tag{4.71}$$

Therefore, we confirm that T_{ij} is the sum of a symmetric isotropic tensor $T^{(s)} = T\delta_{ij}$ and an antisymmetric tensor. The fluctuation-dissipation theorem is valid only when the antisymmetric temperature tensor is identically zero which, in our case, requires that $b_1 = -b_2 = b$, i.e. Γ is a symmetric tensor, corresponding to the case of a particle attached to the origin through a spring and immersed in an elongational flow. In fact, in this case we find:

$$\sigma_{12} = -\frac{kTb}{a_1 a_2 - b^2}; \qquad \sigma_{11} = \frac{kTa_2}{a_1 a_2 - b^2}; \qquad \sigma_{22} = \frac{kTa_1}{a_1 a_2 - b^2}. \qquad (4.72)$$

showing that indeed $\sigma_{ij} = kT\Gamma_{ij}^{-1}$. Therefore, since $T_{12} = 0$, we may conclude that, as expected, both the fluctuation-dissipation theorem and the equipartition theorem are satisfied.

Now, let us consider two additional cases.

- $b_1 = b_2 = b$, corresponding to the case of a particle attached to the origin through a spring and immersed in a rotational flow. Then, if the spring recovering force is isotropic, i.e. $a_1 = a_2 = a$, we find: $\sigma_{12} = 0$ and $\sigma_{11} = \sigma_{22} = \sigma_{33} = kT/a$, showing that the equipartition theorem is satisfied. On the other hand, $T_{12} = -Tb/a$, showing that fluctuation-dissipation is violated. The fact that the equipartition theorem is satisfied is not surprising, as in this case $\Gamma_{ij}^{(s)} \Gamma_{ik}^{(a)} = 0$, i.e. conservative and non conservative forces are perpendicular to each other.
- $b_1 = 0$ and $b_2 = -b$, which corresponds to a particle attached to the origin though a spring and immersed in a simple shear flow along the x_1-direction. Again, for an isotropic spring, with $a_1 = a_2 = a$, we obtain:

$$\sigma_{12} = -\frac{kTb}{2a^2}; \qquad \sigma_{11} = kT\frac{2a^2 + b^2}{2a^3}; \qquad \sigma_{22} = \frac{kT}{a}. \qquad (4.73)$$

Therefore, since $T_{12} = Tb/2a$, neither the fluctuation-dissipation nor the equipartition theorems are satisfied.

4.6 Wiener's Process

The Wiener process takes its name from *Norman Wiener*, who studied it extensively. Basically, it corresponds to the simplest Fokker-Planck equation with no drift and normalized diffusivity, $D = 1/2$, i.e.,

$$\frac{\partial \Pi}{\partial t} = \frac{1}{2} \frac{\partial^2 \Pi}{\partial x^2}; \qquad -\infty < x < \infty, \qquad (4.74)$$

where $\Pi = \Pi(x, t|0)$ is the conditional probability that a random variable X equals x at time t, i.e. $X(t) = x$, assuming that $X(0) = 0$. This equation must be solved with initial condition,

$$\Pi(x, 0|0) = \delta(x), \qquad (4.75)$$

where $\delta(x)$ is the Dirac delta, and we require that $\Pi \to 0$ as $|x| \to \pm\infty$.

Now denote by $\widehat{\Pi}(k, t)$ the Fourier transform of $\Pi(x, t)$, i.e.,

$$\widehat{\Pi}(k, t) = \int_{-\infty}^{\infty} \Pi(x, t) e^{ikx} \, dx, \qquad (4.76)$$

where

$$\Pi(x,t) = \int_{-\infty}^{\infty} \Pi(k,t) e^{-ikx} \frac{dk}{2\pi}. \tag{4.77}$$

Fourier transforming Eq. (4.74) using the definition (4.76) and considering that $\widehat{\Pi}(k,0) = 1$, yields:

$$\widehat{\Pi}(k,t) = \exp\left(-\frac{1}{2}k^2 t\right), \tag{4.78}$$

and anti-transforming through (4.77), we obtain:

$$\Pi(x,t) = \frac{1}{\sqrt{2\pi t}} \exp\left(-\frac{x^2}{2t}\right). \tag{4.79}$$

This represents a Gaussian distribution,[5] with

$$\langle X(t)\rangle = 0; \qquad \langle X^2(t)\rangle = t, \tag{4.80}$$

indicating that an initially sharp distribution spreads in time, with a linearly growing variance. The Wiener process is often called simply Brownian motion, since it describes that process, as we saw in Sect. 3.1.

Again, note that, although the mean value of $X(t)$ is zero, its mean square diverges as $t \to \infty$. That means that the trajectories $X(t)$ are extremely variable. In fact, it can even be shown [2, Sect. 3.8.1] that they are non-differentiable, as the speed of a Brownian particle is almost certainly infinite. We can see that heuristically noting that, as $dX \propto \sqrt{dt}$, then $dX/dt \propto 1/\sqrt{dt}$ and therefore it diverges.

As we saw in Sect. 3.2 [cf. Eq. (3.51], the Wiener process corresponds to the simple Langevin equation,

$$\frac{dX}{dt} = \xi(t), \tag{4.81}$$

where

$$\langle \xi(t)\rangle = 0; \qquad \langle \xi(t)\xi(t')\rangle = \delta(t - t'). \tag{4.82}$$

Since the trajectories are continuous, they must be integrable; therefore, we obtain:

$$X(t) = \int_0^t \xi(s)\,ds, \tag{4.83}$$

or

$$dX(t) = \xi(t)\,dt. \tag{4.84}$$

In fact, since $X(t)$ is continuous, we know that its evolution is described through a Fokker-Planck equation, where the drift and diffusion terms can be determined considering that:

$$\langle X(t)\rangle = \int_0^t \langle \xi(s)\rangle\,ds = 0; \tag{4.85}$$

---

[5]Letting $t \to 0$ in (4.78) and considering the initial condition (4.75), we see that the Dirac delta is a distribution, corresponding to the limit of a series of normalized Gaussian functions as their variance tends to zero. In that limit, as expected, (4.78) shows that its Fourier transform is a constant.

and

$$\langle [X(t)]^2 \rangle = \int_0^t ds \int_0^t ds' \langle \xi(s)\xi(s') \rangle = \int_0^t ds \int_0^t ds' \delta(s-s') = t, \qquad (4.86)$$

so that at the end we obtain a Wiener process, as anticipated.

Thus, the integral of $\xi(t)$ is $X(t)$, which is itself not differentiable, as we have seen above, showing that, mathematically speaking, the Langevin equation does not make any sense. The reason why we were able to use it, obtaining correct results, is that its integral can be interpreted consistently, applying Eq. (4.83). This is the basic idea of stochastic integration, as described in Chap. 5.

From Eq. (4.84) we see intuitively that, as $\xi(t)$ oscillates around its zero mean much more rapidly than $X(t)$, we have:[6]

$$dX(t)^2 = dt. \qquad (4.87)$$

For an n-dimensional Wiener process, this result can be easily generalized as:

$$dX_i(t)\, dX_j(t) = \delta_{ij}\, dt. \qquad (4.88)$$

In practice, this relation simply shows that dX is an infinitesimal of order $1/2$. Accordingly, when random walk is simulated numerically, each increment ΔX_i during time Δt_i can be generated through the expression $\Delta X_i = \xi_i \sqrt{\Delta t_i}$, where ξ_i are random variables with $\langle \xi_i \rangle = 0$ and $\langle \xi_i^2 \rangle = 1$.

Comment The Wiener process is the basis of diffusion. One of its most important properties, due to the fact that the process is Markovian, is the statistical independence of any increment, $\Delta X_i = X(t_i) - X(t_{i-1})$, where $t_{i-1} < t_i$, from all other increments, ΔX_j, with $j \neq i$. Accordingly, the joint probability density for the ΔX_i's is equal to the product among the conditional probabilities (4.79) of each step:

$$\Pi(\Delta X_n; \Delta X_{n-1}; \ldots \Delta X_1|0) = \prod_{i=1}^n \frac{1}{\sqrt{2\pi \Delta t_i}} \exp\left(-\frac{\Delta X_i^2}{2\Delta t_i}\right). \qquad (4.89)$$

The independence of the increments ΔX_i from each other is important in the definition of stochastic integration (see next section).

4.7 Problems

Problem 4.1 Show that the Brownian motion of a free particle can be described through the so-called Kramers equation,

$$\dot{\Pi} + \frac{\partial J_v}{\partial v} = 0; \qquad J_v = -\frac{\zeta}{m} v\Pi - \frac{kT\zeta}{m^2}\frac{\partial \Pi}{\partial v}, \qquad (4.90)$$

where v is the velocity, while $\Pi = \Pi(v, t|v_0)$. Solve it and show that for long times Π tends to a Maxwellian distribution.

---

[6]A formal proof of this statement can be found in [2, Sect. 4.2.5].

Problem 4.2 Show that the phenomenological coefficients M and Q in the Kramers equation (4.90) can be obtained from Eqs. (3.52) and (3.53).

Problem 4.3 Consider the Brownian motion of a particle subjected to a linear external force $F = -Ax$. Show that for long timescales, $t \gg m/\zeta$, we can repeat the same analysis as in Problem 1, obtaining the so-called Smoluchowski equation:

$$\dot{\Pi} + \frac{\partial J_x}{\partial x} = 0; \qquad J_x = -\frac{A}{\zeta} x \Pi - \frac{kT}{\zeta} \frac{\partial \Pi}{\partial x}. \qquad (4.91)$$

Then, taking the limit when $A \to 0$, obtain the Stokes-Einstein result.

References

1. Chandrasekhar, S.: Rev. Mod. Phys. **15**, 1 (1943)
2. Gardiner, C.W.: Handbook of Stochastic Methods. Springer, Berlin (1985)
3. Leporini, D., Mauri, R.: Fluctuations of non-conservative systems. J. Stat. Mech., P03002 (2007)
4. Mauri, R., Leporini, D.: Violation of the fluctuation-dissipation theorem in confined driven colloids. Europhys. Lett. **76**, 1022–1028 (2006)
5. Uhlenbeck, G.E., Ornstein, L.S.: On the theory of the Brownian motion. Phys. Rev. **36**, 823–841 (1930)
6. Wax, N.: Noise and Stochastic Processes. Dover, New York (1954)

Chapter 5
Stochastic Differential Calculus

In this chapter, the basic concepts of stochastic integration are explained in a way that is readily understandable also to a non-mathematician.[1] The fact that Brownian motion, i.e. the Wiener process, is non-differentiable, and therefore requires its own rules of calculus, is explained in Sect. 5.1. In fact, there are two dominating versions of stochastic calculus, each having advantages and disadvantages, namely the Ito stochastic calculus (Sect. 5.2), based on a pre-point discretization rule, named after Kiyoshi Ito, and the Stratonovich stochastic calculus (Sect. 5.3), based on a mid-point discretization rule, developed simultaneously by Ruslan Stratonovich and Donald Fisk. Finally, in Sect. 5.4, we illustrate the main features of Stochastic Mechanics, showing that, by applying the rules of stochastic integration, the evolution of a random variable can be described through the Schrödinger equation of quantum mechanics.

5.1 Introduction

In the previous chapters, we have been concerned only with a very particular type of random process, namely the Ornstein-Uhlenbeck process, where the drift coefficient is linear and the diffusivity is constant. As we saw, that corresponds to the case of a system that has been slightly removed from its equilibrium state, so that it is subjected to a phenomenological linear force, trying to bring it back to equilibrium and, in addition, to the same fluctuating force existing at equilibrium. When we try to generalize these results, though, we run into problems.

Let us consider, for example, the Langevin equation,

$$\frac{dx}{dt} = V(x, t) + B(x, t)\xi(t), \tag{5.1}$$

---

[1]A more formal, yet still understandable, treatment can be found in [4, Sect. 4.3].

R. Mauri, *Non-Equilibrium Thermodynamics in Multiphase Flows*,
Soft and Biological Matter, DOI 10.1007/978-94-007-5461-4_5,
© Springer Science+Business Media Dordrecht 2013

where $V(x, t)$ is a phenomenological velocity, $\xi(t)$ is the normalized Brownian force (4.82), and consequently $B(x, t)$ is a sort of square root of the diffusivity. This equation can be written as:

$$dx = V[x(t), t] dt + B[x(t), t] dW(t), \tag{5.2}$$

where we have considered Eq. (4.84), i.e., $\xi(t)dt = dW(t)$, where $W(t)$ is a Wiener process. Now, integrating (5.2) with initial condition $x(0) = 0$, we obtain:

$$x(t) = \int_0^t V[x(s), s] ds + \int_0^t B[x(s), s] dW(s). \tag{5.3}$$

The first terms on the RHS is well-behaved; the second, however, needs some thinking. Basically, since the Wiener process is a succession of jumps of the random variable x, this equation does not tell us whether we should substitute in $B(x)$ the value of x just before the jump, or at some other instant of time. In fact, Ito's interpretation is that the value of x in $B(x)$ should be taken before the jump, while Stratonovich assumed that it should be taken halfway between its values before and after the jump.[2] Obviously, these two interpretations lead to two different Fokker-Planck equations. Conversely, if we assume that stochastic integration must lead to a given Fokker-Planck equation, describing a known macroscopic convection-diffusion process, then the convective term V must be different in the two interpretations. In any case, it is clear that Eq. (5.1) has no meaning if it is not associated to an integration rule since, as noted by van Kampen [11] "no amount of physical acumen suffices to justify a meaningless string of symbols."

5.2 Ito's Stochastic Calculus

When we discretize the time interval $(0-t)$, such that $t_i = i \Delta t$, with $i = 0, 1, \ldots, N$ and $t_n = t$, the differential equation (5.2) can be interpreted, according to *Ito*, applying the Cauchy-Euler iterative scheme:

$$x_{i+1} = x_i + V(x_i, t_i) \Delta t + B(x_i, t_i) \Delta W_i, \tag{5.4}$$

where $x_i = x(t_i)$ and $\Delta W_i = W(t_{i+1}) - W(t_i)$. Therefore, knowing x_i, we can calculate x_{i+1} by adding to a deterministic, drift term, $V(x_i, t_i) \Delta t$, a stochastic term, $B(x_i, t_i) \Delta W_i$. Here, ΔW_i is the increment of the Wiener process during the time interval $(t_i - t_{i+1})$ and as such, as we saw in Sect. 4.6, it is independent of the value x_i of the random variable at time t_i. Therefore, Ito's iterative solution (5.4) does not "look into the future", meaning that x_i is independent of ΔW_j for $j > i$, so that the

---

[2]A third approach to stochastic integration, based on a post-point discretization rule, was proposed by Peter Hänggi and Yuri Klimontovich to describe relativistic Brownian motion, but it is not considered here for sake of brevity.

problem is well posed and the solution can be formally obtained by letting the mesh size go to zero, i.e. when $\Delta t \to 0$. From that, we obtain the following formula,

$$\mathrm{I} \int_0^t B[x(s), s] \, dW(s) = \lim_{n \to \infty} \sum_{i=0}^n B(x_i) \, \Delta W_i, \qquad (5.5)$$

where the prefix "I" indicates *Ito*'s stochastic integration.

These stochastic integrals have special properties called Ito's isometries, namely:

$$\left\langle \mathrm{I} \int_0^t B[x(s), s] \, dW(s) \right\rangle = 0; \qquad (5.6)$$

and,

$$\left\langle \left(\mathrm{I} \int_0^t B[x(s), s] \, dW(s) \right)^2 \right\rangle = \mathrm{I} \int_0^t \langle B^2[x(s), s] \rangle \, ds. \qquad (5.7)$$

The first Ito's isometry can be easily demonstrated applying the definition (5.5), with:

$$\langle B(x_i, t_i) \, \Delta W_i \rangle = \langle B(x_i, t_i) \rangle \langle \Delta W_i \rangle = 0, \qquad (5.8)$$

where we have considered that ΔW_i, i.e. the Wiener increment during the time interval $(t_i - t_{i+1})$ is independent of B at time t_i. In similar fashion, the second Ito's isometry can be proven considering that $\langle \Delta W_i \Delta W_j \rangle = \delta_{ij} \Delta t$.

Now, we derive an important relation. First, expand an arbitrary function, $f[x(t), t]$, of the random variable $x(t)$, satisfying the Langevin equation (5.2), up to the second order in $dW(t)$,

$$df[x(t), t] = \frac{\partial f}{\partial t} dt + \frac{\partial f}{\partial x} dx(t) + \frac{1}{2} \frac{\partial^2 f}{\partial x^2} dx(t)^2 + \cdots, \qquad (5.9)$$

where $df[x(t), t] = f[x(t + dt), t + dt] - f[x(t), t]$ and the derivatives are taken at $[x(t), t]$. Then substituting (5.2) and considering (4.87), i.e.,

$$dx^2 = B^2 dW^2 + o(dt) = B^2 dt + o(dt), \qquad (5.10)$$

we obtain the so-called *Ito's lemma*:

$$df(x, t) = \left(\frac{\partial f}{\partial t} + V \frac{df}{dx} + \frac{1}{2} B^2 \frac{d^2 f}{dx^2} \right) dt + B \frac{df}{dx} dW. \qquad (5.11)$$

This formula shows that, applying Ito's stochastic differentiation rules, changing variables is not given by ordinary calculus. As a simple example of how this can be deceiving, consider that Ito's lemma (5.11) with $x = W$, $f = x^2$, $V = 0$ and $B = 1$, reduces to:

$$dW^2 = dt + 2W \, dW, \qquad (5.12)$$

so that we obtain by integration:

$$\mathrm{I} \int_0^t W(s)\,dW(s) = \frac{1}{2}\left[W^2(t) - t\right]. \tag{5.13}$$

Here, the second term on the RHS would be absent by the rules of standard calculus. Yet, this term must be present for consistency since, by the first Ito's isometry, the expectation of the left hand-side is zero, i.e. $\langle W\,dW \rangle = 0$. As we saw, this result is a direct consequence of Ito's stochastic integral formula (5.5); in fact, when it is applied to our case, we obtain: $\int W\,dW = \sum W_i \Delta W_i$, where W_i is evaluated at time t_i, while ΔW_i is the increment during the time interval $(t_i - t_{i+1})$ and, as such, being independent of W_i, it has zero mean.

An immediate consequence of Ito's lemma is the derivation of the Fokker-Planck equation. In fact, considering the average,

$$\langle f[x(t)] \rangle = \int_{-\infty}^{\infty} f(x) \Pi(x, t|0)\,dx, \tag{5.14}$$

and taking the time derivative, we obtain:

$$\frac{d}{dt}\langle f[x(t)] \rangle = \int_{-\infty}^{\infty} f(x) \frac{\partial \Pi(x, t|0)}{\partial t}\,dx. \tag{5.15}$$

On the other hand, averaging Eq. (5.11) and considering that $\langle dW \rangle = 0$ we have:

$$\frac{d}{dt}\langle f[x(t)] \rangle = \left\langle V\frac{df}{dx} + \frac{1}{2}B^2 \frac{d^2 f}{dx^2} \right\rangle = \int_{-\infty}^{\infty} \left(V\frac{df}{dx} + \frac{1}{2}B^2 \frac{d^2 f}{dx^2} \right) \Pi(x, t|0)\,dx.$$

Integrating by parts, the integral becomes:

$$\left[Vf\Pi + \frac{1}{2}B^2 \Pi \frac{\partial f}{\partial x} - \frac{1}{2}f \frac{\partial(B^2 \Pi)}{\partial x} \right]_{-\infty}^{\infty} + \int_{-\infty}^{\infty} f\left(-\frac{\partial(V\Pi)}{\partial x} + \frac{1}{2}\frac{\partial^2(B^2 \Pi)}{\partial x^2} \right)dx,$$

where the first term is zero, as Π tends exponentially to zero at infinity. Consequently,

$$\int_{-\infty}^{\infty} f\left(\frac{\partial \Pi}{\partial t} + \frac{\partial(V\Pi)}{\partial x} - \frac{1}{2}\frac{\partial^2(B^2 \Pi)}{\partial x^2} \right)dx = 0, \tag{5.16}$$

and therefore, as $f(x)$ is arbitrary, we obtain the Fokker-Planck equation,[3]

$$\frac{\partial \Pi}{\partial t} = -\frac{\partial(V\Pi)}{\partial x} + \frac{\partial^2(D\Pi)}{\partial x^2}. \tag{5.17}$$

where $D = \frac{1}{2}B^2$ is the diffusivity.

---

[3]This equation is also referred to as the *forward Kolmogorov equation*.

For many variable systems, the stochastic differential equation becomes:

$$dx = \mathbf{V}[\mathbf{x}(t), t]\, dt + \mathbf{B}[\mathbf{x}(t), t] \cdot d\mathbf{W}(t), \qquad (5.18)$$

where $d\mathbf{W}(t)$ describes an n-dimensional Wiener process (4.88), while the Fokker-Planck equation is:

$$\frac{\partial \Pi}{\partial t} + \nabla \cdot \mathbf{J} = 0, \qquad (5.19)$$

where $\mathbf{J}$ is the probability flux,

$$J_i = V_i \Pi - \sum_{j=1}^{n} \nabla_j (D_{ij} \Pi), \qquad (5.20)$$

where

$$D_{ij} = \frac{1}{2} \sum_{k=1}^{n} B_{ik} B_{kj} \qquad (5.21)$$

is the diffusivity tensor. Note that, since we obtain the same Fokker-Planck equation by replacing $\mathbf{B}$ by $\mathbf{B} \cdot \mathbf{U}$, where $\mathbf{U}$ is a unitary tensor with $\sum_l U_{il} U_{lj} = \delta_{ij}$,[4] we can assume without loss of generality that $\mathbf{B}$ is also symmetric, i.e. $B_{ij} = B_{ji}$ [5] and therefore the diffusivity tensor can be assumed to be symmetric. In fact, from (5.18), we see that $dx_i\, dx_j = 2D_{ij}\, dt$, with $D_{ij} = D_{ji}$.

5.3 Stratonovich's Stochastic Calculus

The choice of Ito's stochastic integration is not unique. In fact, *Stratonovich* defined the following alternative definition:

$$S \int_0^t B[x(s), s]\, dW(s) = \lim_{n \to \infty} \sum_{i=0}^{n} B\left(\frac{x_i + x_{i+1}}{2}, t_i \right) \Delta W_i, \qquad (5.22)$$

where the prefix "S" indicates Stratonovich's stochastic integration. Obviously, for well-behaved functions, the two definitions (5.5) and (5.22) would give the same result in the limit of $\Delta t \to 0$. Here, however, this is not so. For example, an obvious difference between the two stochastic integration rules is that, since in (5.22) B is not independent of ΔW, as in Ito's stochastic integration, Ito's isometries are not

---

[4]$\mathbf{U} \cdot \mathbf{v}$ describes a rigid rotation of the n-dimensional vector $\mathbf{v}$ that keeps its length unchanged. The transposed tensor, $\mathbf{U}^+$, represents a rotation in the opposite direction, so that when we apply both $\mathbf{U}$ and $\mathbf{U}^+$, the vector $\mathbf{v}$ returns to its original state.

[5]Actually, we can even find an appropriate unitary transformation so that $\sum_{l,m} U_{il} B_{lm} U_{mj} = \mathrm{diag}\{\lambda_1, \lambda_2, \ldots, \lambda_n\}$.

valid anymore when we follow Stratonovich's integration rules. Also, we should add that Stratonovich's approach is far more problematic than Ito's. In fact, the discretized equation (5.4) in Stratonovich's interpretation would become:

$$x_{i+1} = x_i + V(x_i, t_i)\Delta t + B\left(\frac{x_i + x_{i+1}}{2}, t_i\right)\Delta W_i, \qquad (5.23)$$

showing that B depends on the behavior of the Wiener process in the future and therefore, as it lacks the important property of the Ito integral of not "looking into the future", Stratonovich's integral cannot be used, for example, to predict stock price evolution in financial mathematics. In physics, however, stochastic differential equations are not a direct description of physical reality, but are instead coarse-grained versions of more microscopic models. In that respect, Stratonovich's interpretation appears to be the correct way to model such processes.

The connection between Ito's and Stratonovich's integrals can be derived expanding (5.22) as follows,

$$S\int_0^t B[x(s), s]\, dW(s) = \lim_{n\to\infty} \sum_{i=0}^n \left(B(x_i, t_i)\,\Delta W_i + \frac{1}{2}\frac{dB}{dx}\Delta x\, \Delta W_i\right). \qquad (5.24)$$

Then, considering that at leading order $dx = B\, dW$ and $dW^2 = dt$, we obtain:

$$S\int_0^t B(x, s)\, dW(s) = I\int_0^t B(x, s)\, dW(s) + \frac{1}{2}B(x, s)\frac{dB}{dx}(x, s)\, ds. \qquad (5.25)$$

Therefore, we see that the Stratonovich stochastic differential equation (SDE) $dx = V^S\, dt + B\, dW(t)$ leads to the same Fokker-Planck equation as the Ito SDE, $dx = V\, dt + B\, dW(t)$, provided that the convective terms are related as:

$$V^S = V - \frac{1}{2}B\frac{dB}{dx}. \qquad (5.26)$$

For systems with many variables, the Stratonovich SDE is:

$$d\mathbf{x} = \mathbf{V}^S(\mathbf{x}, t)\, dt + \mathbf{B}(\mathbf{x}, t)\cdot d\mathbf{W}(t), \qquad (5.27)$$

where the Stratonovich velocity $\mathbf{V}^S$ is related to its Ito counterpart through:

$$V_i^S = V_i - \frac{1}{2}\sum_{j,k} B_{kj}\nabla_j B_{ik}, \qquad (5.28)$$

and, as we saw in the previous section, we can assume, without loss of generality, that $\mathbf{B}$ is symmetric, i.e. $B_{ij} = B_{ji}$.

Using these expressions, the Fokker-Planck equation can be rewritten in terms of the Stratonovich velocity, obtaining:

$$\frac{\partial \Pi}{\partial t} = -\frac{\partial(V^S\Pi)}{\partial x} + \frac{1}{2}\frac{\partial}{\partial x}\left(B\frac{\partial}{\partial x}(B\Pi)\right), \qquad (5.29)$$

and for many variable systems,

$$\frac{\partial \Pi}{\partial t} + \nabla \cdot \mathbf{J}^\Pi = 0, \tag{5.30}$$

where

$$J_i^\Pi = V_i^S \Pi - \frac{1}{2} \sum_{j,k=1}^n B_{ik} \nabla_j (B_{kj} \Pi). \tag{5.31}$$

Finally, converting Ito's lemma (5.11), we can show that the rule for a change of variables in Stratonovich SDE is exactly the same as in ordinary calculus, i.e. [4, Sect. 4.3.6],

$$df(x) = \frac{df}{dx} dx = \frac{df}{dx}(x)\left[V^S(x,t) dt + B(x,t) dW(t)\right]. \tag{5.32}$$

Consequently, Stratonovich's stochastic integrals yield physically meaningful results. For example, the integral (5.13) gives:

$$\mathrm{S}\int_0^t W(s) dW(s) = \frac{1}{2} W^2(t), \tag{5.33}$$

as one would expect.

5.4 Stochastic Mechanics

In this section we follow the derivation of stochastic mechanics by E. Nelson [8]. Consider a Brownian particle, subjected to a conservative force $\mathbf{F} = -\nabla V$, and diffusing with diffusion coefficient $D = kT$, where the drag coefficient has been normalized. As we saw, $\mathbf{x}(t)$ is a smooth diffusion process with,

$$\langle dx_i(t) \rangle = F_i dt, \tag{5.34}$$

and

$$dx_i(t) dx_j(t) = 2D\delta_{ij} dt. \tag{5.35}$$

Notice that, in agreement with (4.88), in (5.35) there is no conditional expectation.

One way to describe this motion is to apply the forward Langevin equation,

$$\dot{\mathbf{x}}^+ - \mathbf{v}^+ = \mathbf{f}, \tag{5.36}$$

where $\mathbf{v}^+ = \langle \dot{\mathbf{x}}^+ \rangle$ represents the mean forward particle velocity, $\mathbf{f}$ is the Brownian force, while the forward derivative has been defined as:

$$\dot{\mathbf{x}}^+ = \frac{D^+}{Dt} \mathbf{x}(t) = \lim_{\Delta t \to 0} \frac{\mathbf{x}(t + \Delta t) - \mathbf{x}(t)}{\Delta t}. \tag{5.37}$$

As we saw, another way to describe this process is by defining the probability density Π, satisfying the forward Fokker-Planck equation,

$$\frac{\partial \Pi}{\partial t} + \nabla \cdot \mathbf{J}^{+} = 0; \quad \mathbf{J}^{+} = \mathbf{F}\Pi - D\nabla\Pi. \tag{5.38}$$

Therefore, we see that the force $\mathbf{F}$ coincides with the mean forward velocity (4.7), i.e. $\mathbf{v}^{+} = \mathbf{F}$.

Now, let us consider the backward Langevin equation,

$$\dot{\mathbf{x}}^{-} - \mathbf{v}^{-} = \mathbf{f}, \tag{5.39}$$

where $\mathbf{v}^{-} = \langle \dot{\mathbf{x}}^{-} \rangle$ represents the mean backward particle velocity while the backward derivative has been defined as:

$$\dot{\mathbf{x}}^{-} = \frac{D^{-}}{Dt}\mathbf{x}(t) = \lim_{\Delta t \to 0} \frac{\mathbf{x}(t) - \mathbf{x}(t - \Delta t)}{\Delta t}. \tag{5.40}$$

Clearly, this equation can be obtained from the forward Langevin equation by replacing Δt with $-\Delta t$. Identical results could be obtained starting from the backward Fokker-Planck equation,

$$\frac{\partial \Pi}{\partial t} + \nabla \cdot \left(\mathbf{v}^{-}\Pi + D\nabla\Pi\right) = 0. \tag{5.41}$$

Comparing the above relations, we see that $\mathbf{v}^{-} = \mathbf{F} - 2D\nabla \log \Pi$.

Now, define the current velocity, $\mathbf{v}$, and the osmotic velocity, $\mathbf{u}$, as,

$$\mathbf{v} = \frac{1}{2}\left(\mathbf{v}^{+} + \mathbf{v}^{-}\right); \quad \mathbf{u} = \frac{1}{2}\left(\mathbf{v}^{+} - \mathbf{v}^{-}\right). \tag{5.42}$$

Clearly, that means:

$$\mathbf{v} = \mathbf{F} - D\nabla \log \Pi; \quad \mathbf{u} = D\nabla \log \Pi. \tag{5.43}$$

Substituting the first of the above equations into the forward Fokker-Plank equation (5.38), we see that we obtain the continuity equation,

$$\frac{\partial \Pi}{\partial t} + \nabla \cdot (\mathbf{v}\Pi) = 0. \tag{5.44}$$

In Nelson's notation [7], defining

$$\mathbf{u} = 2D\nabla R; \quad \text{i.e.} \quad \Pi = e^{2R}, \tag{5.45}$$

and

$$\mathbf{v} = 2D\nabla S, \tag{5.46}$$

the continuity equation yields,

$$\frac{\partial R}{\partial t} = -D\nabla^2 S - 2D(\nabla S) \cdot (\nabla R), \tag{5.47}$$

that is, taking the gradient,[6]

$$\frac{\partial \mathbf{u}}{\partial t} = -D\nabla(\nabla \cdot \mathbf{v}) - \nabla(\mathbf{v} \cdot \mathbf{u}). \tag{5.48}$$

Another independent equation was derived by Nelson imposing that $\mathbf{F} = m\mathbf{a}$, where m is the mass of the particle and $\mathbf{a}$ its acceleration, i.e. the second time-derivative of $\mathbf{x}(t)$,

$$\mathbf{a}(t) = \frac{1}{2}\left[\left\langle\frac{D^+}{Dt}\right\rangle\left\langle\frac{D^-}{Dt}\right\rangle + \left\langle\frac{D^-}{Dt}\right\rangle\left\langle\frac{D^+}{Dt}\right\rangle\right]\mathbf{x}(t). \tag{5.49}$$

Here, we have considered that acceleration is time-reversible, and we have defined the mean forward stochastic derivative,

$$\left\langle\frac{D^+}{Dt}\right\rangle = \frac{\partial}{\partial t} + \mathbf{v}^+ \cdot \nabla + D\nabla^2 \tag{5.50}$$

and the mean backward stochastic derivative,

$$\left\langle\frac{D^-}{Dt}\right\rangle = \frac{\partial}{\partial t} + \mathbf{v}^- \cdot \nabla - D\nabla^2. \tag{5.51}$$

Note that, when we apply these operators to the particle position vector $\mathbf{x}$ we find again $\mathbf{v}^+ = \langle\dot{\mathbf{x}}^+\rangle$ and $\mathbf{v}^- = \langle\dot{\mathbf{x}}^-\rangle$, while when we apply the mean backward stochastic derivative and the mean forward stochastic derivative to the probability density Π, we obtain the backward Fokker-Planck equation and the forward Fokker-Planck equation, respectively. So, at the end, we obtain Nelson's definition of acceleration:

$$\mathbf{a} = \frac{1}{2}\left\langle\frac{D^-}{Dt}\right\rangle\mathbf{v}^+ + \frac{1}{2}\left\langle\frac{D^+}{Dt}\right\rangle\mathbf{v}^-, \tag{5.52}$$

i.e.,

$$\mathbf{a} = \frac{1}{2}\left(\frac{\partial}{\partial t} + \mathbf{v}^- \cdot \nabla - D\nabla^2\right)\mathbf{v}^+ + \frac{1}{2}\left(\frac{\partial}{\partial t} + \mathbf{v}^+ \cdot \nabla + D\nabla^2\right)\mathbf{v}^-, \tag{5.53}$$

and finally:

$$\mathbf{a} = \frac{1}{m}\mathbf{F} = \frac{\partial \mathbf{v}}{\partial t} + \mathbf{v} \cdot \nabla\mathbf{v} - \mathbf{u} \cdot \nabla\mathbf{u} - D\nabla^2\mathbf{u}. \tag{5.54}$$

---

[6]Note that, since $\mathbf{v}$ is a potential vector field, $\nabla(\nabla \cdot \mathbf{v}) = \nabla^2\mathbf{v}$.

Now, considering that $\mathbf{F} = -\nabla V$, $\mathbf{v} \cdot \nabla \mathbf{v} = \frac{1}{2} \nabla v^2$ and $\mathbf{u} \cdot \nabla \mathbf{u} = \frac{1}{2} \nabla u^2$, (in fact, $\mathbf{v}$ and $\mathbf{u}$ are potential vector fields) this equation is equivalent to:

$$\frac{\partial S}{\partial t} + D\nabla S \cdot \nabla S - D\nabla R \cdot \nabla R - kT\nabla^2 R = -\frac{V}{2Dm}. \tag{5.55}$$

Here comes the interesting part. Assume that the diffusion coefficient has the following form:

$$D = \frac{\hbar}{2m}, \tag{5.56}$$

where $\hbar = h/2\pi$ and h is the Planck constant. Then, denoting by $\rho = \Pi$ the probability density, define

$$\psi = e^{R+iS} = \sqrt{\rho}e^{iS}, \tag{5.57}$$

and impose that ψ satisfies the Schrödinger equation,

$$i\hbar\frac{\partial \psi}{\partial t} = -\frac{\hbar^2}{2m}\nabla^2\psi + V\psi. \tag{5.58}$$

It is easy to see that R and S satisfy Eqs. (5.47) and (5.55) (and, of course, ρ satisfies the continuity equation).

This result is equivalent to Bohm's formulation of quantum mechanics [2], where Eq. (5.55) can be considered as a Bernoulli (or Hamilton-Jacobi) equation for a frictionless compressible fluid,

$$\frac{\hbar}{m}\frac{\partial S}{\partial t} + \frac{1}{2}v^2 + \frac{1}{m}V - U_B = 0, \tag{5.59}$$

where

$$U_B = \frac{1}{2}u^2 + \frac{\hbar}{2m}\nabla \cdot \mathbf{u} = \frac{\hbar^2}{4m^2}\left[\nabla^2 \ln \rho + \frac{1}{2}(\nabla \rho)^2\right] \tag{5.60}$$

is Bohm's quantum potential energy, equal to the sum of a specific dilatation energy and a diffusion kinetic energy. In Bohm's theory, the quantum potential is the origin of the effect by which the wave function guides the motion of the particle, as was proposed in de Broglie's pilot-wave theory. The claim that Nelson's formulation provides an alternative to Bohm's that is realist but without a dualist ontology rests on the claim that this term arises from the stochastic fluctuations of the particle.

Another way of looking at this problem is to start from Madelung's hydrodynamic description of quantum mechanics [10]. Then, Π plays the role of the fluid number density, $\Pi = \rho$, while $\mathbf{v}$ and $\mathbf{u}$ are its mean and fluctuating velocity, respectively. So, mass conservation yields the continuity equation (5.44),

$$\frac{\partial \rho}{\partial t} + \nabla \cdot (\mathbf{v}\rho) = 0, \tag{5.61}$$

while energy conservation imposes that the total mechanical energy of this friction-less "fluid", $\mathcal{H}$, is conserved, where,

$$\mathcal{H} = \int \rho(\mathbf{x}) \, H(\mathbf{x}) \, d\mathbf{x}, \tag{5.62}$$

is the Hamiltonian, with

$$H = \frac{1}{2} m \left(v^2 + u^2 \right) + V \tag{5.63}$$

denoting the energy of a single particle. Here, the $(\mathbf{v} \cdot \mathbf{u})$ term has been omitted since it is not invariant under time reversal.[7] At this point, imposing that the total energy is conserved, i.e. $\dot{\mathcal{H}} = 0$, we obtain again Eq. (5.55). This approach has been clarified mathematically by Guerra and Morato [5, 6], using a stochastic minimization condition.

We saw that Nelson's stochastic mechanics leads to predictions that agree with those of standard quantum mechanics and are confirmed by experiments. The underlying fundamental assumption is that the interaction with a background field causes the system to undergo a diffusion process, with diffusion coefficient $\hbar/2m$. So, "had the Schrödinger equation been derived from stochastic mechanics, the history and conceptual foundations of modern physics would have been different." (Nelson, 1985) Yet, it is not clear whether quantum mechanics and stochastic mechanics could agree on everything, for example when dealing with the measurement problem [9], the commutability among operators [3] or if a separate quantization condition, as in the old quantum theory, should be added separately [12]. In fact, since Nelson's paper in 1969, many researchers have lost their sleep in trying to understand whether this is all a coincidence or there is something fundamental underneath. A review of this subject can be found in the work by Adler [1].

Comment 5.1 In general, when in statistical mechanics we write Eq. (5.62), we expect that $H(\mathbf{x})$, being the energy of a particle at $\mathbf{x}$, is a property of an individual trajectory, so that the properties of the ensemble are reflected completely in the linear explicit dependence of the Hamiltonian on ρ. This expectation would seem to rule out H being itself a function of ρ, while in stochastic mechanics we saw that H depends on $\mathbf{u}$, which in turn depends on $\ln \rho$. This makes clear that Nelson's is not simply a theory of an ensemble of particles undergoing Brownian motion, as it costs a lot of energy for a particle to be at a point where the gradient of the probability density—in the ensemble of which it is a member—is large. So, perhaps, since the same thing happens in the mean field theory, which is a coarse-grained approximation of a more microscopic physical reality, quantum mechanics could be interpreted in the same way, assuming that the quantum fluid can interact with itself non-locally.

---

[7]In Nelson's formulation the condition of time reversal invariance is imposed in the definition (5.49) of the stochastic acceleration.

Comment 5.2 Nelson [8] writes: "Let me remark that I have no evidence for the background field hypothesis (if I did, I would gladly sacrifice an ox). This hypothesis is in no way used in the mathematical development of stochastic mechanics, but I believe it to be essential for a physical understanding of the theory."

5.5 Problems

Problem 5.1 Show that:

$$I \int_0^t W^3(s)\,dW(s) = \frac{1}{4}W^4(t) - \frac{3}{2}\int_0^t W^2(s)\,ds. \tag{5.64}$$

Problem 5.2 Show that:

$$I \int_0^t W^n(s)\,dW(s) = \frac{1}{n+1}W^{n+1}(t) - \frac{n}{2}\int_0^t W^{n-1}(s)\,ds. \tag{5.65}$$

Problem 5.3 The price $S(t)$ of a share of stock grows at a constant rate, r, so that $dS/dt = rS$, or $dS/S = r\,dt$. Then, to model the volatility of the market, we add a white noise, so that:

$$\frac{dS}{S} = r\,dt + \sigma\,dW,$$

where W is a Wiener process. Solve this differential equation by Ito's rule.

References

1. Adler, L.: Quantum Theory as an Emergent Phenomenon. Cambridge University Press, Cambridge (2004)
2. Bohm, D.: Phys. Rev. **85**, 166 (1952)
3. Correggi, M., Morchio, G.: Ann. Phys. **296**, 371 (2002)
4. Gardiner, G.W.: Handbook of Stochastic Methods. Springer, Berlin (1985)
5. Guerra, F.: Phys. Rep. **77**, 263 (1981)
6. Guerra, F., Morato, L.M.: Phys. Rev. D **27**, 1774 (1983)
7. Nelson, E.: Phys. Rev. **150**, 1079 (1969)
8. Nelson, E.: Quantum Fluctuations. Princeton University Press, Princeton (1985)
9. Pavon, M.: J. Math. Phys. **40**, 5565 (1999)
10. Tessarotto, M., Ellero, M., Nicolini, P.: Phys. Rev. A **75**, 012105 (2007)
11. van Kampen, N.G.: Stochastic Processes in Physics and Chemistry, p. 245. North-Holland, Amsterdam (1981)
12. Wallstrom, T.C.: Phys. Rev. A **49**, 1613 (1994)

Chapter 6
Path Integrals

In the previous chapters we saw that stochastic processes can be described using two equivalent approaches, one Lagrangian, leading to the Langevin equation, the other Eulerian, exemplified in the Fokker-Planck equation. Both descriptions allow to determine the stochastic properties of a system, provided that these properties are known at an earlier time. In addition, though, a third approach exists, where the evolution of a system in time is formulated by writing down the probability of observing a trajectory, or "path", of its macroscopic variables. The first successful attempt to define path integration is due to Norbert Wiener, who in 1921 replaced the classical notion of a single, unique trajectory of a Brownian particle with a sum, or functional integral, over an infinity of possible trajectories, to compute the probability distribution describing the diffusion process [10]. In a second development, this concept was applied to quantum mechanics, first by Paul Dirac [2] and then by Richard Feynman [3], expressing the propagator of the Schrödinger equation in terms of a complex-valued path integral.

The goal of this chapter is to develop Wiener's path integral formulation of stochastic processes, with particular emphasis on its application to non-equilibrium thermodynamics. After deriving in Sect. 6.1 the path integral for the propagator of a free Brownian particle, its extension to a particle immersed in a force field is determined in Sect. 6.2. Then, in Sect. 6.3, we define the minimum path and the quadratic approximation, while, in Sect. 6.4, a few examples of applications are illustrated.

6.1 Free Brownian Motion

The behavior of a Brownian particle is described in terms of the conditional probability $\Pi(\mathbf{x}, t|\mathbf{x}_0)$ that the particle is located on position $\mathbf{x}$ at time t, provided that at time $t = 0$ it was located at $\mathbf{x}_0$. In the absence of any external force field or boundaries, Π satisfies Wiener's diffusion equation (4.74),

$$\frac{\partial \Pi}{\partial t} - D\nabla^2 \Pi = \delta(\mathbf{x} - \mathbf{x}_0)\,\delta(t), \qquad (6.1)$$

R. Mauri, *Non-Equilibrium Thermodynamics in Multiphase Flows*,
Soft and Biological Matter, DOI 10.1007/978-94-007-5461-4_6,
© Springer Science+Business Media Dordrecht 2013

where D is the particle diffusivity, $\nabla = \partial/\partial \mathbf{r}$ and $\nabla^2 = \nabla \cdot \nabla$. The solution of this problem is the Gaussian distribution (4.79),

$$\Pi(\mathbf{x}, t|\mathbf{x}_0) = (4\pi Dt)^{-3/2} \exp\left[-\frac{|\mathbf{x} - \mathbf{x}_0|^2}{4Dt}\right]. \tag{6.2}$$

Now, the Brownian particle travels from $\mathbf{x}_0$ at time 0 to $\mathbf{x}$ at time t through a series of intermediate steps which define a "path" $\mathbf{x}(\tau)$. So, let us divide the time interval $(0 - t)$ into $N + 1$ equal interval of length ϵ, separated by time points $t_1 < t_2 < \cdots < t_N$, with $t_0 = 0$ and $t_{N+1} = t$. Since the diffusion process is Markovian (see Sect. 4.1), the joint probability distribution to find a particle (which starts at $\mathbf{x}_0$ at time 0) at positions $\mathbf{x}_1$ at time t_1, $\mathbf{x}_2$ at time t_2, ..., $\mathbf{x}_N$ at time t_N and $\mathbf{x}$ at time t, is equal to the product of the propagators (6.2) for each step, i.e.

$$\Pi(\mathbf{x}, t; \mathbf{x}_N, t_N; \mathbf{x}_{N-1}, t_{N-1}; \ldots \mathbf{x}_1, t_1|\mathbf{x}_0)$$

$$= \prod_{j=1}^{N+1} \Pi(\mathbf{x}_j, t_j - t_{j-1}|\mathbf{x}_{j-1})$$

$$= (4\pi D\epsilon)^{-3(N+1)/2} \exp\left[-\frac{1}{4D\epsilon}\sum_{j=1}^{N+1}(\mathbf{x}_j - \mathbf{x}_{j-1})^2\right], \tag{6.3}$$

with $\mathbf{x}_{N+1} = \mathbf{x}$. Therefore, when $N \to \infty$, this represents the probability for the particle to follow the particular path $\mathbf{x}(\tau)$ from $\mathbf{x}(0) = \mathbf{x}_0$ to $\mathbf{x}(t) = \mathbf{x}$, which is specified by the intermediate points $\mathbf{x}(t_j) = \mathbf{x}_j$, for $j = 1, 2, \ldots, N$. In that limit, it is customary to write the exponential in (6.3) in the continuous notation,

$$\lim_{\epsilon \to 0} \exp\left[-\frac{1}{4D\epsilon}\sum_{j=1}^{N+1}(\mathbf{x}_j - \mathbf{x}_{j-1})^2\right] = \exp\left[-\frac{1}{4D}\int_0^t \left(\frac{d\mathbf{x}}{d\tau}\right)^2 d\tau\right]. \tag{6.4}$$

Naturally, when we integrate (6.3) over all the possible intermediate coordinates $\mathbf{x}_1, \mathbf{x}_2, \ldots, \mathbf{x}_N$, we must recover the original conditional probability distribution (6.2). In this way, we derive the so-called Wiener (path) integral, i.e.,

$$\Pi(\mathbf{x}, t|\mathbf{x}_0) = \int_{\mathbf{x}_0, 0}^{\mathbf{x}, t} \exp\left[-\frac{1}{4D}\int_0^t \left(\frac{d\mathbf{x}}{d\tau}\right)^2 d\tau\right]\mathcal{D}\mathbf{x}(t), \tag{6.5}$$

where $\Pi(\mathbf{x}, t|\mathbf{x}_0)$ is the Gaussian distribution (6.2). Here, the second member represents integration over all the possible paths connecting $\mathbf{x}_0$ at time $t = 0$ to $\mathbf{x}$ at time t, including multiplication by the normalization factor, that is formally:

$$\int_{\mathbf{x}_0, 0}^{\mathbf{x}, t} \mathcal{D}\mathbf{x}(t) = \lim_{N \to \infty}(4\pi D\epsilon)^{-3(N+1)/2}\int d\mathbf{x}_1 \int d\mathbf{x}_2 \ldots \int d\mathbf{x}_N. \tag{6.6}$$

A simple generalization of the Wiener integral arises in the case of a dilute chemical species that diffuses in a medium where it undergoes a first-order chemical reaction. Then the particle concentration $c(\mathbf{x}, t)$ satisfies the equation,

$$\frac{\partial c}{\partial t} = D\nabla c - kc, \tag{6.7}$$

where $k = k(\mathbf{x})$ is the reaction constant, expressing the probability, per unit time, that the Brownian particle reacts and disappears. Accordingly, the probability that a Brownian particle will survive, without being reabsorbed, as it moves along an arbitrary particle path $\mathbf{x}(\tau)$, with $\mathbf{x}(0) = \mathbf{x}_0$ and $\mathbf{x}(t) = \mathbf{x}$, equals

$$\exp\left[-\int_0^t k(\mathbf{x}(\tau)) \, d\tau\right]. \tag{6.8}$$

The propagator $G(\mathbf{x}, t|\mathbf{x}_0)$ of Eq. (6.7) is therefore the product of (6.5) and (6.8), i.e.,

$$G(\mathbf{x}, t|\mathbf{x}_0) = \int_{\mathbf{x}_0, 0}^{\mathbf{x}, t} \exp\left[-\frac{1}{4D}\int_0^t \left(|\dot{\mathbf{x}}|^2 - 4Dk\right) d\tau\right] \mathcal{D}\mathbf{x}(t). \tag{6.9}$$

Note that, unlike the propagator of the diffusion equation, which coincides, within a constant factor, with the conditional probability, the propagator of the diffusion-reaction equation differs from the related conditional probability, as its volume integral changes with time [1].

6.2 Brownian Motion in a Field of Force

When a Brownian particle is subjected to a force $\mathbf{F}(\mathbf{x})$, the conditional probability $\Pi(\mathbf{x}, t|\mathbf{x}_0)$ satisfies the Fokker-Planck equation (4.22), i.e.,

$$\dot{\Pi} + \nabla \cdot \mathbf{J} = \delta(\mathbf{x} - \mathbf{x}_0)\delta(t), \tag{6.10}$$

where $\mathbf{J}$ is the probability flux,

$$\mathbf{J} = \mathbf{V}\Pi - D\nabla\Pi, \quad \mathbf{V} = \zeta^{-1}\mathbf{F}, \tag{6.11}$$

with ζ denoting the drag coefficient. The most straightforward way to derive a path integral for this process is to consider the following transformation,[1]

$$\Pi(\mathbf{x}, t|\mathbf{x}_0) = G(\mathbf{x}, t|\mathbf{x}_0) \exp\left(\frac{1}{2D}\int_{\mathbf{x}_0, 0}^{\mathbf{x}, t} \mathbf{V} \cdot d\mathbf{x}\right), \tag{6.12}$$

---

[1] See [9, pp. 20–21].

where the line integral follows a path starting at $\mathbf{x}_0$ and ending at $\mathbf{x}$ after a time t. Now, let us confine ourselves to the case of conservative forces, so that $\mathbf{V}$ can be written as the gradient of a scalar potential $\phi(\mathbf{x})$,

$$\mathbf{V} = -\nabla\phi. \tag{6.13}$$

In this case, the line integral in (6.12) reduces to $\phi(\mathbf{x}) - \phi(\mathbf{x}_0)$, i.e. it depends only on the end points of the line integral and not on the specific form of the path. Then we see that $G(\mathbf{x}, t|\mathbf{x}_0)$ satisfies the diffusion-reaction equation (6.7), with:

$$k = \frac{1}{4D}(V^2 + 2D\nabla \cdot \mathbf{V}). \tag{6.14}$$

Therefore, substituting Eqs. (6.9) into (6.12), and considering that along a path $d\mathbf{x} = \dot{\mathbf{x}}\,d\tau$, we obtain:

$$\Pi(\mathbf{x}, t|\mathbf{x}_0) = \int_{\mathbf{x}_0,0}^{\mathbf{x},t} \exp\left[-\frac{1}{4D}\int_0^t (|\dot{\mathbf{x}} - \mathbf{V}|^2 + 2D\nabla \cdot \mathbf{V})\,d\tau\right]\mathcal{D}\mathbf{x}(t). \tag{6.15}$$

As shown by Graham [6, and references therein], this expression remain true even when the external force field is non-potential and can be generalized to the case when the diffusivity is a position-dependent tensor.

The same result can be obtained [4] from the Langevin equation,

$$\zeta_{ij}(\dot{x}_j - V_j) = f_i, \tag{6.16}$$

where ζ_{ij} is the resistance dyadic. Here, the Brownian force $\mathbf{f}$ results from the sum of a large number of collisions of the particle with the surrounding fluid, each occurring randomly and independently of the others, so that:

$$\langle f_i(t)\rangle = 0; \qquad \langle f_i(t)f_j(t+\tau)\rangle = 2kT\zeta_{ij}\delta(\tau). \tag{6.17}$$

Applying the central limit theorem this result can be generalized, obtaining that the probability of observing a certain Brownian force function $\mathbf{f}(t)$ is the following Gaussian distribution,

$$\Pi[\mathbf{f}(t)] \propto \exp\left[-\frac{1}{2}\iint [f_i(t)B_{ij}(\tau)f_j(t+\tau)]\,dt\,d\tau\right], \tag{6.18}$$

where $\mathbf{B}$ is a sort of inverse of the variance of the process,

$$B_{ij}(\tau) = \langle f_i(t)f_j(t+\tau)\rangle^{-1} = \frac{1}{2kT}\zeta_{ij}^{-1}\delta(\tau). \tag{6.19}$$

Now, since $\mathbf{f}(t)$ and $\mathbf{x}(t)$ are linearly related through the Langevin equation, the probability $\Pi[\mathbf{x}(t)]$ that the particle follows the path $\mathbf{x}(t)$ is proportional to $\Pi[\mathbf{f}(t)]$. Consequently, substituting Eqs. (6.16) and (6.19) into (6.18), we obtain:

$$\Pi[\mathbf{x}(t)] = G(\mathbf{x}, t|\mathbf{x}_0)\exp\left(-\frac{1}{4kT}\int_{\mathbf{x}(t)}(\dot{x}_i - V_i)\zeta_{ij}(\dot{x}_j - V_j)\,dt\right), \tag{6.20}$$

where the normalizing term G is a Jacobian, depending only on the end points. When diffusivity is constant, Graham[2] showed that Eq. (6.20) must be interpreted following the Stratonovich rule of stochastic integration (see Sect. 5.3), obtaining:[3]

$$G(\mathbf{x}, t|\mathbf{x}_0) = \exp\left[-\frac{1}{2}\int_0^t (\nabla \cdot \mathbf{V})\, dt\right]. \tag{6.21}$$

Finally, the conditional probability $\Pi(\mathbf{x}, t|\mathbf{x}_0)$ that the particle moves from $\mathbf{x}_0$ at time $t = 0$ to $\mathbf{x}$ at time t will be equal to the sum of the contributions (6.20) of all paths connecting the two events, obtaining

$$\Pi(\mathbf{x}, t|\mathbf{x}_0) = \int \exp\left[-\frac{S[\mathbf{x}(\tau)]}{4kT}\right]\mathcal{D}[\mathbf{x}(\tau)], \tag{6.22}$$

where the integral is taken over all paths[4] such that $\mathbf{x}(0) = \mathbf{x}_0$ and $\mathbf{x}(t) = \mathbf{x}$, with,

$$S[\mathbf{x}(t)] = \int_0^t \mathcal{L}[\mathbf{x}(\tau), \tau]\, d\tau, \tag{6.23}$$

and,

$$\mathcal{L}[\mathbf{x}(\tau), \tau] = \sum_{ij}(\dot{x}_i - V_i)\zeta_{ij}(\dot{x}_j - V_j) + 2kT\sum_i \nabla_i V_i. \tag{6.24}$$

Clearly, in the isotropic case, when $\zeta_{ij} = \zeta\delta_{ij}$ and $D = kT/\zeta$, we find again Eq. (6.15).

When the last term in (6.24) can be neglected, then $\mathcal{L}[\mathbf{x}(\tau), \tau]$ is the rate of energy dissipation at time τ and $S[\mathbf{x}(t)]$ coincides with the energy dissipated along the trajectory $\mathbf{x}(\tau)$ during the time interval t. This is particularly true for linear velocity fields, when $\mathcal{L}$ is referred to as the Onsager-Machlup function and the above relations lead to a principle of least energy dissipation [7].

6.3 Minimum Path

Among all paths, let us denote by $\mathbf{y}(\tau)$ the one that minimizes S. According to the Hamilton-Jacobi formalism of classical mechanics (see Sects. D.3 and D.4), the momentum $\mathbf{p}$ along the minimum path can be defined as

$$p_i = \left[\frac{\partial \mathcal{L}}{\partial \dot{x}_i}\right]_{\mathbf{x} = \mathbf{y}} = 2\sum_j \zeta_{ij}(\dot{y}_j - V_j). \tag{6.25}$$

---

[2]Graham [6] also showed that when diffusivity depends on position, the exponent in Eq. (6.21) contains two more terms, depending on the gradients of ζ_{ij}.

[3]Had we followed Ito's rule of stochastic integration, (see Sect. 5.2), we would find $G = 1$, which is not correct.

[4]The path integral can be defined rigorously even when the paths $\mathbf{x}(\tau)$ are not continuous functions; see [5].

Now, defining the "Hamiltonian" $\mathcal{H}$ (in reality, $\mathcal{H}$ has the units of an energy per unit time) as $\mathcal{H} = \mathbf{p} \cdot \dot{\mathbf{y}} - \mathcal{L}$, we obtain:

$$\mathcal{H} = \sum_{ij} (\dot{y}_i + V_i)\zeta_{ij}(\dot{y}_j - V_j) - 2kT\nabla \cdot \mathbf{V}. \tag{6.26}$$

The minimum path is determined explicitly through the Hamilton equation,

$$\dot{\mathbf{p}} = -\frac{\partial \mathcal{H}}{\partial \mathbf{y}}, \tag{6.27}$$

that is

$$\dot{\mathbf{p}} + (\nabla \mathbf{V}) \cdot \mathbf{p} = 2kT\nabla(\nabla \cdot \mathbf{V}). \tag{6.28}$$

This equation could be obtained directly by applying the Euler-Lagrange equation to (6.23). In the isotropic case, when $\zeta_{ij} = \delta_{ij}$ and $D = kT/\zeta$, Eq. (6.28) can be rewritten in the following simple form,

$$\ddot{\mathbf{y}} = \nabla U + \dot{\mathbf{y}} \times \mathbf{B}, \tag{6.29}$$

where

$$U = \frac{1}{2}V^2 + D\nabla \cdot \mathbf{V}; \qquad \mathbf{B} = -\nabla \times \mathbf{V}. \tag{6.30}$$

Similar results were obtained by Wiegler [9], who studied the motion of Brownian particles in conservative force fields, where $\mathbf{B} = \mathbf{0}$. So, the minimum path describes the trajectory of a particle of unit mass and unit electric charge immersed in an electric field U and a magnetic field $\mathbf{B}$.

It is intriguing that the dissipative motion of a Brownian particle is described in terms of the conservative motion of this "particle", whose "energy" $\mathcal{H}$ is constant. In fact, multiplying Eq. (6.29) by $\dot{\mathbf{y}}$ and considering that $\frac{d}{d\tau} = \dot{\mathbf{y}} \cdot \nabla$, we can see that $d\mathcal{H}/d\tau = 0$, i.e. $\mathcal{H}$ is constant, along the minimum path.

Now, in general, the domain of integration of the path integral is composed of all paths whose distance from the minimum path is of order $\delta \sim D/\tilde{V}$ or less, where $\tilde{V}$ is a typical value of $\mathbf{V}$. Therefore, expressing any path $\mathbf{x}(\tau)$ as the "sum" of the minimum path $\mathbf{y}(\tau)$ and a "fluctuating" part $\tilde{\mathbf{x}}(\tau)$,

$$\mathbf{x}(\tau) = \mathbf{y}(\tau) + \tilde{\mathbf{x}}(\tau), \tag{6.31}$$

where $\tilde{\mathbf{x}}(0) = \tilde{\mathbf{x}}(t) = 0$, then $\mathcal{S}[\mathbf{x}(t)]$ can be expanded formally around $\mathbf{y}(\tau)$ as:

$$\mathcal{S}[\mathbf{x}(t)] = \mathcal{S}_{min} + \frac{1}{2}\tilde{\mathbf{x}}\tilde{\mathbf{x}} : \left[\frac{\partial^2 \mathcal{S}}{\partial \tilde{\mathbf{x}} \partial \tilde{\mathbf{x}}}\right]_{\mathbf{x}=\mathbf{y}} + \cdots, \tag{6.32}$$

with $\mathcal{S}_{min} = \mathcal{S}[\mathbf{y}(t)]$, where we have considered that the first order derivative is identically zero. Accordingly, we see that, if within distances of $O(\delta)$ from the minimum path $\mathbf{F}$ can be approximated as a linear function, then $\mathcal{S}$ is a quadratic functional, and therefore the expansion (6.32) terminates after the second derivative, with the last term being a function of $\tilde{\mathbf{x}}$ only, and not of $\mathbf{y}$.[5] Finally, substituting (6.32) into (6.22)–(6.24) we obtain:

$$\Pi(\mathbf{x}, t|\mathbf{x}_0) = W(t)\exp\left[-\frac{1}{4kT}\int_0^t \mathcal{L}_{min}(\tau)\,d\tau\right], \qquad (6.33)$$

where the normalization function $W(t)$ depends on t only, and is independent of the endpoints, while $\mathcal{L}_{min}$ can be obtained from (6.24) with $\mathbf{x}(\tau) = \mathbf{y}(\tau)$, i.e.,

$$\mathcal{L}_{min} = \sum_{ij}(\dot{y}_i - V_i)\zeta_{ij}(\dot{y}_j - V_j) + 2kT\sum_i \nabla_i V_i. \qquad (6.34)$$

This result shows that under very general conditions the path integral is determined exclusively by the minimum path (6.28), determining the Boltzmann-like distribution (6.33). When compatible with the end points, the minimum path is obviously $\dot{\mathbf{y}} = \mathbf{V}$.

In the case of a conservative force field, $\zeta_{ij}V_j = -\nabla_i\phi$, so that $\dot{y}_i\zeta_{ij}V_j = -\dot{\phi}$ and $\mathcal{L}_{min}(\tau) = \mathcal{L}_{min}(-\tau)$. Consequently, in this case we obtain:

$$\Pi(\mathbf{x}, t|\mathbf{x}_0) = \Pi(\mathbf{x}_0, t|\mathbf{x}), \qquad (6.35)$$

showing that the process is time-reversible, thus justifying, for example, the derivation of the fluctuation-dissipation theorem. In this case, the steady state, equilibrium probability distribution can be obtained considering the reverse path, $\dot{\mathbf{y}} = -\mathbf{V} = \nabla\phi$, starting from the equilibrium point, $\mathbf{x}_0$, with $\phi(\mathbf{x}_0) = 0$, and ending at $\mathbf{x}$. Then, we find $\mathcal{L}_{min} = 4\dot{\mathbf{y}}\cdot\nabla\phi = 4\dot{\phi}$, so that at the end we obtain the usual Boltzmann distribution, $\Pi(\mathbf{x}) = W\exp[-\phi(\mathbf{x})/kT]$.

6.4 Linear Case

In this section we consider several examples of application of the path integral approach to the motion of Brownian particles immersed in linear velocity fields, with drag coefficient $\zeta_{ij} = \zeta\delta_{ij}$ and diffusivity $D = kT/\zeta$. In all these cases, as the external perturbation is linear, the only contribution of the path integral comes from the minimum path. In addition, as the divergence of the velocity field is a constant (zero, in most cases), the calculation of the minimum path simplifies considerably.

---

[5]For a detailed proof of this statement, see [8].

6.4.1 Uniform Flow Field

As a first example, consider the diffusion of a Brownian particle in a uniform velocity field, i.e. $\mathbf{V}$ is uniform. Instead of applying directly Eq. (6.33), here we will perform the calculation explicitly.

As we saw, the conditional probability $\Pi(\mathbf{X}, t|\mathbf{0})$ that the particle moves from the origin, $\mathbf{x} = \mathbf{0}$ at time $t = 0$ to a position $\mathbf{X}$ at time t will be equal to the sum of the contributions of all paths connecting the two points, obtaining from Eqs. (6.22)–(6.24),

$$\Pi(\mathbf{X}, t|\mathbf{0}) = \int \exp\left[-\frac{\mathcal{E}[\mathbf{x}(\tau)]}{4kT}\right] \mathcal{D}[\mathbf{x}(\tau)], \tag{6.36}$$

where the integral is taken over all paths such that $\mathbf{x}(0) = \mathbf{0}$ and $\mathbf{x}(t) = \mathbf{X}$, with $\mathcal{E}$ denoting the energy dissipated along the path,

$$\mathcal{E}[\mathbf{x}(\tau)] = \int_0^t \zeta(\dot{\mathbf{x}} - \mathbf{V})^2 d\tau. \tag{6.37}$$

The minimum path $\mathbf{y}(\tau)$ satisfies the Euler-Lagrange equation (6.29), $\ddot{\mathbf{y}} = 0$, with $\mathbf{y}(0) = \mathbf{0}$ and $\mathbf{y}(t) = \mathbf{X}$, obtaining: $\mathbf{y}(\tau) = \mathbf{X}\tau/t$. Thus, denoting $\mathbf{x} = \mathbf{y} + \tilde{\mathbf{x}}$, Eq. (6.37) becomes:

$$\mathcal{E}[\mathbf{x}(t)] = \frac{\zeta}{t^2} \int_0^t (\mathbf{X} - \mathbf{V}t)^2 d\tau + 2\frac{\zeta}{t} \int_0^t (\mathbf{X} - \mathbf{V}t) \cdot \dot{\tilde{\mathbf{x}}} d\tau + \zeta \int_0^t \dot{\tilde{\mathbf{x}}}^2 d\tau. \tag{6.38}$$

Here, the second integral on the RHS is identically zero since $\tilde{\mathbf{x}}(0) = \tilde{\mathbf{x}}(t) = \mathbf{0}$. Consequently, we see that the path integral is determined only by the contribution of the minimum path and it reduces to the well-known result,

$$\Pi(\mathbf{X}, t|\mathbf{0}) = W(t) \exp\left[-\frac{(\mathbf{X} - \mathbf{V}t)^2}{4Dt}\right], \tag{6.39}$$

where $W(t)$ is a normalization factor, which is independent of the endpoints.

$$W(t) = \int \exp\left[-\frac{1}{4D} \int_0^t \dot{\tilde{\mathbf{x}}}^2 d\tau\right] \mathcal{D}[\tilde{\mathbf{x}}(\tau)]. \tag{6.40}$$

6.4.2 Harmonic Oscillator

Consider a typical Ornstein-Uhlenbeck process, where a Brownian particle immersed in a quiescent fluid is subjected to a linear potential force, (i.e. a harmonic oscillator), attracting the particle towards the origin, with $\mathbf{F} = \zeta\mathbf{V} = -\zeta\mathbf{M} \cdot \mathbf{x}$, where $M_{ij} = M_{ji}$. Accordingly, we can choose a reference frame where the axes coincide with the principal directions of the $\mathbf{M}$ matrix, so that $M_{ij} = M_i \delta_{ij}$, with $M_i > 0$.

Now, the conditional probability function $\Pi(\mathbf{X}, t|\mathbf{0})$ that describes the motion of this Brownian particle is given by Eqs. (6.33)–(6.34). Here, the minimum path equation (6.29) reduces to:

$$\ddot{y}_i = M_i^2 y_i, \tag{6.41}$$

which, coupled to the conditions $y_i(0) = 0$ and $y_i(t) = X_i$, yields:

$$y_i(\tau) = X_i \frac{\sinh(M_i\tau)}{\sinh(M_i t)}. \tag{6.42}$$

Substituting this result into Eq. (6.34), i.e. $\mathcal{L}_{min} = \zeta(\dot{\mathbf{y}} - \mathbf{V})^2$, we obtain:

$$\mathcal{L}_{min} = \zeta \sum_i \frac{M_i^2 X_i^2}{\sinh^2(M_i t)} (\cosh(M_i\tau) + \sinh(M_i\tau))^2 + C, \tag{6.43}$$

where $C = kT \, \mathrm{tr}(\mathbf{M})$ is an irrelevant constant. Finally, from Eq. (6.33) we find the following Gaussian distribution:

$$\Pi(\mathbf{X}, t|\mathbf{0}) = W(t) \exp\left[-\frac{1}{4D} \sum_i M_i X_i^2 \left[1 + \coth(M_i t)\right]\right], \tag{6.44}$$

where $W(t)$ is a normalization factor, which is independent of the endpoints. Therefore, considering that $g_i = \zeta M_i/kT$, we obtain again Eqs. (3.44)–(3.46) and (4.55)–(4.57). In particular, for long times, $t \gg M^{-1}$, this solution tends to the equilibrium distribution,

$$\Pi(\mathbf{X}) = W \exp\left[-\frac{1}{2} \sum_i g_i X_i^2\right]. \tag{6.45}$$

6.4.3 Simple Shear Flow

Consider a Brownian particle immersed in a simple shear flow field, $V_1 = \gamma x_2$; $V_2 = 0$. Following the same procedure as before, we see that the minimum path equation (6.29) reduces to:

$$\dddot{y}_1 - \ddot{y}_2 = 0; \qquad \dddot{y}_2 + \gamma \ddot{y}_1 - \gamma^2 \dot{y}_2 = 0, \tag{6.46}$$

which, coupled to the conditions $y_i(0) = 0$; $y_i(t) = X_i$, yields:

$$y_1(\tau) = C_1\left(\tilde{\tau}^3 - 6\tilde{\tau}\right) + C_2\tilde{\tau}^2; \qquad y_2(\tau) = 3C_1\tilde{\tau}^2 + 2C_2\tilde{\tau}, \tag{6.47}$$

with $\tilde{\tau} = \gamma\tau$, where

$$C_1 = \frac{\tilde{t}X_2 - 2X_1}{\tilde{t}(\tilde{t}^2 + 12)}; \qquad C_2 = \frac{(6 - \tilde{t}^2)X_2 + 3\tilde{t}X_1}{\tilde{t}(\tilde{t}^2 + 12)}, \tag{6.48}$$

and $\tilde{t} = \gamma t$. Substituting this result into Eqs. (6.33) and (6.34), we find the following Gaussian distribution:

$$\Pi(\mathbf{X}, t|0) = W(t)\exp\left[-\frac{3\gamma(2X_1 - \tilde{t}X_2)^2}{4D\tilde{t}(\tilde{t}^2 + 12)} - \frac{X_2^2}{4Dt}\right]. \qquad (6.49)$$

Therefore, the variances of this distribution are:

$$\langle X_1^2 \rangle == 2Dt\left[1 + \frac{1}{3}(\gamma t)^2\right]; \qquad \langle X_1 X_2 \rangle = D\gamma t^2; \qquad \langle X_2^2 \rangle = 2Dt. \qquad (6.50)$$

Here we see that, as expected, the mean free displacement in the flow direction grows like t^3.

Comment It should be stressed that, as we have seen, path integrals give us no dramatic new results in the study of a single Brownian particle (and of quantum particles as well). Indeed, most if not all the simple results in quantum or stochastic mechanics which are obtained applying path integration can be derived with considerably greater ease using the standard formulations, that is by solving the appropriate partial differential equations.[6] Despite that, however, path integration constitutes a very worthwhile contribution to our understanding of stochastic processes (and quantum mechanics, as well), as this way of looking at diffusive processes is, arguably, more intuitive than the usual approaches. Furthermore, the close relation between statistical mechanics and quantum mechanics, or between statistical field theory and quantum field theory, is plainly visible via path integration.

6.5 Problems

Problem 6.1 Find the probability distribution function of a Brownian particle immersed in an elongational incompressible flow field, $V_1 = \gamma x_2$ and $V_2 = \gamma x_1$, which is initially located at the origin.

Problem 6.2 Find the probability distribution function of a Brownian particle immersed in an straining incompressible flow field, $V_1 = \gamma x_1$ and $V_2 = -\gamma x_2$, which is initially located at the origin.

References

1. Chalchian, M., Demichev, A.P., Demichev, A.: Path Integrals in Physics: Stochastic Processes and Quantum Mechanics. Taylor & Francis, New York (2001)

---

[6]For systems with very many degrees of freedom, though, such as in field theory, path integrals turn out to be considerably more useful.

2. Dirac, P.A.M.: Phys. Z. Sowjetunion **3** (1932)
3. Feynman, R.P.: Rev. Mod. Phys. **20**, 367 (1948)
4. Feynman, R.P., Hibbs, A.R.: Quantum Mechanics and Path Integrals. McGraw-Hill, New York (1965)
5. Gelfand, I.M., Yaglom, A.M.: J. Math. Phys. **1**, 48 (1960)
6. Graham, R.: Phys. Rev. Lett. **38**, 51 (1977)
7. Onsager, L., Machlup, S.: Phys. Rev. **91**, 1505 (1953)
8. Schulman, L.S.: Techniques and Applications of Path Integration. Wiley, New York (1981)
9. Wiegel, F.W.: Introduction to Path-Integral Methods in Physics and Polymer Science. Word Scientific, Singapore (1986)
10. Wiener, N.: Proc. Natl. Acad. Sci. USA **7**, 253 (1921), also see p. 294

Chapter 7
Balance Equations

In this chapter (see Sects. 7.1–7.6) we derive the macroscopic conservation laws of matter, momentum, energy and angular momentum for a multicomponent system subjected to conservative external forces and in which chemical reactions may occur. Then, in Sect. 7.7, the time growth of the macroscopic entropy of the system is derived, showing that it can be expressed as the product between thermodynamic fluxes and their conjugated thermodynamic forces. It should be noted that, as shown in Appendix E, these equations are consequences of the fundamental mechanical laws governing the motions of the constituent particles of the system.

7.1 General Balance Equation

Consider a macroscopic system in a fixed volume V, bounded by a closed surface S. Define an arbitrary extensive quantity $F(t)$ as

$$F(t) = \int_V \rho(\mathbf{r}, t) f(\mathbf{r}, t) \, d^3\mathbf{r},\qquad(7.1)$$

where $f(\mathbf{r}, t)$ represents the density of $F(t)$ per unit mass and $\rho(\mathbf{r}, t)$ is the mass per unit volume as a function of position and time. The most general equation describing the variation of $F(t)$ is:

$$\frac{dF(t)}{dt} = \int_V \frac{\partial}{\partial t}(\rho f) \, d^3\mathbf{r} = -\oint_S J_n^{(F)} \, d^2\mathbf{r} + \int_V \sigma^{(F)} \, d^3\mathbf{r},\qquad(7.2)$$

where $\sigma^{(F)}$ is the source density, i.e. the amount of F generated per unit volume and per unit time, while $J_n^{(F)}$ is the flux of F leaving the volume, that is the amount of F crossing the surface S per unit surface and per unit time, having the direction of the unit vector $\mathbf{e}_n$ perpendicular to the surface and directed outward. According to one of the many Cauchy's theorems, $J_n^{(F)}$ can be written as:

$$J_n^{(F)} = \mathbf{e}_n \cdot \mathbf{J}^{(F)},\qquad(7.3)$$

R. Mauri, *Non-Equilibrium Thermodynamics in Multiphase Flows*,
Soft and Biological Matter, DOI 10.1007/978-94-007-5461-4_7,
© Springer Science+Business Media Dordrecht 2013

where $\mathbf{J}^{(F)}$ is a vector. This statement can be formally proved applying Eq. (7.2) to a tetrahedron and letting it shrink to zero, as shown in most of the undergraduate fluid mechanics textbooks. Finally, applying the divergence theorem,

$$\oint_S \mathbf{e}_n \cdot \mathbf{J}^{(F)} d^2\mathbf{r} = \int_V \nabla \cdot \mathbf{J}^{(F)} d^3\mathbf{r}, \tag{7.4}$$

we obtain the differential equation for the local balance of f,

$$\frac{\partial}{\partial t}(\rho f) + \nabla \cdot \mathbf{J}^{(F)} = \sigma^{(F)}. \tag{7.5}$$

Clearly, this equation must be coupled with an expression for the source density and a constitutive relation relating the flux $\mathbf{J}^{(F)}$ to f and its gradients. We will do it for the balance equations for mass, momentum, energy, angular momentum and entropy, showing that irreversible thermodynamics can determine the structure of these relations.

7.2 Conservation of Mass

For mass transport, we set $F = M$ in Eq. (7.1), where M is the mass, so that $f = 1$. The mass flux can be written as

$$\mathbf{J}^{(M)} = \rho \mathbf{v}, \tag{7.6}$$

where $\mathbf{v}$ is the fluid velocity. Equation (7.6) can be seen as a definition of the mean velocity, while the mass flux $\mathbf{J}^{(M)}$ is a momentum density and is therefore a primitive quantity.[1] So, considering that, as mass is conserved, i.e. $\sigma^{(M)} = 0$, Eq. (7.5) becomes:

$$\frac{\partial \rho}{\partial t} + \nabla \cdot (\rho \mathbf{v}) = 0. \tag{7.7}$$

The law of conservation of mass (7.7) can also be written in an alternative form by defining the substantial (also referred to as material or barycentric) derivative,

$$\frac{D}{Dt} = \frac{\partial}{\partial t} + \mathbf{v} \cdot \nabla, \tag{7.8}$$

---

[1] Actually, $\mathbf{v}$ is the mean velocity of the fluid particles that are contained within the physical point-like volume defined in (1.6) (that is a volume large enough so as to neglect thermal fluctuations and yet small enough to neglect the effects of macroscopic gradients). Clearly, when we deal with a single component fluid, the average can be intended as mass or as molar average, as the two quantities are the same. As we will see in the next section, though, in multicomponent flows the two averages are different from each other and we will choose to define $\mathbf{v}$ as the mass-averaged velocity.

indicating the time rate of change measured in a Lagrangian reference frame, that reflects the point of view of an observer that moves together with the fluid, with velocity $\mathbf{v}$. So we obtain:

$$\frac{D\rho}{Dt} = -\rho \nabla \cdot \mathbf{v}, \tag{7.9}$$

or,

$$\frac{1}{\tilde{v}} \frac{D\tilde{v}}{Dt} = \nabla \cdot \mathbf{v}, \tag{7.10}$$

where $\tilde{v} = \rho^{-1}$ is the specific volume.

Before we proceed, consider the following equality:

$$\frac{\partial}{\partial t}(\rho f) = \rho \frac{\partial f}{\partial t} - f \nabla \cdot (\rho \mathbf{v}) = \rho \frac{\partial f}{\partial t} - \nabla \cdot (\rho f \mathbf{v}) + \rho \mathbf{v} \cdot \nabla f, \tag{7.11}$$

that is:

$$\rho \frac{Df}{Dt} = \frac{\partial(f\rho)}{\partial t} + \nabla \cdot (\rho f \mathbf{v}). \tag{7.12}$$

This is the differential version of the integral Reynolds transport theorem,

$$\frac{d}{dt} \int_{V_m(t)} (\rho f) d^3\mathbf{r} = \int_{V_m(t)} \frac{\partial(f\rho)}{\partial t} + \nabla \cdot (\rho f \mathbf{v}) d^3\mathbf{r} = \int_{V_m(t)} \rho \frac{Df}{Dt} d^3\mathbf{r}, \tag{7.13}$$

where $V_m(t)$ is a material volume moving together with the fluid. Obviously, when $f = 1$, this equation reduces to

$$\frac{d}{dt} \int_{V_m(t)} \rho \, d^3\mathbf{r} = 0, \tag{7.14}$$

indicating that mass is conserved.

Substituting (7.12) into (7.5), we can rewrite the general balance equation in a Lagrangian frame as follows,

$$\rho \frac{Df}{Dt} = \frac{\partial(f\rho)}{\partial t} + \nabla \cdot (\rho f \mathbf{v}) = -\nabla \cdot \mathbf{J}_d^{(F)} + \sigma_{(F)}, \tag{7.15}$$

where

$$\mathbf{J}_d^{(F)} = \mathbf{J}^{(F)} - \rho f \mathbf{v}. \tag{7.16}$$

This indicates that, when measured in a reference frame moving with the fluid, the flux $\mathbf{J}^{(F)}$ (which is defined as the flux observed at a fixed point in space) is reduced by an amount $\rho f \mathbf{v}$, i.e. its convective contribution. The remaining part is the flux that takes place *in the absence of convection*, and we will refer to it as the diffusive flux, $\mathbf{J}_d^{(F)}$. The two frameworks, i.e. one moving with the fluid and the other fixed in space, are generally referred to as Lagrangian and Eulerian, respectively.

7.3 Conservation of Chemical Species

Let us consider a mixture composed of n components, among which r chemical reactions are possible. Again, we set $F = M^{(k)}$ in Eq. (7.1), where $M^{(k)}$ is the mass of species k, so that $f = \phi^{(k)} = \rho^{(k)}/\rho$ is the mass fraction, defined as the ratio between the mass concentration (mass per unit volume) of species k and the density of the mixture. Clearly, here we have:

$$\sum_{k=1}^{n} \rho^{(k)} = \rho, \qquad \sum_{k=1}^{n} \phi^{(k)} = 1. \tag{7.17}$$

The rate of change of the mass of component k can be determined applying Eq. (7.5), where the mass flux of component k is $\mathbf{J}^{(k)} = \rho^{(k)}\mathbf{v}^{(k)}$, with $\mathbf{v}^{(k)}$ denoting the velocity of species k. In addition, unlike the total mass, $M^{(k)}$ is not conserved, as in the j-th chemical reaction a mass $v^{(kj)} \mathcal{J}^{(j)}$ of species k is generated per unit volume and unit time. Here, $v^{(kj)}$ divided by the molecular mass of k is proportional to the stoichiometric coefficient with which k appears in the j-th chemical reaction (the sign is positive or negative, depending on whether it appears on the right or the left side of the reaction), while $\mathcal{J}^{(j)}$ is the chemical reaction rate of reaction j (mole per unit time and per unit volume). Finally, we obtain:

$$\frac{\partial(\rho\phi^{(k)})}{\partial t} + \nabla \cdot \left(\rho\phi^{(k)}\mathbf{v}^{(k)}\right) = \sum_{j=1}^{r} v^{(kj)} \mathcal{J}^{(j)} \quad (k = 1, 2, \ldots, n). \tag{7.18}$$

Since mass is conserved in each separate reaction, we have:

$$\sum_{k=1}^{n} v^{(kj)} = 0 \quad (j = 1, 2, \ldots, r). \tag{7.19}$$

Summing Eq. (7.18) over all k's and considering Eq. (7.19), we obtain the law of conservation of mass, Eq. (7.7), where ρ is the total density, while $\mathbf{v}$ is the mass-averaged velocity,

$$\mathbf{v} = \frac{1}{\rho} \sum_{k=1}^{n} \rho^{(k)}\mathbf{v}^{(k)} = \sum_{k=1}^{n} \phi^{(k)}\mathbf{v}^{(k)}. \tag{7.20}$$

$\mathbf{v}$ is also referred to as the center of mass, or barycentric, velocity.

The balance equation of chemical species k can also be written in a Lagrangian frame, applying Eq. (7.15) as follows:

$$\rho\frac{D\phi^{(k)}}{Dt} = -\nabla \cdot \mathbf{J}_d^{(k)} + \sum_{j=1}^{r} v^{(kj)} \mathcal{J}^{(j)} \quad (k = 1, 2, \ldots, n) \tag{7.21}$$

where the diffusive mass flux,

$$\mathbf{J}_d^{(k)} = \mathbf{J}^{(k)} - \rho\phi^{(k)}\mathbf{v} = \rho\phi^{(k)}\left(\mathbf{v}^{(k)} - \mathbf{v}\right) \tag{7.22}$$

is defined in terms of the relative velocity of the species k with respect to the barycentric velocity.

Note that from (7.20) and (7.22) we see that

$$\sum_{k=1}^{n} \mathbf{J}_d^{(k)} = 0, \tag{7.23}$$

which means that only $(n-1)$ of the n diffusive fluxes are independent. Similarly, only $(n-1)$ of the n equations (7.21) or (7.18) are independent. In fact, by summing Eqs. (7.21) over all k, both members vanish identically as a consequence of (7.17), (7.19) and (7.23). The n-th independent equation describing the change of mass density within the system is Eq. (7.7) or (7.9).

Comment From thermodynamics, we know that the specific volume of a mixture, $\tilde{v} = \rho^{-1}$, is given by:

$$\tilde{v} = \sum \phi^{(k)} \tilde{v}_k + \Delta \tilde{v}_{mix}, \tag{7.24}$$

where $\tilde{v}_k$ is the specific volume of component k alone, i.e. *outside* the mixture, while $\Delta \tilde{v}_{mix}$ is the volume of mixing. The latter accounts for the fact that when two fluids mix they generally loose some volume; so, for example, if we mix 1 liter of water and 1 liter of ethanol, the resulting mixture occupies approximately only 1.9 liters. In an ideal case, for so-called regular mixtures,[2] volumes are additive, so that $\Delta \tilde{v}_{mix} = 0$, In no case, however, densities are additive, since assuming $\rho = \sum \phi^{(k)} \rho_k$, where ρ_k is the density of component k alone, is equivalent to introducing an *ad hoc* volume of mixing.

7.4   Conservation of Momentum

The momentum balance equation is the equation of motion of continuum mechanics, which in Lagrangian formulation can be derived from Eq. (7.15) with $f = \mathbf{v}$ as

$$\rho \frac{D\mathbf{v}}{Dt} = -\nabla \cdot \mathbf{P} + \rho \mathbf{F}, \tag{7.25}$$

where $\mathbf{P}$ is the momentum diffusive flux, which is generally referred to as the pressure tensor (it is equal to the stress tensor, with opposite sign), while the momentum source term, $\rho \mathbf{F}$, expressing the momentum transferred to the fluid per unit time and per unit volume, is a body force, that is a force per unit volume, with $\mathbf{F}$ denoting the external force per unit mass. In turn, the body force can be written as the sum of the

---

[2]Regular mixtures are composed of species that are rather similar to each other, so that they behave in some extent like mixtures of ideal gases, i.e. volume and enthalpy are additive quantities.

separate forces exerted on each chemical species, i.e.,

$$\rho \mathbf{F} = \sum_{k=1}^{n} \rho^{(k)} \mathbf{F}^{(k)}. \tag{7.26}$$

Here, we will restrict the discussion to the case of conservative forces, which can be derived from a potential $\psi^{(k)}$ (energy per unit mass), independent of time,

$$\mathbf{F}^{(k)} = -\nabla \psi^{(k)}; \qquad \frac{\partial \psi^{(k)}}{\partial t} = 0. \tag{7.27}$$

Accordingly, we can define a total potential energy density (energy per unit volume),

$$\rho \psi = \sum_{k=1}^{n} \rho^{(k)} \psi^{(k)}. \tag{7.28}$$

From a microscopic point of view, ψ derives from long-range forces, while the pressure tensor $\mathbf{P}$ results from short-range interactions.

The equation of motion can also be written in an Eulerian framework, i.e. Eq. (7.5),

$$\frac{\partial (\rho \mathbf{v})}{\partial t} + \nabla \cdot (\rho \mathbf{v} \mathbf{v} + \mathbf{P}) = \rho \mathbf{F}, \tag{7.29}$$

where $\rho \mathbf{v} \mathbf{v}$ is a dyadic, expressing the convective part of the momentum flux.

From the equation of motion it is possible to derive a balance equation for the kinetic energy. In fact, multiplying Eq. (7.25) by $\mathbf{v}$ we obtain:

$$\rho \mathbf{v} \cdot \frac{D\mathbf{v}}{Dt} = \rho \frac{D\frac{1}{2}v^2}{Dt} = -\mathbf{v} \cdot (\nabla \cdot \mathbf{P}) + \rho \mathbf{F} \cdot \mathbf{v}. \tag{7.30}$$

Now, considering that $\mathbf{v} \cdot (\nabla \cdot \mathbf{P}) = \nabla \cdot (\mathbf{P} \cdot \mathbf{v}) - (\mathbf{P} : \nabla \mathbf{v})$, where $\mathbf{A} : \mathbf{B} = A_{ij} B_{ij}$, and applying Eq. (7.12), we obtain the kinetic energy balance equation,

$$\frac{\partial (\frac{1}{2} \rho v^2)}{\partial t} + \nabla \cdot \left(\frac{1}{2} \rho v^2 \mathbf{v} + \mathbf{P} \cdot \mathbf{v} \right) = \mathbf{P} : \nabla \mathbf{v} + \rho \mathbf{F} \cdot \mathbf{v}. \tag{7.31}$$

So, we see that the kinetic energy flux is the sum of a convective part, $\frac{1}{2} \rho v^2 \mathbf{v}$, and a diffusive term, $\mathbf{P} \cdot \mathbf{v}$, while the sources of kinetic energy involve the power density (i.e. the work done per unit volume and per unit time) done by the pressure tensor and by the external force.

Now we will derive an equivalent balance equation for the potential energy, $\rho \psi$, with $\psi = \sum \phi^{(k)} \psi^{(k)}$. First, apply Eq. (7.12) to obtain:

$$\rho \frac{D\psi}{Dt} = \rho \sum_{k=1}^{n} \left[\phi^{(k)} \frac{D\psi^{(k)}}{Dt} + \psi^{(k)} \frac{D\phi^{(k)}}{Dt} \right] = \frac{\partial (\rho \psi)}{\partial t} + \nabla \cdot (\mathbf{v} \rho \psi). \tag{7.32}$$

Now, considering that $\psi^{(k)}$ does not depend explicitly on time and applying Eq. (7.21) we obtain:

$$\frac{\partial(\rho\psi)}{\partial t} + \nabla \cdot (\mathbf{v}\rho\psi) = \sum_{k=1}^{n} \rho\phi^{(k)}\mathbf{v} \cdot \nabla\psi^{(k)} - \sum_{k=1}^{n} \rho\psi^{(k)}\nabla \cdot \mathbf{J}_d^{(k)}$$

$$+ \sum_{k=1}^{n}\sum_{j=1}^{r} \rho\psi^{(k)}\nu^{(kj)}\mathcal{J}^{(j)}. \tag{7.33}$$

The last term vanishes in most cases, as the potential energy remains unchanged in a chemical reaction, i.e.,

$$\sum_{k=1}^{n} \psi^{(k)}\nu^{(kj)} = 0. \tag{7.34}$$

In fact, the properties that are responsible for the interactions with the force field are conserved, as it happens with the mass and the electric charge of the chemically reacting particles. Equation (7.33) then reduces to:

$$\frac{\partial(\rho\psi)}{\partial t} + \nabla \cdot \left(\mathbf{v}\rho\psi + \sum_{k=1}^{n} \psi^{(k)}\mathbf{J}_d^{(k)}\right) = -\rho\mathbf{F} \cdot \mathbf{v} - \sum_{k=1}^{n} \mathbf{J}_d^{(k)} \cdot \mathbf{F}^{(k)}. \tag{7.35}$$

Here, we see that the potential energy flux is the sum of a convective part, $\mathbf{v}\rho\psi$, and a diffusive term, representing the transport of energy due to mass diffusion; in fact, trivially, if all species had the same energy, i.e. $\psi^{(k)} = \psi$, this term would vanish, as the sum of all diffusive fluxes is equal to zero. The source terms of potential energy involve the works done by the external forces on the mean convective flow and on the diffusive flows. The first is a sink term and represents the conversion of potential energy into kinetic energy, as an equal but opposite term appears in the kinetic energy equation as well; the second term represents instead the conversion of potential energy into internal energy by diffusion.

Let us add the two equation (7.31) and (7.35) for the rate of change of mechanical (i.e. kinetic plus potential) energy:

$$\frac{\partial}{\partial t}\left[\rho\left(\frac{1}{2}v^2 + \psi\right)\right] + \nabla \cdot \left[\mathbf{v}\rho\left(\frac{1}{2}v^2 + \psi\right) + \mathbf{J}_d^{ME}\right] = \sigma^{ME}, \tag{7.36}$$

where

$$\mathbf{J}_d^{(ME)} = \mathbf{P} \cdot \mathbf{v} + \sum_{k=1}^{n} \psi^{(k)}\mathbf{J}_d^{(k)}, \tag{7.37}$$

is the diffusive mechanical energy flux, while,

$$\sigma^{(ME)} = \mathbf{P}:\nabla\mathbf{v} + \sum_{k=1}^{n} \mathbf{J}_d^{(k)} \cdot \nabla\psi^{(k)}, \tag{7.38}$$

is the mechanical energy source, showing that mechanical energy is (obviously) not conserved.

7.5 Conservation of Energy

The total energy per unit volume, ρe, of the mixture is the sum of kinetic, potential and internal energy,

$$\rho e = \frac{1}{2}\rho v^2 + \rho \psi + \rho u, \tag{7.39}$$

where u is the thermodynamic internal energy, which includes the energies of thermal agitation and short-range molecular interactions. In general, the internal energy will also satisfy a balance equation (7.5),

$$\frac{\partial}{\partial t}(\rho u) + \nabla \cdot \mathbf{J}^{(U)} = \dot{q}, \tag{7.40}$$

where $\mathbf{J}^{(U)}$ is the internal energy flux, while $\dot{q} = \sigma^{(U)}$ is the internal energy source, which is generally referred to as heat source. The internal energy flux can be written as the sum of a convective and a diffusive component, with the latter generally called the heat flux, $\mathbf{J}^{(q)}$, i.e.,

$$\mathbf{J}^{(U)} = \rho u \mathbf{v} + \mathbf{J}^{(q)}. \tag{7.41}$$

Since the total energy is conserved, it satisfies the balance equation (7.5), with no energy source, i.e. $\sigma^{(E)} = 0$,

$$\frac{\partial}{\partial t}(\rho e) + \nabla \cdot \mathbf{J}^{(E)} = 0, \tag{7.42}$$

where $\mathbf{J}^{(E)} = \mathbf{J}^{(ME)} + \mathbf{J}^{(U)}$ is the energy flux, equal to the sum of the mechanical energy flux and the internal energy flux, while $\sigma^{(E)} = \sigma^{(ME)} + \dot{q} = 0$ is the energy source, which is identically zero. Therefore, from Eq. (7.38) we obtain:

$$\dot{q} = -\mathbf{P} : \nabla \mathbf{v} - \sum_{k=1}^{n} \mathbf{J}_d^{(k)} \cdot \nabla \psi^{(k)}, \tag{7.43}$$

showing that the heat source, i.e. the conversion of mechanical energy into heat, is due to momentum and mass diffusion. In addition, from Eqs. (7.37), (7.40) and (7.42) we obtain,

$$\mathbf{J}^{(E)} = \rho e \mathbf{v} + \mathbf{J}_d^{(E)}, \quad \mathbf{J}_d^{(E)} = \mathbf{J}^{(q)} + \mathbf{P} \cdot \mathbf{v} + \sum_{k=1}^{n} \psi^{(k)} \mathbf{J}_d^{(k)}, \tag{7.44}$$

showing that the diffusive energy flux is the sum of, respectively, the heat flux, the work dissipated by viscous forces, and the net potential energy loss. As we saw in the previous section, this last term describes the fact that when species k diffuses out of the material volume, it carries an energy $\psi^{(k)}$, so that, in general, in a diffusive process there is a net energy flow even if there is no net mass flow.

Naturally, the heat flux, $\mathbf{J}^{(q)}$, as well as the pressure tensor, $\mathbf{P}$, and the diffusive mass fluxes, $\mathbf{J}_d^{(k)}$, must be expressed in terms of constitutive equations, as we will see in the next section. Also, note that, since $h = u + p/\rho$, the internal energy u can be replaced by the enthalpy h in the energy equation, as the dp/dt term can be neglected in any low Mach number process.

The internal energy balance can also be formulated in an alternate form substituting Eq. (7.21) into the last term of the heat source (7.43), obtaining,

$$\sum_{k=1}^{n} \mathbf{J}_d^{(k)} \cdot \nabla \psi^{(k)} = \sum_{k=1}^{n} \nabla \cdot \left(\mathbf{J}_d^{(k)} \psi^{(k)} \right) + \rho \sum_{k=1}^{n} \psi^{(k)} \frac{D\phi^{(k)}}{dt} - \sum_{k=1}^{n} \sum_{j=1}^{r} \psi^{(k)} \nu^{(kj)} \mathcal{J}^{(j)},$$

where the last term is identically zero [cf. Eq. (7.34)]. Accordingly, the heat flux and the heat source can be written as:

$$\dot{q}' = -\mathbf{P} : \nabla \mathbf{v} - \rho \sum_{k=1}^{n} \psi^{(k)} \frac{D\phi^{(k)}}{Dt}, \tag{7.45}$$

and

$$\mathbf{J}^{(q)'} = \mathbf{J}^{(q)} + \sum_{k=1}^{n} \psi^{(k)} \mathbf{J}_d^{(k)}. \tag{7.46}$$

The minus sign on the last term in the RHS of Eq. (7.45) is due to the fact that heat is drawn to increase the chemical energy of the mixture, i.e. when both $\psi^{(k)}$ and $\dot{\phi}^{(k)}$ are positive. In addition, (7.46) shows that the flux of the internal energy for multi-component mixtures can be written as the sum of a heat diffusive term and a heat transport term due to each diffusing species.

Finally, it is useful to note that Eq. (7.40) is simply the first law of thermodynamics. In fact, defining the heat per unit mass q so that $\rho Dq/Dt + \nabla \cdot \mathbf{J}^{(q)} = 0$, we obtain:

$$\frac{Du}{Dt} = \frac{Dq}{Dt} - p \frac{D\tilde{v}}{Dt} - \tilde{v} \widetilde{\mathbf{P}} : \nabla \mathbf{v} - \tilde{v} \sum_{k=1}^{n} \mathbf{J}_d^{(k)} \cdot \nabla \psi^{(k)}, \tag{7.47}$$

where $\tilde{v} = \rho^{-1}$ is the specific volume, p is the scalar hydrostatic (i.e. thermodynamic) pressure, with $\mathbf{P} = p\mathbf{I} + \widetilde{\mathbf{P}}$, and $\mathbf{I}$ denoting the unit tensor. Here, first we have considered that $\mathbf{I} : \nabla \mathbf{v} = \nabla \cdot \mathbf{v}$ and then we have applied Eq. (7.10). It should be stressed that, in general, the pressure tensor can be written as the sum of an isotropic part and a so-called deviatoric (i.e. trace-free) part,

$$\mathbf{P} = P\mathbf{I} + \check{\mathbf{P}} \tag{7.48}$$

where

$$P = \frac{1}{3}\mathbf{I}{:}\mathbf{P} \tag{7.49}$$

is referred to as the dynamic pressure, which is generally different than the hydrostatic pressure p, while $\mathbf{\check{P}}$ denotes the deviators part of $\mathbf{P}$ (and of $\mathbf{\tilde{P}}$ as well). That means that $\mathbf{\tilde{P}}$ is not the deviatoric part of the pressure tensor, that is $\mathbf{\tilde{P}}$ is not divergence-free, as P, not p, is the trace of the pressure tensor. In general, we can write:

$$P = p + p', \tag{7.50}$$

where p' is a non-thermodynamic pressure-like term that will be determined in Sect. 8.2 applying Onsager's relations.

7.6 Conservation of Angular Momentum

The angular momentum per unit mass can be written as an axial, or pseudo, vector $\mathbf{g}$. Unlike absolute tensors, like momentum (and pressure, velocity and stress tensor as well), that are unaffected by whether the coordinate system is right or left-handed, pseudo-vectors change sign if the basis set is changed from one that is right-handed to one that is left-handed. Familiar examples are the cross products of any two absolute vectors, or the curl of an absolute vector field, such as the vorticity or, in this case, the angular momentum. Performing a second such operation on a pseudo-vector, such as taking a curl, will yield an absolute entity. Also, if the cross product of a pseudovector and an absolute vector is formed, the result with be an absolute vector.[3]

Alternatively, we may define the antisymmetric tensor $\Gamma_{ij} = -\Gamma_{ji}$, with Cartesian components $\Gamma_{12} = -\Gamma_{21} = g_3$ (cycl.). Defining Ricci's third-order antisymmetric tensor ϵ_{ijk} as $\epsilon_{123} = 1$, $\epsilon_{321} = -1$ (cycl.) and $\epsilon_{ijk} = 0$ otherwise,[4] that means $\Gamma_{ij} = \epsilon_{ijk} g_k$ and, viceversa, $g_i = \frac{1}{2}\epsilon_{ijk}\Gamma_{jk}$. Similarly to Eq. (7.25), the angular momentum balance yields:

$$\rho\frac{D\mathbf{g}}{Dt} = -\nabla\cdot(\mathbf{r}\times\mathbf{P}), \tag{7.51}$$

that is,

$$\rho\frac{Dg_k}{Dt} = -\sum_{l,m,n=1}^{3}\epsilon_{klm}\frac{\partial}{\partial x_n}(x_m P_{nl}), \tag{7.52}$$

---

[3] See discussion in [2].

[4] Using Ricci's tensor, the cross product between two vector $\mathbf{a}$ and $\mathbf{b}$ can be written as: $(\mathbf{a}\times\mathbf{b})_i = \sum_{j,k}\epsilon_{ijk}a_j b_k$. Note that $\sum_m\epsilon_{ijm}\epsilon_{mkl} = \delta_{ik}\delta_{jl} - \delta_{il}\delta_{jk}$ and $\sum_{jk}\epsilon_{ijk}\epsilon_{jkl} = 2\delta_{il}$.

where the RHS represents the torque exerted on a mass element by the pressure tensor. Now, multiplying both members by ϵ_{ijk} and considering that $\sum_m \epsilon_{ijk}\epsilon_{klm} = \delta_{il}\delta_{km} - \delta_{im}\delta_{jl}$, we obtain:

$$\rho\frac{D\Gamma_{ij}}{Dt} = -\sum_{k=1}^{3}\frac{\partial}{\partial x_k}(x_i P_{kj} - x_j P_{ki}). \tag{7.53}$$

The angular momentum $\mathbf{g}$ can be split into two parts,

$$\mathbf{g} = \mathbf{g}^{(e)} + \mathbf{g}^{(i)}, \quad \text{i.e.} \quad \boldsymbol{\Gamma} = \boldsymbol{\Gamma}^{(e)} + \boldsymbol{\Gamma}^{(i)}, \tag{7.54}$$

where $\mathbf{g}^{(e)} = \mathbf{r} \times \mathbf{v}$ and $\boldsymbol{\Gamma}^{(e)}$ is the usual (i.e. external) angular momentum density due to the fluid motion, with $\Gamma_{ij}^{(e)} = (x_i v_j - x_j v_i)$, while $\mathbf{g}^{(i)} = \mathbf{s}$ (with $\boldsymbol{\Gamma}^{(i)} = \mathbf{S} = \boldsymbol{\epsilon} \cdot \mathbf{s}$) is an internal angular momentum (spin) per unit mass, arising as a consequence of the possible rotational motion of the microstructure. From Eq. (7.25), considering that, since $v_i = Dx_i/Dt$,

$$\rho\frac{D\Gamma_{ij}^{(e)}}{Dt} = \rho\left(x_i\frac{Dv_j}{Dt} - x_j\frac{Dv_i}{Dt}\right), \tag{7.55}$$

we obtain:

$$\rho\frac{D\Gamma_{ij}^{(e)}}{Dt} = -\sum_{k}\frac{\partial}{\partial x_k}(x_i P_{kj} - x_j P_{ki}) + (P_{ij} - P_{ji}). \tag{7.56}$$

Subtracting (7.56) from (7.53) we obtain:

$$\rho\frac{DS_{ij}}{Dt} = -(P_{ij} - P_{ji}), \quad \text{i.e.} \quad \rho\frac{D\mathbf{S}}{Dt} = -2\mathbf{P}^{(a)}, \tag{7.57}$$

where $P_{ij}^{(a)} = \frac{1}{2}(P_{ij} - P_{ji})$ is the antisymmetric part of the total pressure tensor $\mathbf{P}$. The same relation holds also between the spin vector $\mathbf{s} = \frac{1}{2}\boldsymbol{\epsilon}{:}\mathbf{S}$ and the internal torque density (i.e. per unit mass) $\mathbf{t} = \frac{1}{2}\boldsymbol{\epsilon}{:}\mathbf{P}$, where we have used Gibbs' notation: $\mathbf{A}{:}\mathbf{B} = A_{ij}B_{ij}$,

$$\rho\frac{D\mathbf{s}}{Dt} = -2\mathbf{t}. \tag{7.58}$$

From the equations above it is clear that if the system possesses no intrinsic internal motion, i.e. when $\mathbf{S} = 0$, then $\boldsymbol{\Gamma} = \boldsymbol{\Gamma}^{(e)}$ and Eqs. (7.58) and (7.57) yield:

$$P_{ij} = P_{ji}, \quad \text{i.e.} \quad \mathbf{P}^{(a)} = 0, \quad \mathbf{t} = \mathbf{0}. \tag{7.59}$$

that is the pressure tensor is symmetric and there is no internal body torque.

In the following, the internal body torque will be assumed to be zero, so that the pressure tensor is symmetric.

7.7 Entropy Equation

From thermodynamics we know that at equilibrium the entropy per unit mass, s, is a known function of the variables that are necessary to define the thermodynamic state of the mixture. Assuming that we have a single phase, with n reacting chemical species, we need to define $n + 1$ variables, i.e. $s = s(u, \tilde{v}, \phi_k)$. In fact, the Gibbs relation gives,

$$T \, ds = du + p \, d\tilde{v} - \sum_{k=1}^{n} \mu^{(k)} \, d\phi^{(k)}, \tag{7.60}$$

where $\mu^{(k)}$ is the chemical potential (i.e. the partial specific Gibbs free energy) of component k. Assuming local equilibrium, as we saw in Sect. 1.1, within a small mass element we may assume that the entropy of the system, and its variation, has the same dependence on all thermodynamic variables as in real equilibrium. In particular, Eq. (7.60) remains valid for a mass element that we follow along its center of mass:

$$T \frac{Ds}{Dt} = \frac{Du}{Dt} + p \frac{Dv}{Dt} - \sum_{k=1}^{n} \mu^{(k)} \frac{D\phi^{(k)}}{Dt}. \tag{7.61}$$

Substituting Eqs. (7.47) and (7.21), we obtain:

$$T\rho \frac{Ds}{Dt} = -\nabla \cdot \left(\mathbf{J}^{(q)} - \sum_{k=1}^{n} \mathbf{J}_d^{(k)} \mu^{(k)} \right) - \tilde{\mathbf{P}} {:} \nabla \mathbf{v} - \sum_{k=1}^{n} \mathbf{J}_d^{(k)} \cdot \nabla \tilde{\mu}^{(k)} - \sum_{j=1}^{r} \mathcal{J}^{(j)} A^{(j)},$$

where we have introduced the so called chemical affinities,

$$A^{(j)} = \sum_{k=1}^{n} v^{(kj)} \mu^{(k)}, \tag{7.62}$$

while,

$$\tilde{\mu}^{(k)} = \mu^{(k)} + \psi^{(k)} \tag{7.63}$$

is the total (i.e. chemical plus non-chemical) potential. For example, in the presence of an electrostatic field, $\tilde{\mu}^{(k)}$ is the electrochemical potential, with $\psi^{(k)} = z^{(k)}\phi$, where $z^{(k)}$ is the charge per unit mass of component k, and ϕ the electrostatic potential.

Rearranging and casting this equation in the form (7.15), we have:

$$\rho \frac{Ds}{Dt} + \nabla \cdot \mathbf{J}_d^{(S)} = \sigma^{(S)}, \tag{7.64}$$

where

$$\mathbf{J}_d^{(S)} = \frac{1}{T} \left(\mathbf{J}^{(q)} - \sum_{k=1}^{n} \mathbf{J}_d^{(k)} \mu^{(k)} \right) \tag{7.65}$$

is the diffusive entropy flux, while

$$T\sigma^{(S)} = -\frac{1}{T}\mathbf{J}^{(q)} \cdot \nabla T - \sum_{k=1}^{n} \mathbf{J}_d^{(k)} \cdot \left[T\nabla\left(\frac{\mu^{(k)}}{T}\right) + \nabla\psi^{(k)} \right] - \tilde{\mathbf{P}}{:}\nabla\mathbf{v} - \sum_{j=1}^{r} \mathcal{J}^{(j)} A^{(j)},$$

(7.66)

is the entropy production.

Different, albeit equivalent, expressions can also be found [1], considering that

$$\nabla\left(\frac{\mu^{(k)}}{T}\right) = \frac{1}{T}[\nabla\mu^{(k)}]_T + \left(\frac{\partial(\mu^{(k)}/T)}{\partial T}\right)\nabla T,$$

(7.67)

where the subscript "T" indicates that the derivative must be taken at constant temperature, Then, taking into account that $h^{(k)} = \mu^{(k)} - T\partial\mu^{(k)}/\partial T$, where $h^{(k)}$ is the partial molar enthalpy, we obtain:

$$T\nabla\left(\frac{\mu^{(k)}}{T}\right) = [\nabla\mu^{(k)}]_T - \frac{h^{(k)}}{T}\nabla T.$$

(7.68)

Finally, we obtain the following alternative expressions for the heat flux and the entropy production,

$$\mathbf{J}^{(q)'} = \mathbf{J}^{(q)} - \sum_{k=1}^{n} \mathbf{J}_d^{(k)} h^{(k)},$$

(7.69)

and,

$$T\sigma^{(S)} = -\frac{1}{T}\mathbf{J}^{(q)'} \cdot \nabla T - \sum_{k=1}^{n} \mathbf{J}_d^{(k)} \cdot \left([\nabla\mu^{(k)}]_T + \nabla\psi^{(k)}\right) - \tilde{\mathbf{P}}{:}\nabla\mathbf{v} - \sum_{j=1}^{r} \mathcal{J}^{(j)} A^{(j)}.$$

(7.70)

Still, another form of the entropy production can be obtained using the equality,

$$T\nabla\left(\frac{\mu^{(k)}}{T}\right) = \nabla\mu^{(k)} - \frac{\mu^{(k)}}{T}\nabla T,$$

(7.71)

and the definition (7.65) of the entropy diffusive flux, obtaining,

$$T\sigma^{(S)} = -\mathbf{J}_d^{(S)} \cdot \nabla T - \sum_{k=1}^{n} \mathbf{J}_d^{(k)} \cdot \nabla\tilde{\mu}^{(k)} - \tilde{\mathbf{P}}{:}\nabla\mathbf{v} - \sum_{j=1}^{r} \mathcal{J}^{(j)} A^{(j)}.$$

(7.72)

7.8 Problems

Problem 7.1 Derive the equation of conservation of the electrical charge.

Problem 7.2 Show that when some of the chemical species are electrically charged in the entropy equation the following additional term is included:

$$T\sigma^{(S)} = \sum_{k=1}^{n} z^{(k)} \mathbf{J}_d^{(k)} \cdot (\mathbf{E} + \mathbf{v} \times \mathbf{B}),$$ (7.73)

where $z^{(k)}$ and $\mathbf{J}_d^{(k)}$ are the charge per unit mass and the diffusive flux of component k, respectively, while $\mathbf{E}$ and $\mathbf{B}$ are the electric and magnetic fields.

References

1. de Groot, S.R., Mazur, P.: Non-Equilibrium Thermodynamics. Dover, New York (1984), Chap. III.3
2. Leal, L.G.: Laminar Flow and Convective Transport Processes, pp. 178–179. Butterworth, Stoneham (1992)

Chapter 8
Constitutive Relations

Assuming linear relations between the thermodynamic fluxes and the thermodynamic forces derived in Chap. 7, in this chapter we apply Onsager's reciprocal relations to determine the most general symmetry conditions that are satisfied by the phenomenological coefficients (Sects. 8.1 and 8.2). Then, in Sect. 8.3, these conditions are applied to particular cases, obtaining the constitutive relations of pure fluids, multicomponent mixtures, metals and non-isotropic media. Finally, in Sect. 8.4, we see how these same symmetry conditions can be applied to find the cross correlation properties of the fluctuating thermodynamic fluxes.

8.1 Introduction

In the previous chapter, we found that the entropy production rate can be written as the sum of products between thermodynamic fluxes and forces [cfr. Eq. (7.72)]. Now, using the notation of Sect. 2.3, we should distinguish between x- and y-variables, these latter being defined as even and odd functions of the particle velocities, respectively, i.e. [cfr. Eq. (2.34)]:

$$\frac{1}{k}\sigma^{(S)} = \frac{1}{k}\frac{D\Delta S}{Dt} = \sum_{i=1}^{n} J_i^{(x)} X_i + \sum_{i=1}^{n} J_i^{(y)} Y_i. \qquad (8.1)$$

Here,

$$J_i^{(x)} = \frac{Dx_i}{Dt}, \qquad J_i^{(y)} = \frac{Dy_i}{Dt} \qquad (8.2)$$

are diffusive fluxes, being the material derivatives of the independent x- and y-variables, while,

$$X_i = \frac{1}{k}\frac{\partial \sigma^{(S)}}{\partial J_i^{(x)}} = \frac{1}{k}\frac{\partial \Delta S}{\partial x_i}; \qquad Y_i = \frac{1}{k}\frac{\partial \sigma^{(S)}}{\partial J_i^{(y)}} = \frac{1}{k}\frac{\partial \Delta S}{\partial y_i} \qquad (8.3)$$

are the thermodynamic forces defined in (1.16).

R. Mauri, *Non-Equilibrium Thermodynamics in Multiphase Flows*,
Soft and Biological Matter, DOI 10.1007/978-94-007-5461-4_8,
© Springer Science+Business Media Dordrecht 2013

Now we restrict ourselves to the linear regime, near the equilibrium point, and therefore we assume that fluxes and forces are related through linear relations,[1]

$$J_i^{(x)} = \sum_{j=1}^{n} L_{ij}^{(xx)} X_j + \sum_{j=1}^{n} L_{ij}^{(xy)} Y_j, \qquad (8.4)$$

$$J_i^{(y)} = \sum_{j=1}^{n} L_{ij}^{(yx)} X_j + \sum_{j=1}^{n} L_{ij}^{(yy)} Y_j. \qquad (8.5)$$

Under this hypothesis, the phenomenological coefficients L_{ik} satisfy the Onsager-Casimir reciprocity relations,

$$L_{ij}^{(xx)} = L_{ij}^{(xx)}; \qquad L_{ij}^{(xy)} = -L_{ji}^{(yx)}; \qquad L_{ij}^{(yy)} = L_{ij}^{(yy)}. \qquad (8.6)$$

In the following, Eq. (8.6) will be applied to find the constitutive relations of single- and multi-phase fluids, as well as anisotropic solids. Then, at the end, we will apply the fluctuation-dissipation theorem to find the fluctuating part of the thermodynamic fluxes.

Comment 8.1 A fundamental objection against applying Onsager's reciprocal relation to any $J_i X_i$ term in the entropy production expression (8.1) was raised by Coleman and Truesdell [3], who stated that "merely to exhibit a bilinear form for the production of entropy and to assume its entries J_i and X_i linearly related through (8.4)–(8.5) does not imply that the J_i are time derivatives of a thermodynamic variables x or that the forces X_i are determined by Eq. (1.16)."

This objection can be partially answered considering that $J_i = x v_i^{(d)} = (dx/dt)_i$ and $X_i \propto \nabla_i x = (\partial S/\partial x)_i$, indicating that the diffusive flux and the conjugated force associated with a variable x equal, respectively, the time variation of x and its associated entropy gradient as we move along the i-direction with diffusive velocity $v_i^{(d)}$. So, the Onsager relations should be interpreted as referred to a coordinate reference system moving with mean velocity.

Comment 8.2 The Onsager relations can be simply obtained applying the linear phenomenological relations (8.4)–(8.5) to the Betti-Maxwell-Lorentz reciprocity relations, which establish that, for any two non-equilibrium configurations (J_i', X_i') and (J_i'', X_i''), we have: $\sum_i J_i' X_i'' = \sum_i J_i'' X_i'$ (see Sect. F.3.1 in Appendix F).

The complete relations (8.6) can then be obtained simply imposing that we cannot have any $X_i Y_j$ terms in the expression (8.1), since σ_S must be invariant to rotations [2]. So, it seems that the arguments based on statistical fluctuations and the principle of microscopic reversibility are not essential in deriving Onsager's reciprocity relations and therefore they play no fundamental role in the thermodynamic

---

[1]Examples are Fourier's law of heat transport, Fick's law of mass transport and Newton's law of momentum transport.

theory of irreversible processes [1]. On the other hand, one could also argue that the macroscopic balance equations, together with the above reciprocity relations, are but a reflection of more fundamental microscopic balance relations which, as shown in Appendix E, have exactly the same structure as their macroscopic counterparts.

8.2 Applying the Reciprocity Relations

First of all, before we can apply Eq. (8.6), we must rewrite Eq. (7.72) in terms of independent variables. Accordingly, considering that among the n material diffusive fluxes only $n - 1$ are independent [cfr. Eq. (7.23)], we have:

$$\sigma^{(S)} = -\frac{1}{T}\mathbf{J}_d^{(S)} \cdot \nabla T - \sum_{k=1}^{n-1}\frac{1}{T}\mathbf{J}_d^{(k)} \cdot \nabla\widetilde{\mu}^{(kn)} - \frac{1}{T}\widetilde{\mathbf{P}}{:}\nabla\mathbf{v} - \sum_{j=1}^{r}\frac{1}{T}\mathcal{J}^{(j)}A^{(j)}, \quad (8.7)$$

where $\widetilde{\mu}^{(kn)} = \widetilde{\mu}^{(k)} - \widetilde{\mu}^{(n)}$. Physically, this stresses that diffusion needs at least two components, as it consists of the transport of one component into another. Now, comparing Eq. (8.7) with Eq. (8.1), we could identify $\mathbf{J}_d^{S}$, $\mathbf{J}_d^{(k)}$, $\widetilde{\mathbf{P}}$ and $\mathcal{J}^{(j)}$ as fluxes, and $-\nabla T/kT$, $-\nabla\widetilde{\mu}^{(kn)}/kT$, $-\nabla\mathbf{v}/kT$ and $-A^{(j)}/kT$ as the respective thermodynamic forces, the third ones being y-variables (i.e. velocities), while the others are x-variables. At this point, we assume that near equilibrium forces and fluxes are related through linear phenomenological equations (8.4) and (8.5), so that any Cartesian component of a flux can be a linear function of the Cartesian components of all thermodynamic forces. In reality, however, according to the Curie symmetry principle, some of these relations cannot exist when the fluxes and their related thermodynamic forces do not have the same tensorial character, due to the fact that the constitutive relations (8.4)–(8.5) must be invariant under reflection and under rotation. In particular, in an isotropic medium, the Curie principle states that fluxes and forces of different tensorial characters do not couple.[2] This is particularly relevant in our case, as in (8.7) some fluxes (and their related forces) are scalar, some are vectors and one is a second-rank tensor.

Before we proceed, some care must be taken in interpreting Curie's principle, because any second-rank tensor $\mathbf{A}$ may be decomposed as:

$$A_{ij} = A\delta_{ij} + \epsilon_{ijk}a_k + \breve{A}_{ij}^{(s)}, \quad (8.8)$$

where A is one third of the trace,

$$A = \frac{1}{3}A_{ij}\delta_{ij} = \frac{1}{3}(A_{11} + A_{22} + A_{33}), \quad (8.9)$$

---

[2]In certain cases, this fact is trivial. For example, the phenomenological coefficient relating a vectorial flux and a second-order tensorial force must be a third-order tensor, and there is no isotropic third-order tensor. For details, see [5].

a is an axial vector deriving from the antisymmetric part of **A**,

$$a_i = \frac{1}{2}\epsilon_{ijk}A^{(a)}_{jk}; \qquad A^{(a)}_{ij} = \frac{1}{2}(A_{ij} - A_{ji}), \tag{8.10}$$

while $\check{\mathbf{A}}^{(s)}$ is the symmetric, deviatoric (i.e. trace-free) part of **A**,

$$\check{A}^{(s)}_{ij} = \frac{1}{2}(A_{ij} + A_{ji}) - A\delta_{ij}. \tag{8.11}$$

Here, we have used Ricci's third-order antisymmetric tensor ϵ_{ijk}, defined in Sect. 7.6 as $\epsilon_{123} = 1$, $\epsilon_{321} = -1$ (cycl.) and $\epsilon_{ijk} = 0$ otherwise. Using Ricci's tensor, the cross product between two vectors, **a** and **b**, can be written as: $(\mathbf{a} \times \mathbf{b})_i = \sum_{j,k}\epsilon_{ijk}a_jb_k$.[3]

In our case, when $\mathbf{A} = \widetilde{\mathbf{P}}$, as the deviatoric part of $\widetilde{\mathbf{P}}$ and that of **P** coincide with each other, we see that the axial vector, **t**, is the internal torque per unit mass, $\mathbf{t} = \frac{1}{2}\boldsymbol{\epsilon}:\mathbf{P}^{(a)}$.[4] When $\mathbf{A} = \nabla\mathbf{v}$, the trace is the velocity divergence, $\nabla \cdot \mathbf{v}$, while the axial vector is the angular velocity, $\boldsymbol{\Omega}$, i.e. one half of the vorticity, $\boldsymbol{\Omega} = \frac{1}{2}\nabla \times \mathbf{v}$. Therefore:

$$\widetilde{\mathbf{P}} = p'\mathbf{I} + \boldsymbol{\epsilon} \cdot \mathbf{t} + \check{\mathbf{P}}^{(s)}; \qquad \nabla\mathbf{v} = \frac{1}{3}(\nabla \cdot \mathbf{v})\mathbf{I} + \boldsymbol{\epsilon} \cdot \boldsymbol{\Omega} + \check{\mathbf{S}}, \tag{8.12}$$

where p' is the non-thermodynamic pressure-like term defined in (7.50), while the symmetric and traceless part of the velocity gradient,

$$\check{\mathbf{S}} = \frac{1}{2}(\nabla\mathbf{v} + \nabla\mathbf{v}^{\top}) - \frac{1}{3}(\nabla \cdot \mathbf{v})\mathbf{I}. \tag{8.13}$$

is generally referred to as the shear rate tensor.

Using these results the third term on the RHS of Eq. (8.7) becomes:

$$\widetilde{\mathbf{P}}:(\nabla\mathbf{v}) = p'\nabla \cdot \mathbf{v} + \check{\mathbf{P}}^{(s)}:\check{\mathbf{S}}, \tag{8.14}$$

where we have considered that $\mathbf{I}:\mathbf{I} = 3$, the pressure tensor is symmetric (i.e. $\mathbf{t} = \mathbf{0}$) and that the doubly contracted product of a symmetric and an antisymmetric tensor is identically zero, i.e. $\mathbf{A}^{(s)}:\mathbf{B}^{(a)} = 0$. The fact that in isotropic media without any internal body torque the angular velocity does not influence the entropy production rate is an obvious result since the properties of the system should be rotation invariant.

Considering all the above relations, the entropy production rate (8.7) can be written as the sum of three separate contributions, each containing fluxes and thermodynamic forces of the same tensorial character, as fluxes and forces of different

---

[3]Note that $\sum_m \epsilon_{ijm}\epsilon_{mkl} = \delta_{ik}\delta_{jl} - \delta_{il}\delta_{jk}$ and $\sum_{jk}\epsilon_{ijk}\epsilon_{jkl} = 2\delta_{il}$.

[4]Here we have used Gibbs' notation: $\mathbf{A}:\mathbf{B} = A_{ij}B_{ij}$.

tensorial character do not interfere with one another,

$$\sigma_0^{(S)} = -\frac{1}{T} p' \nabla \cdot \mathbf{v} - \frac{1}{T} \sum_{j=1}^{r} \mathcal{J}^{(j)} A^{(j)}; \tag{8.15}$$

$$\sigma_1^{(S)} = -\frac{1}{T} \mathbf{J}_d^{(S)} \cdot \nabla T - \frac{1}{T} \sum_{k=1}^{n-1} \mathbf{J}_d^{(k)} \cdot \nabla \tilde{\mu}^{(kn)}; \tag{8.16}$$

$$\sigma_2^{(S)} = -\frac{1}{T} \check{\mathbf{P}}^{(s)} : \check{\mathbf{S}}, \tag{8.17}$$

which are each separately positive definite.

At this point, the phenomenological equations relating fluxes and thermodynamic forces of the same tensorial order are:

$$p' = -l^{(vv)} \frac{1}{kT} \nabla \cdot \mathbf{v} - \sum_{j=1}^{r} l^{(vj)} \frac{1}{kT} A^{(j)}; \tag{8.18}$$

$$\mathcal{J}^{(i)} = -l^{(iv)} \frac{1}{kT} \nabla \cdot \mathbf{v} - \sum_{j=1}^{r} l^{(ij)} \frac{1}{kT} A^{(j)}; \tag{8.19}$$

$$\mathbf{J}_d^S = -L^{(qq)} \frac{1}{kT} \nabla T - \sum_{k=1}^{n-1} L^{(qk)} \frac{1}{kT} \nabla \tilde{\mu}^{(kn)}; \tag{8.20}$$

$$\mathbf{J}_d^i = -L^{(iq)} \frac{1}{kT} \nabla T - \sum_{k=1}^{n-1} L^{(ik)} \frac{1}{kT} \nabla \tilde{\mu}^{(kn)}; \tag{8.21}$$

$$\check{\mathbf{P}}^{(s)} = -L \frac{1}{kT} \check{\mathbf{S}} \tag{8.22}$$

and the following reciprocity relations can be established:

$$-l^{(vj)} = l^{(jv)}; \qquad l^{(ij)} = l^{(ji)}; \tag{8.23}$$

$$L^{(iq)} = L^{(qi)}; \qquad L^{(ij)} = L^{(ji)}. \tag{8.24}$$

The first of these relations describes a cross effect between an x- and a y-variable and therefore has a minus sign, while all the other relations describe cross effects between x-x or y-y variables.

8.3 Constitutive Relations

8.3.1 Single-Phase Fluid

First, let us consider the simplest case of a single-phase fluid. In this case, there is no reaction or diffusion terms, so that Eq. (8.18)–(8.22) reduce to:

$$p' = -\zeta \nabla \cdot \mathbf{v}; \tag{8.25}$$

$$\mathbf{J}_d^{(S)} = -\kappa \nabla T / T; \tag{8.26}$$

$$\check{\mathbf{P}}^{(s)} = -2\eta \check{\mathbf{S}} \tag{8.27}$$

where $\zeta = l^{(vv)}/kT$ is the bulk viscosity (or second viscosity), $\kappa = L^{(qq)}/k$ is the heat conductivity, while $\eta = L/2kT$ is the shear viscosity. Therefore, considering the definition (7.65) of diffusive entropy flux, we obtain the following constitutive equation for the heat flux and the pressure tensor:

$$\mathbf{J}^{(q)} = -\kappa \nabla T, \tag{8.28}$$

$$\mathbf{P} = \left[p - \zeta (\nabla \cdot \mathbf{v}) \right] \mathbf{I} - \eta \left[\nabla \mathbf{v} + \nabla \mathbf{v}^+ - \frac{2}{3} (\nabla \cdot \mathbf{v}) \right]. \tag{8.29}$$

Sometime this last relation is expressed in the following equivalent form:

$$\mathbf{P} = \left[p - \lambda (\nabla \cdot \mathbf{v}) \right] \mathbf{I} - \eta \left(\nabla \mathbf{v} + \nabla \mathbf{v}^+ \right), \tag{8.30}$$

where $\lambda = \zeta - 2\eta/3$.

Another way to obtain this constitutive relation is to consider that the proportionality term between two second-order tensors is a forth-rank tensor., i.e. $P_{ij} = \upsilon_{ijkl} \nabla_k \upsilon_l$. For isotropic fluids, υ_{ijkl} must be an isotropic tensor and therefore it must be of the form: $\upsilon_{ijkl} = \eta_1 \delta_{ik} \delta_{jl} + \eta_2 \delta_{il} \delta_{jk} - \lambda \delta_{ij} \delta_{kl}$. Now, imposing that $\upsilon_{ijkl} = \upsilon_{jikl}$ since $\mathbf{P}$ is symmetric, we find $\eta_1 = \eta_2$, therefore obtaining the constitutive relation (8.30).

8.3.2 Binary Mixtures: Thermo-Diffusion

Let us consider a non-reactive binary mixture. Compared to the single-component case, here the entropy production rates $\sigma^{(0)}$ and $\sigma^{(2)}$ remain unchanged, so that, at the end, we obtain the same constitutive relation (8.29) for the pressure tensor. On the other hand, $\sigma^{(1)}$ has one more term, describing the diffusion of one component into the other, i.e.

$$\sigma_1^{(S)} = -\frac{1}{T^2} \left(\mathbf{J}^{(q)} - \mathbf{J}_d^{(1)} \mu^{(12)} \right) \cdot \nabla T - \frac{1}{T} \mathbf{J}_d^{(1)} \cdot \nabla \tilde{\mu}^{(12)}, \tag{8.31}$$

where $\mu^{(12)} = \mu^{(1)} - \mu^{(2)}$ and $\widetilde{\mu}^{(12)} = \widetilde{\mu}^{(1)} - \widetilde{\mu}^{(2)}$ are the differences between the (chemical and mechanical) potentials of the two components. Note that (8.31) is invariant to interchanging the two components, as $\mathbf{J}_d^{(2)} = -\mathbf{J}_d^{(1)}$.

For purely diffusive processes, it is more convenient to use the expression (7.70) for the entropy production rate. That means considering that $\nabla \mu^{(1)} = [\nabla \mu^{(1)}]_T + s^{(1)} \nabla T$, and $h^{(1)} = \mu^{(1)} - s^{(1)} T$, where the subscript "$T$" means that the gradient is taken at constant temperature, while $s^{(1)}$ and $h^{(1)}$ are the partial entropy and partial enthalpy of component 1, respectively. At the end, we obtain:

$$\sigma_1^{(S)} = -\frac{1}{T^2} \left(\mathbf{J}^{(q)} - \mathbf{J}_d^{(1)} h^{(12)} \right) \cdot \nabla T - \frac{1}{T} \mathbf{J}_d^{(1)} \cdot \left[\nabla \widetilde{\mu}^{(12)} \right]_T, \tag{8.32}$$

with $h^{(12)} = h^{(1)} - h^{(2)}$, where we have assumed that the potential $\psi^{(k)}$ is not an explicit function of the temperature.

At this point, the phenomenological equations relating fluxes and thermodynamic forces are:

$$\mathbf{J}^{(q)'} = -L^{(qq)} \frac{1}{kT^2} \nabla T - L^{(q1)} \frac{1}{kT} \left[\nabla \widetilde{\mu}^{(12)} \right]_T; \tag{8.33}$$

$$\mathbf{J}_d^{(1)} = -L^{(1q)} \frac{1}{kT} \nabla T - L^{(11)} \frac{1}{kT} \left[\nabla \widetilde{\mu}^{(12)} \right]_T; \tag{8.34}$$

where [cf. Eq. (7.69)],

$$\mathbf{J}^{(q)'} = \mathbf{J}^{(q)} - \mathbf{J}_d^{(1)} h^{(12)}, \tag{8.35}$$

so that the following reciprocity relations can be established:

$$L^{(1q)} = L^{(q1)}. \tag{8.36}$$

In particular, assuming that no external forces are present (i.e. $\psi^{(k)} = 0$ and $\widetilde{\mu}^{(k)} = \mu^{(k)}$) and that viscous effects can be neglected, so that pressure is uniform, then $[\mu^{(12)}]_T$ is a function of the composition $\phi^{(1)}$ only.[5] Accordingly, applying the Gibbs-Duhem relation, $\phi^{(1)} \nabla \mu^{(1)} + \phi^{(2)} \nabla \mu^{(2)} = 0$, we obtain:

$$\left[\nabla \widetilde{\mu}^{(12)} \right]_T = \left(1 + \frac{\phi^{(1)}}{\phi^{(2)}} \right) \left[\nabla \mu^{(1)} \right]_T = \frac{\partial \mu^{(1)} / \partial \phi^{(1)}}{\phi^{(2)}} \nabla \phi^{(1)}. \tag{8.37}$$

Using these results, the phenomenological equations (8.33) and (8.34) can be rewritten as:

$$\mathbf{J}^{(q)'} = -\kappa \nabla T - \rho \phi^{(1)} T \frac{\partial \mu^{(1)}}{\partial \phi^{(1)}} D'' \nabla \phi^{(1)}, \tag{8.38}$$

$$\mathbf{J}_d^{(1)} = -\rho \phi^{(1)} \phi^{(2)} D' \nabla T - \rho D \nabla \phi^{(1)}. \tag{8.39}$$

---

[5]Clearly, this is always true in liquid systems.

Here, κ is the heat conductivity,

$$\kappa = \frac{L^{(qq)}}{kT^2},$$
(8.40)

D'' is the *Dufour coefficient*,

$$D'' = \frac{L^{(q1)}}{\rho\phi^{(1)}\phi^{(2)}kT^2},$$
(8.41)

D' is the *thermal diffusion* coefficient,

$$D' = \frac{L^{(1q)}}{\rho\phi^{(1)}\phi^{(2)}kT^2},$$
(8.42)

and D is the diffusion coefficient,

$$D = \frac{L^{(11)}}{\rho\phi^{(2)}kT} \frac{\partial\mu^{(1)}}{\partial\phi^{(1)}}.$$
(8.43)

Here, D' characterizes the thermo-diffusion phenomenon, that is the flow of matter caused by a temperature gradient, while D'' describes the opposite, so-called Dufour effect, that is a heat flow caused by concentration gradients in isothermal conditions. According to the Onsager reciprocal relation (8.36), we have,

$$D'' = D',$$
(8.44)

showing that thermo-diffusion and Dufour effects are strictly related.

The thermo-diffusion phenomenon is observed at the scale of one millimeter or less.[6] It is labeled "positive" when particles move from a hot to cold region and "negative" when the reverse is true. Typically the heavier/larger species in a mixture exhibits positive thermophoretic behavior while the lighter/smaller species exhibit negative behavior. The phenomenon of thermophoresis in liquid mixtures is generally called *Soret* effect, as it differs from that in gas mixtures and is not as well understood, in general.

Let us see how the thermo-diffusion coefficient could be measured. Consider a system enclosed in a reservoir, where convection phenomena can be neglected. Furthermore, assume that our system consists of a binary mixture, whose components are liquids of approximately the same density, so that the overall density ρ is roughly uniform. For measuring the thermal diffusion coefficient, D', let us fix the temperature difference between two walls. At steady state, applying (8.38), where the heat flux is constant and the Dufour term can be neglected compared to heat

---

[6]A simple example of thermo-diffusion is when the hot rod of an electric heater is surrounded by tobacco smoke: as the small particles of air nearest the hot rod are heated, they create a fast flow away from the rod, down the temperature gradient, thereby carrying with them the slower-moving particles of the tobacco smoke.

conducting term, we see that the temperature gradient will be uniform. With this result, applying (8.39) with $\mathbf{J}_d^{(1)} = \mathbf{0}$, we obtain the so-called *Soret coefficient*, s_T,

$$s_T = \frac{D'}{D} = -\frac{1}{\phi^{(1)}\phi^{(2)}}\frac{\nabla\phi^{(1)}}{\nabla T}, \tag{8.45}$$

showing that the thermo-diffusion coefficient D' can be determined by measuring the resulting concentration gradient.[7] Experimentally, s_T turns out to be of the order of 10^{-4}–10^{-2} K^{-1}, both for liquids and gaseous mixtures [6]. Accordingly, for liquids $D \approx 10^{-5}$ cm^2/s and $D' = 10^{-7}$–10^{-9} cm^2/s K, while for gases $D \approx 10^{-1}$ cm^2/s and $D' = 10^{-3}$–10^{-5} cm^2/s K.[8]

8.3.3 Metals: Thermo-Electricity

A metal can be considered as a binary system, where the first component is formed by the electrons and the second by the positive ion lattice. Since there is no viscous dissipation and no chemical reactions, the entropy production rate (7.72) becomes:

$$T\sigma^{(S)} = -\mathbf{J}_d^{(S)} \cdot \nabla T - \sum_{k=1}^{2}\mathbf{J}_d^{(k)} \cdot \nabla\widetilde{\mu}^{(k)}, \tag{8.46}$$

where the electrochemical potential,

$$\widetilde{\mu}^{(k)} = \mu^{(k)} + \psi^{(k)} \tag{8.47}$$

is the sum of the chemical potential, $\mu^{(k)}$, and the product $\psi^{(k)} = z^{(k)}\phi$, where $z^{(k)}$ is the charge per unit mass of component k, and ϕ is the electrostatic potential, so that $\mathbf{E} = -\nabla\phi$ is the electric field. Now, when the diffusive fluxes are measured with respect to the ion lattice, the barycentric velocity is zero, $\mathbf{v} = \mathbf{0}$, so that a) ion diffusion can be neglected; b) the magnetic field appearing in Eq. (7.73) does not play any role.[9] Accordingly, Eq. (8.46) becomes:

$$T\sigma^{(S)} = -\mathbf{J}_d^{(S)} \cdot \nabla T - \mathbf{J}^{(e)} \cdot \nabla\phi^{(e)}, \tag{8.48}$$

with the superscript "e" indicating the electron phase, where

$$\mathbf{J}^{(e)} = z^{(e)}\mathbf{J}_d^{(e)} \tag{8.49}$$

[7]Care should be paid to implement this experiment, as the final stationary state is reached after a time L^2/D, which may be very long.

[8]Sometimes, it is preferred to use the thermal diffusion factor, $k_T = \phi^{(1)}\phi^{(2)}T D'/D$, with values laying between 0.01 and 1.

[9]For details, see [4].

is the total electric current, while

$$\phi^{(e)} = \phi + \frac{\mu^{(e)}}{z^{(e)}} \qquad (8.50)$$

is an effective electrostatic potential.[10]

Using this expression for the entropy production rate, wee see that the phenomenological equations relating fluxes and thermodynamic forces are:

$$\mathbf{J}_d^{(S)} = -L^{(qq)}\nabla T - L^{(qe)}\nabla\phi^{(e)}; \qquad (8.51)$$

$$\mathbf{J}^{(e)} = -L^{(eq)}\nabla T - L^{(ee)}\nabla\phi^{(e)}, \qquad (8.52)$$

where the following reciprocity relations can be established:

$$L^{(qe)} = L^{(eq)}. \qquad (8.53)$$

Sometimes, it is more useful to write these equations as:

$$\mathbf{J}_d^{(S)} = -\frac{k}{T}\nabla T + \pi\mathbf{J}^{(e)}; \qquad (8.54)$$

$$\nabla\phi^{(e)} = -\eta_S\nabla T - R\mathbf{J}^{(e)}, \qquad (8.55)$$

where k is the heat conductivity at zero electrical current, R is the isothermal resistivity of the medium, $\pi = L^{(qe)}/L^{(ee)}$ is the *Peltier* coefficient, while $\eta_S = L^{(eq)}/L^{(ee)}$ is the thermal electricity, or *Seebeck*, coefficient. Then, the reciprocity relation (8.53) becomes,

$$\eta_S = \pi. \qquad (8.56)$$

These constitutive relations describe the thermo-electric effects, that is the direct conversion of temperature differences into electric voltage and vice-versa. The former is the Seebeck effect: in the absence of any electrical current, an electric field can be induced by a temperature gradient. Its most important application is the thermocouple, that is a device consisting of two wires made of different metals, A and B, connecting two heat reservoirs having a ΔT temperature difference. In addition, the thermocouple is an open circuit, e.g. it is connected to a balanced potentiometer, so that there will be no electric current. Therefore, since the Seebeck coefficient is a function of the nature of the substance (and of temperature, as well), it is easily seen from (8.55) that, between the capacitor plates, a potential difference $\Delta\phi^{(e)} = \pi_{AB}\Delta T$ will be induced, where $\pi_{AB} = (\pi_A - \pi_B)$.

The opposite phenomena describes the Peltier effect, where a heat flux is induced by an electric current in isothermal conditions. Therefore, when a unit electric current traverses a junction of two different conductors at uniform temperature, a certain heat must be supplied, or withdrawn, over the Joule heat, to keep the junction at

---

[10]Note that, consequently, $\mathbf{E}' = -\nabla\phi^{(e)} = \mathbf{E} - \nabla\mu^{(e)}/z^{(e)}$ is an effective electric field.

constant temperature. This is the Peltier heat, which, from (8.54), can be expresses as $\mathbf{J}_d^{(S)}/\mathbf{J}^{(e)} = \pi_{AB}$.

Therefore, the Onsager relation establishes the so-called second Thomson relation between the Peltier heat and the Seebeck thermoelectric potential as follows:

$$\pi_{AB} = \frac{d\phi^{(e)}}{dT}. \tag{8.57}$$

A third thermo-electric phenomenon is the Thomson effect, that is the heating or cooling of a current-carrying conductor in the presence of a temperature gradient.[11] In fact, substituting the constitutive relations (8.54)–(8.55) into the entropy generation term (8.46) and the entropy equation (7.64) we obtain, after rearranging:

$$\rho\frac{Ds}{Dt} = \frac{1}{T}\nabla\cdot(k\nabla T) - \nabla\cdot\left(\pi\mathbf{J}^{(e)}\right) + \frac{1}{T}R|\mathbf{J}^{(e)}|^2, \tag{8.58}$$

where we have applied the Onsager relation (8.56). Here the first term on the RHS represents the entropy change due to heat conduction, while the last term is the Joule heat; the second term is the most interesting, as it is related to both Peltier and Thomson heats. In fact, considering that $\nabla\cdot\mathbf{J}^{(e)} = 0$ due to electroneutrality, we have:

$$-\nabla\cdot\left(\pi\mathbf{J}^{(e)}\right) = -\frac{1}{T}\mathbf{J}^{(e)}\cdot\nabla\pi, \quad \text{with } \nabla\pi = (\nabla\pi)_T + \frac{\partial\pi}{\partial T}\nabla T, \tag{8.59}$$

and so:

$$-\nabla\cdot\left(\pi\mathbf{J}^{(e)}\right) = -\frac{1}{T}\mathbf{J}^{(e)}\cdot(\nabla\pi)_T + \frac{1}{T}\sigma\mathbf{J}^{(e)}\cdot\nabla T, \tag{8.60}$$

where,

$$\sigma = -T\frac{\partial\pi}{\partial T} \tag{8.61}$$

is the *Thomson coefficient*. The first term in (8.60) describes the Peltier heat, which is induced by changes of the Peltier coefficient at constant temperature, due, for example, to the non-homogeneity of the system. The second term, on the other hand, describes the so-called Thomson heat, that is a heat effect due to the combined action of an electric current and a temperature gradient. Equation (8.61) is known as Thomson's first relation.

---

[11] See the beautiful description in [8].

8.3.4 Diffusion in Multicomponent Mixtures

Consider isothermal and isobaric diffusion in isotropic, non reactive mixtures in the absence of external forces. The entropy production rate can be written as:

$$\sigma^{(S)} = -\frac{1}{T} \sum_{k=1}^{n-1} \mathbf{J}_d^{(k)} \cdot \nabla \mu^{(kn)}, \tag{8.62}$$

with the following phenomenological equations:

$$\mathbf{J}_d^{(i)} = -\sum_{j=1}^{n-1} L^{(ij)} \frac{1}{kT} \nabla \mu^{(jn)}; \tag{8.63}$$

where $L^{(ij)} = L^{(ji)}$.

First, let us see what happens for binary mixtures. In this case, as we saw in the previous section, we obtain the following constitutive relation:

$$\mathbf{J}_d^{(1)} = -\rho D \nabla \phi^{(1)}, \tag{8.64}$$

where the diffusion coefficient D is related to the Onsager coefficient $L^{(11)}$ through the Eq. (8.43), namely,

$$D = \frac{L^{(11)}}{\rho \phi^{(2)} kT} \frac{\partial \mu^{(1)}}{\partial \phi^{(1)}}. \tag{8.65}$$

The chemical potential $\mu^{(1)}$ can be written as [7]:

$$\mu^{(1)} = \mu_0^{(1)} + \frac{RT}{M_w^{(1)}} \ln \left(f^{(1)} x^{(1)} \right), \tag{8.66}$$

where $M_w^{(1)}$ is the molecular mass of component 1, $f^{(1)}$ is the activity coefficient, which is equal to unity for ideal mixtures, while $x^{(1)}$ is the mole fraction, which is related to the mass fraction through the following relation,

$$\phi^{(1)} = \frac{M_w^{(1)} x^{(1)}}{M_w^{(1)} x^{(1)} + M_w^{(2)} x^{(2)}}. \tag{8.67}$$

Accordingly, $\mu_0^{(1)}$ is the chemical potential of a hypothetical one-molar ideal solution. Therefore, considering that from this last expression we obtain,

$$\frac{\partial x^{(1)}}{\partial \phi^{(1)}} = \frac{x^{(1)} x^{(2)}}{\phi^{(1)} \phi^{(1)}}, \tag{8.68}$$

we see that Eq. (8.43) becomes:

$$D = L^{(11)} \frac{N_A x^{(2)}}{\rho \phi^{(1)} [\phi^{(2)}]^2 M_w^{(1)}} \left(1 + \frac{\partial \ln f^{(1)}}{\partial \ln x^{(1)}} \right), \tag{8.69}$$

where the last factor is missing for ideal mixtures.

Now, let us consider the general, multicomponent case. As we did in the previous section, apply the Gibbs-Duhem relation,

$$\sum_{i=1}^{n} \phi^{(i)} \nabla \mu^{(i)} = 0, \tag{8.70}$$

obtaining:

$$\sigma^{(S)} = k \sum_{k=1}^{n-1} \mathbf{J}_d^{(i)} \cdot \mathbf{X}^{(i)}, \tag{8.71}$$

where

$$\mathbf{X}^{(i)} = -\frac{1}{kT} \sum_{j=1}^{n-1} A^{(jk)} \nabla \mu^{(k)}; \quad A^{(jk)} = \delta_{jk} + \frac{\phi^{(k)}}{\phi^{(n)}}. \tag{8.72}$$

Now, express the chemical potential in terms of composition as

$$\nabla \mu^{(k)} = \sum_{\ell=1}^{n-1} M^{(k\ell)} \nabla \phi^{(\ell)}; \quad M^{(k\ell)} = \frac{\partial \mu^{(k)}}{\partial \phi^{(\ell)}}, \tag{8.73}$$

so that the phenomenological equations can be written as:

$$\mathbf{J}_d^{(i)} = -\sum_{\ell=1}^{n-1} D^{(i\ell)} \nabla \tilde{\phi}^{(\ell)}, \tag{8.74}$$

where $D^{(i\ell)}$ are the diffusion coefficients,

$$D^{(i\ell)} = \sum_{j=1}^{n-1} L^{(ij)} G^{(j\ell)}; \quad G^{(j\ell)} = \frac{1}{kT} \sum_{j,k=1}^{n-1} A^{(jk)} M^{(kl)}. \tag{8.75}$$

Now, inverting Eq. (8.75), we obtain:

$$L^{(ij)} = \sum_{k=1}^{n-1} D^{(ik)} \left[G^{(kj)} \right]^{-1}. \tag{8.76}$$

These results can be simplified considering that $G^{(ij)}$ is a symmetric matrix. In fact, from

$$[dg]_{T,P} = \sum_{i=1}^{n} \mu^{(i)} d\phi^{(i)} = \sum_{i=1}^{n-1} \mu^{(in)} d\phi^{(i)}, \quad \mu^{(in)} = \mu^{(i)} - \mu^{(n)}, \tag{8.77}$$

we easily find the Maxwell relation,

$$\frac{\partial \mu^{(in)}}{\partial \phi^{(j)}} = \frac{\partial \mu^{(jn)}}{\partial \phi^{(i)}}. \tag{8.78}$$

At this point, if we eliminate $\mu^{(n)}$ with the help of the Gibbs-Duhem relations (8.70), these expressions become:

$$\sum_{k=1}^{n-1} A^{(ik)} M^{(kj)} = \sum_{k=1}^{n-1} A^{(jk)} M^{(ki)}, \tag{8.79}$$

that is $G^{(ij)} = G^{(ji)}$.

Considering this symmetry result and applying Onsager's reciprocal relation $L^{(ij)} = L^{(ji)}$ to (8.76), we finally obtain:

$$\sum_{k=1}^{n-1} D^{(ik)} \left[G^{(kj)} \right]^{-1} = \sum_{k=1}^{n-1} D^{(jk)} \left[G^{(ik)} \right]^{-1}; \quad \text{i.e.} \quad \mathbf{D} \cdot \mathbf{G}^{-1} = \mathbf{G}^{-1} \cdot \mathbf{D}^+, \tag{8.80}$$

or equivalently, multiplying by $\mathbf{G}$ in front and in the back on both members,

$$\sum_{k=1}^{n-1} G^{(ik)} D^{(kj)} = \sum_{k=1}^{n-1} D^{(ki)} G^{(kj)}; \quad \text{i.e.} \quad \mathbf{G} \cdot \mathbf{D} = \mathbf{D}^+ \cdot \mathbf{G}. \tag{8.81}$$

These are $\frac{1}{2}(n-1)(n-2)$ relations, which reduce the number of $(n-1)^2$ diffusion coefficients in (8.74) to $\frac{1}{2}n(n-1)$ independent coefficients.

For ternary mixtures, Eqs. (8.81) consist of a single relation, which reduces the independent diffusion coefficients from 4 to 3. Written explicitly, Eq. (8.81) is:

$$G^{(11)} D^{(12)} + G^{(12)} D^{(22)} = D^{(11)} G^{(12)} + D^{(21)} G^{(22)}. \tag{8.82}$$

The simplest case is that of an ideal mixture, where all components have the same molecular weight, M_w, so that mass and molar fractions are equal to each other. Then,

$$\mu^{(i)} = \frac{RT}{M_w} \ln \phi^{(i)}, \qquad M^{(ik)} = \frac{\partial \mu^{(i)}}{\partial \phi^{(k)}} = \frac{RT}{M_w} \frac{\delta^{(ik)}}{\phi^{(i)}}. \tag{8.83}$$

Therefore:

$$G^{(ik)} = \frac{N_A}{M_w} \left(\frac{\delta^{(ik)}}{\phi^{(i)}} + \frac{1}{\phi^{(n)}} \right). \tag{8.84}$$

At the end we obtain:

$$\phi^{(2)} \left(1 - \phi^{(2)} \right) D^{(12)} - \phi^{(1)} \left(1 - \phi^{(1)} \right) D^{(21)} = \phi^{(1)} \phi^{(2)} \left(D^{(11)} - D^{(22)} \right). \tag{8.85}$$

As you can see, even in this simplest case, the result is rather elaborate!!

8.3.5 Non Isotropic Media

Consider first the heat conduction in a non isotropic matrix. In the absence of any mass transport, the entropy production rate can be written simply as

$$\sigma^{(S)} = -\frac{1}{T^2} \mathbf{J}^{(q)} \cdot \nabla T, \tag{8.86}$$

with phenomenological equations,

$$\mathbf{J}^{(q)} = -\mathbf{L}^{(qq)} \cdot \frac{1}{kT^2} \nabla T = -\boldsymbol{\kappa} \cdot \nabla T, \tag{8.87}$$

where, according to Onsager's reciprocal relation, $\kappa = \mathbf{L}^{(qq)}/kT^2$ is a symmetric tensor, i.e.

$$\kappa_{ij} = \kappa_{ji}. \tag{8.88}$$

This relation was first proposed by Maxwell.

Identical relations exist in the following cases:

- Electric conduction. Here, as it appears from Eq. (8.55) the flux is the electrical current density, $\mathbf{J}^{(e)}$, the thermodynamic force is the gradient of the effective electrostatic potential, $\nabla \phi^{(e)}$, and the phenomenological coefficient is the inverse of the resistance tensor, $\mathbf{R}^{-1}$. The symmetry of the resistance tensor can also be seen as a consequence of Maxwell's reciprocal theorem (see Sect. F.3.1).
- Flow through porous media. Here the flux is the mean fluid velocity, $\mathbf{v}$, the thermodynamic force is the pressure gradient, ∇p, and the phenomenological coefficient is the Darcy permeability tensor, $\mathbf{k}$ (see Sect. 10.3). The symmetry of the permeability tensor can also be seen as a consequence of the Lorentz reciprocal theorem of slow viscous flows (see Sect. F.3.1).
- Elasticity. Here the flux is the deformation vector, $\boldsymbol{\xi}$ the thermodynamic force is the load, i.e. a force per unit volume $\mathbf{F}$, and the phenomenological coefficient is the flexibility tensor, $\mathbf{f}$. The symmetry of the flexibility tensor is generally referred to as the Betti-Maxwell relation.

Similar considerations can also apply to momentum transport, leading to symmetry relations of the viscosity forth-rank tensor. In fact, the constitutive relation (8.27) can be generalized as:

$$\check{P}_{ij}^{(s)} = -2\eta_{ijk\ell} \check{S}_{k\ell}, \tag{8.89}$$

where, by construction, considering that both $\check{\mathbf{P}}^{(s)}$ and $\check{\mathbf{S}}$ are symmetric and trace-free, we have:

$$\eta_{ijk\ell} = \eta_{jik\ell} = \eta_{ij\ell k}; \qquad \delta_{ij}\eta_{ijk\ell} = \eta_{ijk\ell}\delta_{k\ell} = 0. \tag{8.90}$$

In addition, the Onsager reciprocal relation gives the additional following symmetry relation:

$$\eta_{ijk\ell} = \eta_{k\ell ij}. \tag{8.91}$$

Righi-Leduc Effect In the presence of a magnetic field, the Onsager relation (8.88) can be written:

$$\kappa_{ij}(\mathbf{B}) = \kappa_{ji}(-\mathbf{B}). \tag{8.92}$$

Decomposing κ as the sum of a symmetric and an antisymmetric tensor [see Eq. (8.8)–(8.10)], we obtain:

$$\kappa_{ij}^s(\mathbf{B}) = \kappa_{ij}^s(-\mathbf{B}); \qquad \kappa_{ij}^a(\mathbf{B}) = -\kappa_{ij}^a(-\mathbf{B}). \tag{8.93}$$

In addition, applying the decomposition (8.12), we obtain:

$$\kappa = \kappa^s + \boldsymbol{\epsilon} \cdot \mathbf{k}', \tag{8.94}$$

where $\mathbf{k}' = \frac{1}{2}\boldsymbol{\epsilon}{:}\kappa^a$ is the axial vector that can be obtained from the antisymmetric part of κ. Clearly, $\mathbf{k}'(\mathbf{B}) = -\mathbf{k}'(-\mathbf{B})$, and therefore it is non-zero only in the presence of a magnetic field. Substituting (8.94) into (8.37), we obtain:

$$\mathbf{J}_q = -\kappa^s \cdot \nabla T + \mathbf{k}' \times \nabla T. \tag{8.95}$$

Let us consider the simplest case of a system that is isotropic in the absence of a magnetic field. If $\mathbf{B}$ is applied along the z-axis, the heat conduction tensor has the form:

$$\kappa = \begin{pmatrix} \kappa_{xx} & \kappa_{xy} & 0 \\ -\kappa_{xy} & \kappa_{xx} & 0 \\ 0 & 0 & \kappa_{zz} \end{pmatrix} \tag{8.96}$$

where, according to (8.93),

$$\kappa_{xx}(\mathbf{B}) = \kappa_{xx}(-\mathbf{B}); \qquad \kappa_{zz}(\mathbf{B}) = \kappa_{zz}(-\mathbf{B}); \qquad k_{xy}(\mathbf{B}) = -k_{xy}(-\mathbf{B}). \tag{8.97}$$

Note that both the form (8.96) of the heat conductivity and the reciprocity relations (8.97) could be easily derived imposing that the phenomenological equation (8.87) is invariant for rotations around the z-axis. As such, the Onsager relations are already satisfied owing to spatial symmetry.

This phenomenon shows that, in the presence of a magnetic field, heat flow in an isotropic medium can have a different direction than the temperature gradient. It is called the Righi-Leduc effect, and is the thermal analogue of the Hall effect, which arises when electric conduction occurs in a magnetic field.

8.4 Fluctuations of Thermodynamic Fluxes

In Chap. 3 [cf. Eq. (3.43)] we have seen that a flux, $\dot{x}_i$, is the sum of a phenomenological part and a fluctuation part, $\tilde{J}_i$,

$$\dot{x}_i = L_{ik} X_k + \tilde{J}_i, \tag{8.98}$$

where $X_i = k^{-1} \partial \dot{S} / \partial \dot{x}_i$ is the thermodynamic force conjugated to the flux $\dot{x}_i$, while
L_{ik} are phenomenological transport coefficients, which are related to each other[12]
through the Onsager reciprocity relation, $L_{ik} = L_{ki}$. In addition, the fluctuating
fluxes satisfy the relation (3.50):

$$\langle \tilde{J}_i(t_1) \tilde{J}_k(t_2) \rangle = 2 L_{ik} \delta(t_1 - t_2), \tag{8.99}$$

showing that, according to the fluctuation-dissipation theorem, their intensity is proportional to the related transport coefficients at the same time.

Now, let us consider the case of a single isotropic fluid. We saw that the total
entropy production equals:

$$\dot{S} = \int_V \sigma^{(S)} \, dV = \lim_{\Delta V \to 0} \sum \sigma^{(S)} \Delta V, \tag{8.100}$$

where

$$\sigma^{(S)} = -\frac{1}{T^2} J_i^{(q)} \nabla_i T - \frac{1}{T} \tilde{P}_{ij} \nabla_1 v_j \tag{8.101}$$

is the entropy production per unit volume. Therefore, we see that if we identify
$\mathbf{J}^{(q)}$ and $\tilde{\mathbf{P}}$ as fluxes, the respective thermodynamic forces are $-\frac{1}{kT^2} \nabla T \Delta V$ and
$-\frac{1}{kT} \nabla v \Delta V$ and the Langevin equation can be written as:

$$J_i^{(q)} = -L_{ik}^{(qq)} \frac{1}{kT^2} \nabla T \Delta V + \tilde{J}_i^{(q)}, \tag{8.102}$$

and

$$\tilde{P}_{ij} = -L_{ijk\ell}^{(vv)} \frac{1}{kT} \nabla_k v_\ell \Delta V + \tilde{J}_{ij}^{(v)}, \tag{8.103}$$

where we have taken into account the fact that, according to Curie's principle, heat
and momentum fluxes cannot be cross correlated, since they have different tensorial
order. Comparing Eqs. (8.102) and (8.103) with the constitutive relations (8.28) and
(8.29), we see that

$$L_{ik}^{(qq)} = \frac{kT^2}{\Delta V} \kappa \, \delta_{ij}, \tag{8.104}$$

---

[12]Here we assume that the variables are all of the x-type, i.e. they are invariant to time reversal
transformation. Generalization to y-type variables is straightforward.

and

$$L_{ijk\ell}^{(vv)} = \frac{kT}{\Delta V}\left[\eta(\delta_{ik}\delta_{j\ell} + \delta_{i\ell}\delta_{jk}) + \left(\zeta - \frac{2}{3}\eta\right)\delta_{ij}\delta_{k\ell}\right]. \qquad (8.105)$$

Note that in the constitutive equations above the fluxes are taken in the same location (as well as at the same time) as the forces. That means that there are no non-local effects, i.e. the response of the system in a certain point (and at a certain time) depends only on its configuration at the same point. Accordingly, we see that (a) fluctuations of the shear stress are not correlated with those of the heat flux; (b) fluctuations of the shear stress in two different volumes ΔV are not correlated with each other, and likewise for the heat flux. Accordingly,

$$\langle \tilde{J}_i^{(q)}(\mathbf{r}_1, t_1)\tilde{J}_{k\ell}^{(v)}(\mathbf{r}_2, t_2)\rangle = 0; \qquad (8.106)$$

$$\langle \tilde{J}_i^{(q)}(\mathbf{r}_1, t_1)\tilde{J}_k^{(q)}(\mathbf{r}_2, t_2)\rangle = 0 \quad \text{if } \mathbf{r}_1 \neq \mathbf{r}_2, \qquad (8.107)$$

$$\langle \tilde{J}_{ij}^{(v)}(\mathbf{r}_1, t_1)\tilde{J}_{k\ell}^{(v)}(\mathbf{r}_2, t_2)\rangle = 0 \quad \text{if } \mathbf{r}_1 \neq \mathbf{r}_2. \qquad (8.108)$$

On the other hand, when the fluctuating fluxes are evaluated at the same location we find for the heat flux:

$$\langle \tilde{J}_i^{(q)}(\mathbf{r}, t_1)\tilde{J}_k^{(q)}(\mathbf{r}, t_2)\rangle = \frac{2\kappa kT^2}{\Delta V}\delta_{ik}\delta(t_1 - t_2). \qquad (8.109)$$

Then, going to the limit $\Delta V \to 0$, we obtain:

$$\langle \tilde{J}_i^{(q)}(\mathbf{r}_1, t_1)\tilde{J}_k^{(q)}(\mathbf{r}_2, t_2)\rangle = 2\kappa kT^2\delta_{ik}\delta(t_1 - t_2)\delta(\mathbf{r}_1 - \mathbf{r}_2). \qquad (8.110)$$

Proceeding in the same way for the shear stresses we obtain:

$$\langle \tilde{J}_{ij}^{(v)}(\mathbf{r}_1, t_1)\tilde{J}_{k\ell}^{(v)}(\mathbf{r}_2, t_2)\rangle$$
$$= 2kT\left[\eta(\delta_{ik}\delta_{j\ell} + \delta_{i\ell}\delta_{jk}) + \left(\zeta - \frac{2}{3}\eta\right)\delta_{ij}\delta_{k\ell}\right]\delta(t_1 - t_2)\delta(\mathbf{r}_1 - \mathbf{r}_2). \qquad (8.111)$$

The same procedure can be applied to evaluate the fluctuations of the diffusive mass flux in isothermal binary mixtures. In that case, the Langevin equation becomes:

$$\mathbf{J}_d^{(1)} = -\rho D\nabla\phi^{(1)} + \tilde{\mathbf{J}}^{(1)}, \qquad (8.112)$$

where the diffusion coefficient D is related to the Onsager coefficient $L^{(11)}$ through the following equation [cf. Eq. (8.43)],

$$D = \frac{L^{(11)}}{\rho\phi^{(2)}kT}\frac{\partial\mu^{(1)}}{\partial\phi^{(1)}}, \qquad (8.113)$$

while the fluctuating flux is characterized by:

$$\langle \tilde{J}_i^{(1)}(\mathbf{r}_1, t_1)\tilde{J}_k^{(1)}(\mathbf{r}_2, t_2)\rangle = 2L^{(11)}\delta_{ik}\delta(t_1 - t_2)\delta(\mathbf{r}_1 - \mathbf{r}_2). \qquad (8.114)$$

In particular, in the dilute limit we find [cf. Eq. (8.69) when $x^{(2)} = \phi^{(2)} = 1$ and $f^{(1)} = 1$, since a dilute mixture is always ideal]:

$$\langle \tilde{J}_i^{(1)}(\mathbf{r}_1, t_1) \tilde{J}_k^{(1)}(\mathbf{r}_2, t_2) \rangle = 2m_1^2 n^{(1)} D \delta_{ik} \delta(t_1 - t_2) \delta(\mathbf{r}_1 - \mathbf{r}_2), \qquad (8.115)$$

where m_1 is the mass of a single particle of the component 1 and $n^{(1)} = \rho^{(1)}/m_1$ is the number density.

8.5 Problems

Problem 8.1 Defining the instantaneous viscosity tensor as $\breve{P}_{ij}^{(s)} = -2\eta_{ijk\ell}\breve{S}_{k\ell}$, find the symmetry relations satisfied by $\eta_{ijk\ell}$.

Problem 8.2 Show that in the dilute limit Eq. (8.115) reduces to (G.32).

References

1. Beretta, G.P., Gyftopoulos, E.P.: J. Chem. Phys. **121**, 2718 (2004)
2. Brenner, H.: Physica A **388**, 3391 (2009)
3. Coleman, B.D., Truesdell, C.: J. Chem. Phys. **33**, 28 (1960)
4. de Groot, S.R., Mazur, P.: Non-Equilibrium Thermodynamics. Dover, New York (1962), Chap. XIII
5. de Groot, S.R., Mazur, P.: Non-Equilibrium Thermodynamics. Dover, New York (1984), Chap. VI.2
6. Legros, et al.: Phys. Rev. A **32**, 1903 (1985)
7. Sandler, L.S.: Chemical and Engineering Thermodynamics, 3rd edn. Wiley, New York (1999), Chap. 7
8. Zemansky, M.W.: Heat and Thermodynamics. McGraw Hill, New York (1957), Sects. 13–10

Chapter 9
Multiphase Flows

In this chapter we derive the equations of motion of multiphase fluids. In the classical theory of multiphase flow, each phase is associated with its own conservation equations (of mass, momentum, energy and chemical species), assuming that it is at local equilibrium and separated from the other phases by zero-thickness interfaces, with appropriate boundary conditions. Instead, here we describe the so-called diffuse interface, or phase field, model, assuming that interfaces have a non-zero thickness, i.e. they are "diffuse", as it is more fundamental than the classical, sharp interface theory and is therefore more suitable to be coupled to all non-equilibrium thermodynamics results. After describing van der Waals' theory of coexisting phases at equilibrium (Sect. 9.2), in Sect. 9.3 we illustrate the main idea of the diffuse interface model, leading to the definition of generalized chemical potentials, where the non uniformity of the composition field is accounted for. Then, in Sect. 9.4, the equations of motion are derived by applying the principle of minimum action, showing that an additional, so called, Korteweg, reversible force appears in the momentum conservation equation. This force is proportional (with a minus sign) to the gradient of the generalized chemical potential and therefore tends to restore the equilibrium conditions (where chemical potentials are uniform). Finally, in Sect. 9.5, we show that for incompressible and symmetric binary mixtures the governing equations simplify considerably.

9.1 Introduction

The transport of momentum, heat and mass in two-phase systems occurs frequently in nature and plays an important role in many areas of science and technology. Familiar examples include: (a) the flow of suspensions through pipes; (b) heat conduction in a composite material; (c) mass transfer from a solid surface to a flowing suspension; (d) bubbly flows through conduits; (e) the flow of granular materials, and many, many more. In fact, it is safe to claim that multiphase transport processes far overshadow the analogous and more commonly investigated single-phase operations in terms of their significance in practical applications.

R. Mauri, *Non-Equilibrium Thermodynamics in Multiphase Flows*,
Soft and Biological Matter, DOI 10.1007/978-94-007-5461-4_9,
© Springer Science+Business Media Dordrecht 2013

The theory of multiphase systems was developed at the beginning of the 19th century by Young, Laplace and Gauss, assuming that different phases are separated by an interface, that is a surface of zero thickness. In this approach, physical properties such as density and concentration, may change discontinuously across the interface and the magnitude of these jumps can be determined by imposing equilibrium conditions at the interface. For example, imposing that the sum of all forces applied to an infinitesimal curved interface must vanish leads to the Young-Laplace equation, stating that the difference in pressure between the two sides of the interface (where each phase is assumed to be at equilibrium) equals the product of surface tension and curvature. Later, this approach was generalized by defining surface thermodynamical properties, such as surface energy and entropy, and surface transport quantities, such as surface viscosity and heat conductivity, thus formulating the thermodynamics and transport phenomena of multiphase systems. At the end of the 19th century, though, another, so-called, diffuse interface (D.I.) approach was proposed, assuming that interfaces have a non-zero thickness, i.e. they are "diffuse." Actually, the basic idea was not new, as it dated back to Maxwell, Poisson and Leibnitz or even Lucretius, who wrote that "a body is never wholly full nor void."[1] Concretely, in a seminal article published in 1893, van der Waals [30] used his equation of state to predict the thickness of the interface, showing that it becomes infinite as the critical point is approached. Later, Korteweg [13] continued this work and proposed an expression for the capillary stresses, which are generally referred to as Korteweg stresses, showing that they reduce to surface tension when the region where density changes from one to the other equilibrium value collapses into a sharp interface.[2]

In the first half of the 20th century, van der Waals' D.I. theory of critical phenomena was generalized by Ginzburg and Landau [17], leading to a general, so-called mean field theory of second-order phase transition, and thereby describing phenomena such as ferromagnetism, superfluidity and superconductivity. Then, at mid 1900, Cahn and Hilliard [6] applied van der Waals' diffuse interface (D.I.) approach to binary mixtures and then used it to describe nucleation and spinodal decomposition [4]. This approach was later extended to model phase separation of polymer blends and alloys [10]. Concomitantly, in the mid 1970s, the D.I. approach was coupled to hydrodynamics, developing a set of conservation equations, that were reviewed by Hohenberg and Halperin [11]. Finally, recent developments in computing technology have stimulated a resurgence of the D.I. approach, above all in the study of systems with complex morphologies and topological changes. A detailed discussion about D.I. theory coupled with hydrodynamics can be found in Antanovskii [2, 3], Lowengrub and Truskinovsky [19], Anderson et al. [1] and, more recently, in Onuki [25], Thiele et al. [21, 29] and Mauri [22]. In order to better understand the basic idea underlying the D.I. theory, let us remind briefly the classical approach to multiphase flow that is used in fluid mechanics. There, the equations

---

[1]T.C. Lucretius [20], "Corpus inani distinctum, quoniam nec plenum naviter extat nec porro vacuum."

[2]For a review of the theory of capillarity, see [28].

of conservation of mass, momentum, energy and chemical species are written separately for each phase, assuming that temperature, pressure, density and composition of each phase are equal to their equilibrium values. Accordingly, these equations are supplemented by appropriate boundary conditions at the interfaces [9]. For example, for the momentum transport we have:

$$\|\mathbf{P}\|_{+}^{-} \cdot \mathbf{n} = \kappa \sigma \mathbf{n} + (\mathbf{I} - \mathbf{nn}) \cdot \nabla \sigma, \qquad \|\mathbf{v}\|_{+}^{-} = 0, \tag{9.1}$$

with $\mathbf{n}$ denoting the normal at the interface, stating that the jump of the momentum flux, or pressure, tensor, $\mathbf{P}$, at the interface is related to the curvature κ, the surface tension σ and its gradient, while velocity $\mathbf{v}$ is continuous. Similar boundary conditions exist also for the transport of heat and mass,

$$\left\|\mathbf{J}^{(q)}\right\|_{+}^{-} \cdot \mathbf{n} = 0; \qquad \|T\|_{+}^{-} = 0, \tag{9.2}$$

and

$$\left\|\mathbf{J}^{(A)}\right\|_{+}^{-} \cdot \mathbf{n} = 0; \qquad \left\|c^{(A)}\right\|_{+}^{-} = (K - 1) c^{(A)}, \tag{9.3}$$

stating that heat flux, $\mathbf{J}^{(q)}$, temperature, T, and the flux of any chemical species A, $\mathbf{J}^{(A)}$, are continuous across the interface, while the concentration, $c^{(A)}$, can undergo a jump, depending on a partition coefficient K, given by thermodynamics. Naturally, this results in a free boundary problem, which means that one of the main problems of this approach is to determine the position of the interface. To that extent, many interface tracking methods have been developed, which have proved very successful in a wide range of situations. However, interface tracking breaks down whenever the interface thickness is comparable to the length scale of the phenomenon that is being studied, such as (a) in near-critical fluids or partially miscible mixtures, as the interface thickness diverges at the critical point, and the morphology of the systems presents self-intersecting free boundaries; (b) near the contact line along a solid surface, in the breakup/coalescence of liquid droplets and, in general, in microfluidics, as the related physical processes act on length scales that are comparable to the interface thickness. In front of these difficulties, the D.I. method offers an alternative approach. Quantities that in the free boundary approach are localized in the interfacial surface, here are assumed to be distributed within the interfacial volume. For example, surface tension is the result of distributed stresses within the interfacial region, which are often called capillary, or Korteweg, stresses. In general, the interphase boundaries are considered as mesoscopic structures, so that any material property varies smoothly at macroscopic distances along the interface, while the gradients in the normal direction are steep. Accordingly, the main characteristic of the D.I. method is the use of an *order parameter*, or *phase field*, which undergoes a rapid but continuous variation across the interphase boundary, while it varies smoothly in each bulk phase, where it can even assume constant equilibrium values. For a single-component system, the phase field is the fluid density ρ, for a liquid binary mixture it is the molar (or mass) fraction ϕ, while in other cases it can be any other parameter, not necessarily with any physical meaning, that allows to reformulate free boundary problems. In all these cases, the D.I. model must include a characteristic interface thickness, over which the phase field changes. In fact, in the asymptotic limit of vanishing interfacial width, the diffuse interface model reduces to the classical free boundary problem.

Based on the above considerations, multiphase flows can be readily modeled using the diffuse interface approach which, being more fundamental than the classical, sharp-interface model, is also more suitable to be coupled to all the non-equilibrium thermodynamics results that we have seen in the previous chapters.

9.2 Equilibrium Conditions

9.2.1 Free Energy and van der Waals' Equation

All thermodynamical properties can be determined from the Helmholtz free energy (B.29). This, in turn, depends on the intermolecular forces which, in a dense fluid, are a combination of weak and strong forces. Fortunately, strong interactions nearly balance each other, so that the net forces acting on each molecule are weak and long-range. In addition, mean field approximation is assumed to be applicable, meaning that molecular interactions are smeared out and can be replaced by the action of a continuous effective medium. Based on these assumptions, the case of dense fluids can be treated as that of nearly ideal gases described in Appendix B.3, so that, allowing for variable density, the molar Helmholtz free energy at constant temperature T can be written as [cf. Eq. (B.54)]

$$f[\rho(\mathbf{x})] = f_{id} + \frac{1}{2} RT N_A \int \left(1 - e^{-\psi(r)/kT}\right) \rho(\mathbf{x} + \mathbf{r}) d^3\mathbf{r}, \qquad (9.4)$$

where k is Boltzmann's constant, $R = N_A k$ is the gas constant, with N_A the Avogadro number, ψ is the pair interaction potential, which depends on the distance $r = |\mathbf{r}|$, ρ is the molar density, while the factor $1/2$ compensates counting twice the interacting molecules. The first term on the RHS,

$$f_{id} = RT \ln \rho, \qquad (9.5)$$

is the molar free energy of an ideal gas (where molecules do not interact). Now, we assume that the interaction potential consists of a long-range term, decaying as r^{-6} (like in the Lennard-Jones potential), while the short-range term is replaced by a hard-core repulsion, i.e.

$$\psi(r) = \begin{cases} -U_0 (r/l)^{-6} & (r > d), \\ \infty & (r < d) \end{cases} \qquad (9.6)$$

where d is the nominal hard-core molecular diameter, l is a typical intermolecular interaction distance, and the non-dimensional constant U_0 represents the strength of the intermolecular potential. When the density is constant, Eq. (9.4) gives the thermodynamic free energy, f_{Th},

$$f_{Th}(T, \rho) = f_{id}(T, \rho) + f_{ex}(T, \rho), \tag{9.7}$$

where

$$f_{ex}(T, \rho) = RT\rho B(T), \tag{9.8}$$

is the excess (i.e. the non ideal part) of the free energy, with

$$B(T) = \frac{1}{2} N_A \int_0^\infty (1 - e^{-\psi(r)/kT}) 4\pi r^2 \, dr \tag{9.9}$$

denoting the first virial coefficient. This integral can be solved as

$$B(T) = 2\pi N_A \int_0^d r^2 \, dr + 2\pi N_A \int_d^\infty (1 - e^{\frac{U_0}{kT}(r/l)^6}) r^2 \, dr = c_2 - \frac{c_1}{RT}, \tag{9.10}$$

where

$$c_1 = \frac{2}{3}\pi U_0 N_A^2 l^6 / d^3 \quad \text{and} \quad c_2 = \frac{2}{3}\pi d^3 N_A \tag{9.11}$$

are the pressure adding term and the excluded molar volume, respectively. Finally we obtain:

$$f_{Th}(\rho, T) = f_{id} + RT c_2 \rho - c_1 \rho \approx RT \ln\left(\frac{\rho}{1 - c_2 \rho}\right) - c_1 \rho, \tag{9.12}$$

that is

$$f_{Th}(\rho, T) = -RT \ln(v - c_2) - \frac{c_1}{v}, \tag{9.13}$$

where $v = \rho^{-1}$ is the molar volume. At this point, applying the thermodynamic equality $P = -(\partial f/\partial v)_{N,T}$, we obtain the van der Waals equation of state,

$$\left(P + \frac{c_1}{v^2}\right) = \frac{RT}{v - c_2}. \tag{9.14}$$

This equation of state could be considerably improved if the term $RT/(v - c_2)$, which is exact in one dimension, is replaced by a more accurate representation of the pressure for a hard-sphere fluid in three dimensions [32].

9.2.2 Critical Point

In the $P - T$ diagram, the vapor-liquid equilibrium curve stops at the critical point, characterized by a critical temperature T_C and a critical pressure P_C. At higher temperatures, $T > T_C$, and pressures, $P > P_C$, the differences between liquid and vapor phases vanish altogether and we cannot even speak of two different phases. In particular, as the critical point is approached, the difference between the specific volume of the vapor phase and that of the liquid phase decreases, until it vanishes at the critical point. Accordingly, near the critical point, since the specific volumes of the two phases, v and $v + \delta v$, are near to each other, we obtain:

$$P(T, v) = P(T, v + \delta v) = P(T, v) + \left(\frac{\partial P}{\partial v}\right)_T \delta v + \frac{1}{2}\left(\frac{\partial^2 P}{\partial v^2}\right)_T (\delta v)^2 + \cdots,$$

where we have considered that the two phases at equilibrium have the same pressure, in addition to having the same temperature. At this point, dividing by v and letting $\delta v \to 0$, we see that at the critical point we have:

$$\left(\frac{\partial P}{\partial v}\right)_T = 0, \quad \text{that is,} \quad \kappa_T \to \infty \text{ as } T \to T_C, \tag{9.15}$$

where κ_T is the isothermal compressibility. Note that this condition is the limit case of the inequality $(\partial P/\partial v)_T \leq 0$, which manifests the internal stability of any single-phase system. In addition, since near an equilibrium point, $\delta f + P \delta v > 0$, expanding δf in a power series of δv, with constant T, we obtain:

$$\delta f_{Th} = \left(\frac{\partial f_{Th}}{\partial v}\right)_T (\delta v) + \frac{1}{2!}\left(\frac{\partial^2 f_{Th}}{\partial v^2}\right)_T (\delta v)^2 + \frac{1}{3!}\left(\frac{\partial^3 f_{Th}}{\partial v^3}\right)_T (\delta v)^3 + \cdots.$$

Finally, considering that $(\partial f_{Th}/\partial v)_T = -P$ and that at the critical point $(\partial^2 f_{Th}/\partial v^2)_T = 0$, we obtain:

$$\frac{1}{3!}\left(\frac{\partial^2 P}{\partial v^2}\right)_{T_C} (\delta v)^3 + \frac{1}{4!}\left(\frac{\partial^3 P}{\partial v^3}\right)_{T_C} (\delta v)^4 + \cdots < 0.$$

Since this equality must be valid for any value (albeit small) of δv (both positive and negative), we obtain:

$$\left(\frac{\partial^2 P}{\partial v^2}\right)_{T_C} = 0, \quad \left(\frac{\partial^3 P}{\partial v^3}\right)_{T_C} < 0. \tag{9.16}$$

Therefore, the critical point corresponds to a horizontal inflection point in the $P - v$ diagram, which means that, since $P = -(\partial f_{Th}/\partial v)_T$,

$$\left(\frac{\partial^2 f_{Th}}{\partial v^2}\right)_{T_C} = 0, \quad \left(\frac{\partial^3 f_{Th}}{\partial v^3}\right)_{T_C} = 0. \tag{9.17}$$

Imposing that at the critical point the $P - v$ curve has a horizontal inflection point, we can determine the constant c_1 and c_2 in the van der Waals equation (the same is

Fig. 9.1 Phase diagram
(P vs. v or μ vs. ϕ)

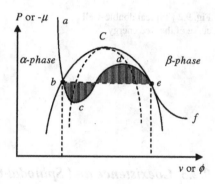

true for any two-parameter cubic equation of state) in terms of the critical constant T_C and P_C, finding:[3]

$$c_1 = \frac{9}{8}RT_C v_C = \frac{27}{64}\frac{(RT_C)^2}{P_C} \quad \text{and} \quad c_2 = \frac{1}{3}v_C = \frac{1}{8}\frac{RT_C}{P_C}. \tag{9.18}$$

Viceversa, the critical pressure, temperature and volume can be determined as functions of c_1 and c_2 as follows:

$$P_C = \frac{1}{27}\frac{c_1}{c_2^2}, \qquad T_C = \frac{8}{27}\frac{c_1}{Rc_2}, \qquad v_C = 3c_2. \tag{9.19}$$

Using these expressions, the van der Waals equation can be written in terms of the reduced coordinates as:

$$\left(P_r + \frac{3}{v_r^2}\right)(3v_r - 1) = 8T_r, \qquad P_r = \frac{P}{P_C}, \qquad v_r = \frac{v}{v_C}, \qquad T_r = \frac{T}{T_C}. \tag{9.20}$$

This equation represents a family of isotherms in the $P_r - v_r$ plane describing the state of any substance, which is the basis of the law of corresponding states. As expected, when $T_r > 1$ the isotherms are monotonically decreasing, in agreement with the stability condition $(\partial P/\partial v)_T < 0$, while when $T_r < 1$ each isotherm has a maximum and a minimum point and between them we have an instability interval, with $(\partial P/\partial v)_T > 0$, corresponding to the two-phase region (see Fig. 9.1).

Note that, considering that $P_C v_C = (3/8)RT_C$ and substituting the expressions for c_1 and c_2 in terms of the intermolecular potential, we obtain the following relation:

$$\left(\frac{l}{d}\right)^2 = \frac{3}{2}\left(\frac{kT_C}{U_0}\right)^{1/3}. \tag{9.21}$$

---

[3]See [18].

Fig. 9.2 Typical double-well
curve of the free energy

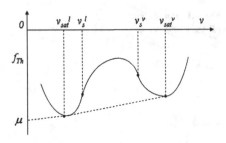

9.2.3 Coexistence and Spinodal Curves

Let us consider a one-component system at equilibrium, whose pressure and temperature are below their critical values, so that it is separated into two coexisting phases, say α and β. According to the Gibbs phase rule, these two phases have the same pressure and temperature and therefore, defining the Gibbs molar free energy $g_{Th} = f_{Th} + Pv$, with $dg_{Th} = -sdT + vdP$, the corresponding equilibrium, or saturation, pressure P_{sat} at a given temperature can be easily determined from the equilibrium condition, stating that at equilibrium the Gibbs molar free energies of the two phases must be equal to each other. So we obtain:

$$g_{Th}^{\beta} - g_{Th}^{\alpha} = \int_{b}^{e} dg_{Th} = 0 \quad \Longrightarrow \quad \int_{b}^{e} v \, dP = [vP]_{b}^{e} - \int_{b}^{e} P \, dv = 0, \quad (9.22)$$

where $P = P(v)$ represents an isotherm transformation. From a geometrical point of view, this relation manifests the equality between the shaded area of Fig. 9.1 (Maxwell's rule), where the points b and e correspond to the equilibrium, or saturation, points of the liquid and vapor phases at that temperature at equilibrium, respectively, with specific volumes v_{e}^{α} and v_{e}^{β}. Conversely, the specific volumes of the two phases at equilibrium could also be determined from the molar free energy f_{Th}, rewriting Eq. (9.20) in terms of reduced coordinates as

$$\frac{f_{Th}}{RT_C} = -T_r \ln(v_C) - T_r \ln\left(v_r - \frac{1}{3}\right) - \frac{9}{8v_r}. \quad (9.23)$$

When $T_r < 1$ a typical curve of the free energy is represented in Fig. 9.2. Now, keeping T_r fixed and considering that the two phases at equilibrium have the same pressure, using the relation $P = -(\partial f_{Th}/\partial v)_T$, we obtain:

$$P^{\alpha} = P^{\beta} \quad \Longrightarrow \quad \left(\frac{\partial f_{th}}{\partial v}\right)_T^{\alpha} = \left(\frac{\partial f_{Th}}{\partial v}\right)_T^{\beta}, \quad (9.24)$$

which, in Fig. 9.2, represents the fact that the two equilibrium points have the same tangent. From this relation we can determine the specific volumes of the two phases at equilibrium, v_{e}^{α} and v_{e}^{β}. This relation can also be obtained considering that the

Fig. 9.3 Phase diagram $(T - v)$ of a single component fluid and $(T - \phi)$ of a binary mixture

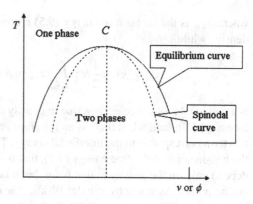

specific volumes of the two phases at equilibrium minimize the total free energy, i.e.,

$$F_{Th} = \int \hat{f}_{Th}(\rho) \, d^3\mathbf{x} = min, \qquad (9.25)$$

where $\hat{f} = \rho f$ is the free energy per unit volume,

$$\hat{f}_{Th} = \rho f_{Th} = \hat{f}_{id} + \hat{f}_{ex} = RT\left[\rho \ln \rho + \rho^2 B(T)\right]. \qquad (9.26)$$

This minimization is carried out in Sect. 9.3.2. In Fig. 9.3, besides the equilibrium curve, we have represented in a $T - v$ diagram the, so called, spinodal curve, defined as the locus of all points (like c and d) satisfying $(\partial P/\partial v)_T = 0$. All points lying outside the region encompassing the equilibrium curve are stable and represent homogeneous, single-phase systems; all points lying inside the region within the bell-shaped spinodal curve are unstable and represent systems that will separate into two phases (one liquid and another vapor, in this case); the region sandwiched between the equilibrium and the spinodal curves represents metastable systems, that is overheated liquid and undercooled vapor. The spinodal points can be also determined using the relation $(\partial P/\partial v)_T = 0$, obtaining:

$$\left(\frac{\partial^2 f_{Th}}{\partial v^2}\right)_T = 0, \qquad (9.27)$$

determining the spinodal specific volumes $\tilde{v}_s^\alpha$ and $\tilde{v}_s^\beta$.

9.3 Diffuse Interfaces

9.3.1 Interfacial Regions

Suppose now that the molar density of the system is not constant. Accordingly, when $U_0 \ll kT$, Eq. (9.4) can be rewritten as

$$f(\mathbf{x}) = f_{Th}(\mathbf{x}) + \Delta f_{NL}(\mathbf{x}), \qquad (9.28)$$

where f_{Th} is the molar free energy (9.5) corresponding to a system with constant density, while

$$\Delta f_{NL}(\mathbf{x}) = \frac{1}{2} N_A^2 \int_{r>d} U(r) \big[\rho(\mathbf{x}+\mathbf{r}) - \rho(\mathbf{x}) \big] d^3 \mathbf{r} \tag{9.29}$$

is a non local molar free energy, due to density changes, typical of the diffuse interface model. In fact, when there is an interface separating two phases at equilibrium, this term corresponds to the interfacial energy. This result is a direct consequence of the non-locality of the free energy (9.4), that is its value at any given point does not depend only on the density at that point, but it depends also on the density at neighboring points. As stated by van der Waals, "the error that we commit in assuming a dependence on the density only at the point considered vanishes completely when the state of equilibrium is that of a homogeneous distribution of the substance. If, however, the state of equilibrium is one where there is a change of density throughout the vessel, as in a substance under the action of gravity, then the error becomes general, however feeble it may be" [30]. Now, in (9.29) the density can be expanded as

$$\rho(\mathbf{x}+\mathbf{r}) = \rho(\mathbf{x}) + \mathbf{r} \cdot \nabla \rho + \frac{1}{2} \mathbf{r}\mathbf{r} : \nabla\nabla\rho + \cdots . \tag{9.30}$$

As we have tacitly assumed that the system is isotropic, we see that the contribution of the linear term vanishes, so that, at leading order, we obtain [26],

$$\Delta f_{NL}(\mathbf{x}) = -\frac{1}{2} RT K \nabla^2 \rho(\mathbf{x}), \tag{9.31}$$

with

$$K = \frac{2\pi}{3} \frac{N_A U_0}{kT} \frac{l^6}{d} = \frac{9\pi}{4} \frac{T_C}{T} N_A d^5, \tag{9.32}$$

where we have substituted Eqs. (9.6), (9.11) and (9.19). Note that, defining a non-dimensional molar density, $\tilde{\rho} = N_A d^3 \rho$, the non local free energy can be rewritten as

$$\Delta f_{NL}(\mathbf{x}) = -\frac{1}{2} RT a^2 \nabla^2 \tilde{\rho}(\mathbf{x}), \tag{9.33}$$

where

$$a = \sqrt{\frac{K}{N_A d^3}} = \sqrt{\frac{9\pi}{4} \frac{T_C}{T}} d \tag{9.34}$$

is the characteristic length. Therefore, in the bulk, the total free energy is:

$$\int_V \widehat{f} d^3 \mathbf{x} = \int_V \rho \left(f_{Th} - \frac{1}{2} RT K \nabla^2 \rho \right) d^3 \mathbf{x}, \tag{9.35}$$

where $\widehat{f} = \rho f$ is the free energy per unit volume.

Comment In the previous analysis, we have truncated the expansion (9.30) after the second-order term, neglecting the next significant (i.e. the forth-order) term. That means assuming that $\epsilon = a^2/\lambda^2 \ll 1$, where λ denotes the thickness of the interfacial region. Now, although this assumption is satisfied near the critical point, where λ diverges as $T \to T_C$ [cf. Eq. (9.52)], far from the critical point, ϵ is not too large. For example, at a water-vapor or an oil-water interface, $\epsilon \approx 0.1$, so that we expect our results to be correct within a 10 % error.

At the wall, the non local free energy (9.29) has an additional contribution of the form

$$\oint \left\{ \rho_s(\mathbf{x}) \left[\int U(r)\rho(\mathbf{x}+\mathbf{r}) d^3\mathbf{r} \right] \right\} d^2\mathbf{x} = \oint f_w(\rho(\mathbf{x})) d^2\mathbf{x}, \qquad (9.36)$$

where the integration is carried out on the surface. Here, ρ_s is the solid density and f_w is the wall free energy per unit surface, that we assume to be a function of the fluid density at the wall only. Now, observing that, integrating by parts,

$$\int \rho(\mathbf{x})\nabla^2\rho(\mathbf{x}) d^3\mathbf{x} = \oint \mathbf{n} \cdot (\rho\nabla\rho) d^2\mathbf{x} - \int |\nabla\rho(\mathbf{x})|^2 d^3\mathbf{x}, \qquad (9.37)$$

we see that the total free energy is the sum of a bulk and a surface free energies, i.e.,

$$F = F_b + F_w, \qquad (9.38)$$

where F_b is the bulk free energy,

$$F_b = \int \widehat{f}(\rho, \nabla\rho, T) d^3\mathbf{x} = RT \int \left[\rho \widetilde{f}_{Th}(\rho, T) + \frac{1}{2}K(T)(\nabla\rho)^2 \right] d^3\mathbf{x}, \qquad (9.39)$$

where $\widetilde{f}_{Th} = f_{Th}/RT = \ln\rho + \rho B(T)$, while F_w is the wall free energy,

$$F_w = RT \oint \left[-\frac{1}{2}K\mathbf{n} \cdot (\rho\nabla\rho) + \widetilde{f}_w(\rho) \right] d^2\mathbf{x}, \qquad (9.40)$$

where $\widetilde{f}_w = f_w/RT$.

At equilibrium, keeping the temperature T constant, the total free energy F will be minimized, subjected to the constraint of mass conservation,

$$\int \rho d^3\mathbf{x} = M = const. \qquad (9.41)$$

Accordingly, introducing a Lagrange multiplier, $RT\widetilde{\mu}$, the minimization condition is:

$$\delta \int \left[\rho(\widetilde{f}_{Th}(\rho) - \widetilde{\mu}) + \frac{1}{2}K|\nabla\rho|^2 \right] d^3\mathbf{x}$$

$$+ \delta \oint \left[-\frac{1}{2}RT K\mathbf{n} \cdot (\rho\nabla\rho) + f_w(\rho) \right] d^2\mathbf{x} = 0, \qquad (9.42)$$

for any arbitrary variation $\delta\rho$ of the density field. Now, consider that, for any function $h(\rho, \nabla\rho)$, we have:

$$\delta h = \frac{\partial h}{\partial \rho}\delta\rho + \frac{\partial h}{\partial(\nabla_i \rho)}\delta(\nabla_i \rho), \quad \text{with } \delta(\nabla_i \rho) = \nabla_i(\delta\rho), \quad (9.43)$$

and

$$\int_V \frac{\partial h}{\partial \nabla_i \rho}\nabla_i(\delta\rho)\,d^3\mathbf{x}\,dt = \oint_S n_i \frac{\partial h}{\partial \nabla_i \rho}\delta\rho\,d^2\mathbf{x} - \int_V \nabla_i\left(\frac{\partial h}{\partial \nabla_i \rho}\right)\delta\rho\,d^3\mathbf{x}. \quad (9.44)$$

Applying these two equalities to Eq. (9.42) we obtain:

$$\int\left[\frac{\partial \widehat{f}}{\partial \rho} - \nabla_i\left(\frac{\partial \widehat{f}}{\partial \nabla_i \rho}\right) - (RT\tilde{\mu})\right]\delta\rho\,d^3\mathbf{x}$$

$$+ \oint_S \mathbf{n}\cdot\left[\frac{1}{2}RTK\nabla\rho + \frac{df_w}{d\rho}\right]\delta\rho\,d^3\mathbf{x} = 0, \quad (9.45)$$

where we have considered that $\delta(\rho\nabla\rho) = \delta\rho\,\nabla\rho + \rho\nabla\delta\rho$ and assumed that $\mathbf{n}\cdot\nabla\delta\rho = 0$ at the boundary.

9.3.2 Generalized Chemical Potential

Choosing $\delta\rho = 0$ at the boundary, Eq. (9.45) reduces to minimizing the bulk free energy. So, predictably, we obtain the Euler-Lagrange equation:

$$\tilde{\mu} = \frac{1}{RT}\left[\frac{\partial \widehat{f}}{\partial \rho} - \nabla_i\left(\frac{\partial \widehat{f}}{\partial \nabla_i \rho}\right)\right], \quad (9.46)$$

that is, substituting (9.39),

$$\tilde{\mu} = \frac{d(\rho\widehat{f}_{Th})}{d\rho} - K\nabla^2\rho. \quad (9.47)$$

This defines the Lagrange multiplier $\tilde{\mu}$, associated with mass conservation. Now, by definition, the first term on the RHS is the Gibbs free energy, which, in a one-component system, coincides with the chemical potential. In fact,

$$RT\tilde{\mu}_{Th} = \frac{d(\rho f_{Th})}{d\rho} = f_{Th} - v\frac{df_{Th}}{dv}, \quad (9.48)$$

where $df_{Th}/dv = -P$. This (apart from the dimensional constant RT) is the equation of the straight line represented in Fig. 9.2, stating that two phases at mutual equilibrium have the same chemical potential. Therefore, Eq. (9.46) can be rewritten as

$$\tilde{\mu}(\rho, \nabla\rho) = \tilde{\mu}_{Th}(\rho) - K\nabla^2\rho, \quad (9.49)$$

showing that at equilibrium, when ρ is non-uniform, it is $\tilde{\mu}$, and not $\tilde{\mu}_{Th}$, that remains uniform and so $\tilde{\mu}$ can be interpreted as a generalized chemical potential. Note that the thermodynamic chemical potential, $\tilde{\mu}_{Th}$, can be determined from the solvability condition of Eq. (9.48), that is,

$$\tilde{\mu}_{Th} = \frac{\rho^\alpha \tilde{f}_{Th}^\alpha - \rho^\beta \tilde{f}_{Th}^\beta}{\rho^\alpha - \rho^\beta} = \frac{v^\alpha \tilde{f}_{Th}^\beta - v^\beta \tilde{f}_{Th}^\alpha}{v^\alpha - v^\beta}, \tag{9.50}$$

as it can also be seen geometrically from Fig. 9.2, stating that the chemical potential equals the intercept of the tangent line on the $v = 0$ vertical axis. When two phases are coexisting at equilibrium, separated by a planar interfacial region centered on $z = 0$, Eq. (9.46) can be solved once the equilibrium molar free energy f is known, imposing that, far from the interface region, the density is constant and equal to its equilibrium value, so that the generalized chemical potential is equal to its thermodynamic value (9.48). In particular, in the vicinity of the critical point, considering that the chemical potential vanishes and expanding the free energy (9.23) as a power series of $\tilde{v} = (v - v_C)/v_C$ and $\tilde{t} = (T - T_C)/T_C$, we obtain at leading order the following equation:

$$\frac{d^2\tilde{v}}{d\tilde{z}^2} - 6\tilde{t}\tilde{v} - \frac{3}{2}\tilde{v}^3 = 0, \tag{9.51}$$

where $\tilde{z} = z/\lambda$ is based on the characteristic length

$$\lambda = \sqrt{\frac{1}{8(-\tilde{t})}} d. \tag{9.52}$$

Equation (9.51) must be solved imposing that

$$\tilde{v}(\tilde{z} \to \pm\infty) = \pm\tilde{v}_e = \pm 2\sqrt{-\tilde{t}}. \tag{9.53}$$

The solution, due (again!) to van der Waals, is:

$$\tilde{v}(\tilde{z}) = \tilde{v}_e \tanh \tilde{z}, \tag{9.54}$$

showing that λ is a typical interfacial thickness. As expected, the interfacial thickness diverges like $(-\tilde{t})^{-1/2}$ as we approach the critical point, while far from the critical point it is of $O(d)$.

9.3.3 Surface Tension

In Sect. 9.3.1 we have seen that the total free energy is the sum of a thermodynamical, constant density, part, and a non local contribution (9.31). When the system is composed of two phases at equilibrium, separated by a plane interfacial region, we may define the surface tension as the energy per unit area stored in this region.

This quantity can be calculated by integrating the specific free energy (9.31) along a coordinate z perpendicular to the interface:

$$\sigma = -\frac{1}{2}RTK \int_{-\infty}^{\infty} \rho \frac{d^2\rho}{dz^2}\, dz = \frac{1}{2}RTK \int_{-\infty}^{\infty} \left(\frac{d\rho}{dz}\right)^2 dz, \qquad (9.55)$$

where we have integrated by parts and considered that, outside the interfacial region, the integrand is identically zero as density is constant. We see that, near the critical point,

$$\sigma \approx RT_C K(\Delta\rho_e)^2 / \lambda \approx \frac{kT_C}{d^2}(-\tilde{t})^{3/2}, \qquad (9.56)$$

where we have considered Eqs. (9.32) and (9.52). In fact, using the density profile (9.54), Eq. (9.55) yields, for a van der Waals system:

$$\sigma = \frac{2}{3}RT_C(-a_c\tilde{t})^{3/2}\left(\frac{K}{v_c^3}\right) = C\frac{kT_C}{d^2}(-\tilde{t})^{3/2}, \qquad (9.57)$$

with $C = 3^{3/2}/(2^{3/2}\pi)$, where we have used Eq. (9.20). These results show that the surface tension decreases as we approach criticality, until it vanishes at the critical point.

Now, following van der Waals, we will show that in a curved interface region there arises a net force, which is compensated by a pressure term, thus obtaining the Young-Laplace equation. To see that, let us denote the position of the interface by $z = h(\xi)$, where ξ is the 2D vector in the support plane, and assume that $|\nabla_\xi h| \ll 1$, where $|\nabla_\xi h|$ is the 2D gradient [26]. Now, the free energy increment due to the interface curvature can be written as:

$$\Delta F = \sigma \int \left(\sqrt{1 + |\nabla_\xi h|^2} - 1\right) d^2\xi \approx \frac{1}{2}\sigma \int |\nabla_\xi h|^2\, d^2\xi. \qquad (9.58)$$

This increment in the free energy induces an increment in the pressure,

$$\Delta P = \delta F / \delta h = -\sigma \nabla^2 h = -\kappa\sigma, \qquad (9.59)$$

where $\kappa = \nabla^2 h$ is the curvature of a weakly curved interface. Applying a rigorous regular perturbation approach to Eq. (9.46), Pismen and Pomeau [27] derived both the Young-Laplace equation (9.59) and the Gibbs-Thomson law, relating the equilibrium temperature or pressure to the interfacial curvature.

From a different perspective, one could say that at equilibrium the two coexisting phases must have the same generalized pressure, defined as

$$\widetilde{P} = -\left[\frac{\partial \widehat{f}}{\partial v} - \nabla_i\left(\frac{\partial \widehat{f}}{\partial \nabla_i v}\right)\right] = P - \rho^2 RTK\nabla^2\rho. \qquad (9.60)$$

9.3.4 Boundary Conditions

As previously noted, the equilibrium state of an unconfined van der Waals fluid can be determined using the generalized chemical potential in the bulk. In general, however, for confined systems surface wettability effects are present and must be taken into account. In our D.I. approach such effects can be accounted for by introducing the simplest additional surface contribution to the free energy functional, which is based on the assumption of local equilibrium, so that wettability is a local quantity, depending on the composition of the mixture at the wall. Accordingly, choosing $\delta\rho = 0$ in the bulk, Eq. (9.45) reduces to minimizing the surface integral, and so we obtain the following boundary condition [12],

$$\frac{1}{2}RT K \mathbf{n} \cdot \nabla\rho = -\frac{df_w}{d\rho}(\rho), \qquad (9.61)$$

where $f_w(\phi)$ is the surface energy at the wall. Assume a linear dependence,

$$f_w(\phi) = \sigma_{Bw} + \phi\Delta\sigma_w, \qquad (9.62)$$

where $\phi = (\rho - \rho_\alpha)/(\rho_\beta - \rho_\alpha)$, ρ_α and ρ_β are the densities of pure phases α and β, respectively, while $\Delta\sigma_w = \sigma_{w\beta} - \sigma_{w\alpha}$ expresses the affinity of the wall to phase β, as compared to phase α. Therefore, considering that $\sigma \cong a\rho RT/M_w$ is the surface tension between the two fluid phases at equilibrium, this boundary condition can be rewritten as

$$\sigma a \mathbf{n} \cdot \nabla\phi = -\Delta\sigma_w, \qquad (9.63)$$

which is generally referred to as the Cahn boundary condition [5]. In the sharp interface limit, $\mathbf{n} \cdot \nabla\phi = \cos\theta$, where θ is the contact angle, and therefore the Cahn boundary condition reduces to the Young-Laplace formula, $\cos\theta = -\Delta\sigma_w/\sigma$. From here we see that, when $\sigma_{Aw} = \sigma_{Bw}$ or when σ_{Aw} and σ_{Bw} are both $\ll \sigma$, then $\theta = \pi/2$; instead, when $\sigma_{Aw} \gg \sigma_{Bw}$ or $\sigma_{Aw} \ll \sigma_{Bw}$, then $\theta = \pi$ and $\theta = 0$, respectively.

9.4 Equations of Motion

9.4.1 Minimum Action Principle

In this section, we confine ourselves to study the reversible motion of a dissipation-free fluid. Now, if $\mathbf{x}(t, \mathbf{x}_0)$ denotes the trajectory of a material particle which is located at $\mathbf{x}_0$ at time $t = 0$, i.e. with $\mathbf{x}_0 = \mathbf{x}(0, \mathbf{x}_0)$, then the fluid velocity field is $\mathbf{v}(\mathbf{x}, t) = \dot{\mathbf{x}}(t, \mathbf{x}_0)$, where the dot denotes time derivative at constant $\mathbf{x}_0$. According to the Hamilton, minimum action, principle, the motion of any conservative system minimizes the following action functional:

$$S = \int_0^t \int_V \mathcal{L}(\mathbf{v}, \phi, \nabla\phi)\, d^3\mathbf{x}\, dt, \qquad (9.64)$$

where

$$\mathcal{L} = \mathcal{L}(\mathbf{v}, \rho, \nabla\rho) = \frac{1}{2}\rho v^2 - \widehat{u}, \qquad (9.65)$$

is the Lagrangian of the system, with $\widehat{u} = \rho u$ denoting the internal energy per unit volume. The minimization must be carried over with the constraint of mass conservation which, as we saw in Sect. 9.3.2, results into the appearance of the Lagrange multiplier $\tilde{\mu}$.

Applying the minimum condition, let us give a virtual displacement δx_i, corresponding to an infinitesimal change of the fluid flow. Let us assume that the virtual displacement is solenoidal, so that no density variation is involved, i.e. $\delta\rho = 0$, and the constraint of mass conservation is identically satisfied. In addition, considering that $f = u - Ts$, since we are considering isoentropic variation, we see that $\delta\widehat{u} = \delta\widehat{f}$. Accordingly, minimization of the action S in (9.64) yields:

$$\delta S = \int_0^t \int_V \left[\rho v_i \delta v_i - \delta(\widehat{f} - RT\tilde{\mu}\rho) \right] d^3\mathbf{x}\, dt = 0, \qquad (9.66)$$

where $\tilde{\mu}$ is given by Eq. (9.46).

Considering that $\delta(dx_i) = d(\delta x_i)$, the first integral term on the RHS of Eq. (9.66) gives, after integrating by parts:

$$\int_0^t \int_V \rho v_i \delta v_i\, d^3\mathbf{x}\, dt = \int_0^t \int_V \rho v_i \delta \frac{dx_i}{dt}\, d^3\mathbf{x}\, dt$$

$$= \int_0^t \int_V \rho v_i \frac{d}{dt}(\delta x_i)\, d^3\mathbf{x}$$

$$= \int_V [\rho v_i \delta x_i]_{t_1}^{t_2}\, d^3\mathbf{x} - \int_0^t \int_V \rho \frac{dv_i}{dt} \delta x_i\, d^3\mathbf{x}\, dt, \quad (9.67)$$

i.e.

$$\int_0^t \int_V \rho v_i \delta v_i\, d^3\mathbf{x}\, dt = - \int_0^t \int_V \rho \frac{dv_i}{dt} \delta x_i\, d^3\mathbf{x}\, dt. \qquad (9.68)$$

Here we have considered that the virtual displacement is equal to zero at the beginning and at the end, i.e. when $t = t_1$ and $t = t_2$, as well as on the boundary, S, of the volume V of integration.

The second integral term on the RHS of Eq. (9.66) gives, considering that $\widehat{f} = \widehat{f}(\rho, \nabla\rho)$:

$$\int_0^t \int_V \delta(\widehat{f} - RT\tilde{\mu}\rho)\, d^3\mathbf{x}\, dt = \int_0^t \int_V \left[\left(\frac{\partial\widehat{f}}{\partial\rho} - RT\tilde{\mu} \right) \delta\rho + \frac{\partial\widehat{f}}{\partial\nabla_j\rho} \delta(\nabla_j\rho) \right] d^3\mathbf{x}\, dt. \qquad (9.69)$$

Proceeding as in Sect. 9.3.1 we obtain [cf. Eq. (9.45)],

$$\int_0^t \int_V \delta(\widehat{f} - RT\tilde{\mu}\rho)\, d^3\mathbf{x}\, dt = \int \left[\frac{\partial\widehat{f}}{\partial\rho} - \nabla_i \left(\frac{\partial\widehat{f}}{\partial\nabla_i\rho} \right) - (RT\tilde{\mu}) \right] \delta\rho\, d^3\mathbf{x}. \qquad (9.70)$$

Now, considering that $\delta\rho = (\nabla_i\rho)(\delta x_i)$, together with the equality,

$$\nabla_i \widehat{f} = \frac{\partial \widehat{f}}{\partial \rho}\nabla_i\rho + \frac{\partial \widehat{f}}{\partial \nabla_j\rho}\nabla_i\nabla_j\rho = \left[\frac{\partial \widehat{f}}{\partial \rho} - \nabla_j\frac{\partial \widehat{f}}{\partial \nabla_j\rho}\right]\nabla_i\rho + \nabla_j\left[\frac{\partial \widehat{f}}{\partial \nabla_j\rho}\nabla_i\rho\right],$$

(9.71)

we obtain:

$$\int_0^t\int_V \delta(\widehat{f} - RT\tilde{\mu}\rho)\,d^3\mathbf{x}\,dt = \int_0^t\int_V \nabla_i(\widehat{f} - RT\tilde{\mu}\rho)$$

$$- \nabla_j\left(\frac{\partial \widehat{f}}{\partial(\nabla_j\rho)}\nabla_i\rho\right)\delta x_i\,d^3\mathbf{x}\,dt. \quad (9.72)$$

Concluding, substituting (9.68) and (9.72) into Eq. (9.66) gives:

$$\int_0^t\int_V \left(\rho\frac{dv_i}{dt} - \nabla_i(\widehat{f} - RT\tilde{\mu}\rho) - \nabla_j P_{ji}\right)\delta x_i\,d^3\mathbf{x}\,dt = 0, \quad (9.73)$$

where

$$P_{ij}^K = -\frac{\partial \widehat{f}}{\partial(\nabla_i\rho)}\nabla_j\rho \quad (9.74)$$

is the Korteweg stress tensor, first derived by Korteweg [13] in 1901.

Now, considering the arbitrariness of the virtual displacement δx_i and applying Reynolds theorem, we finally obtain the linear momentum equation,

$$\rho\frac{dv_i}{dt} + \nabla_i\tilde{p} = F_i^K = \nabla_j P_{ji}^K, \quad (9.75)$$

with d/dt denoting the material derivative and where,

$$\tilde{p} = RT\tilde{\mu}\rho - \widehat{f}, \quad (9.76)$$

is a pressure term. The result could be more easily determined by applying Noether's theorem D.35 (see Sect. D.2).

This equation must be coupled to the continuity condition,

$$\frac{d\rho}{dt} + \rho\nabla\cdot\mathbf{v} = 0, \quad (9.77)$$

and the internal energy equation, which, for non-dissipative systems, includes only the convective term, that is,

$$\rho\frac{du}{dt} = \frac{\partial(\rho u)}{\partial t} + \nabla\cdot(\rho u\mathbf{v}) = 0. \quad (9.78)$$

9.4.2 Korteweg Stresses

Using the expression (9.39) for the free energy, i.e., $\widehat{f} = \rho f_{Th}(\rho) + \frac{1}{2} RT K (\nabla \rho)^2$, the Korteweg stress tensor $\mathbf{P}^K$ becomes:

$$P_{ij}^K = -RT K (\nabla_i \rho)(\nabla_j \rho). \tag{9.79}$$

Note that the Korteweg stress and force depend only on the non local part of the free energy, even when, as in Eqs. (9.84) and (9.87) below, this is not explicitly indicated.

In addition, taking into account the expressions (9.46)–(9.48) for the chemical potential, the pressure term become,

$$\widetilde{p} = RT \widetilde{\mu} \rho - \widehat{f} = P - RT K \left[\rho \nabla^2 \rho + \frac{1}{2} (\nabla \rho)^2 \right], \tag{9.80}$$

where we have considered that $P = \rho^2 df_{Th}/d\rho$ is the thermodynamic pressure. So, finally, the governing equation Eq. (9.75) can be rewritten as:

$$\rho \frac{dv_i}{dt} = \nabla_j T_{ji}, \tag{9.81}$$

where T_{ij} is the stress tensor (i.e. the opposite of the momentum flux tensor),

$$T_{ij} = \left[-P + RT K \left(\rho \nabla^2 \rho + \frac{1}{2} |\nabla \rho|^2 \right) \right] \delta_{ij} - RT K (\nabla_i \rho)(\nabla_j \rho). \tag{9.82}$$

The most important feature of Eq. (9.75) is the appearance of the Korteweg body force, $\mathbf{F}^K$, which can be rewritten as:

$$F_i^K = \nabla_j P_{ji}^K = - \left[\nabla_j \left(\frac{\partial \widehat{f}}{\partial \nabla_j \rho} \right) \nabla_i \rho + \frac{\partial \widehat{f}}{\partial \nabla_j \rho} \nabla_i \nabla_j \rho + \frac{\partial \widehat{f}}{\partial \rho} \nabla_i \rho - \frac{\partial \widehat{f}}{\partial \rho} \nabla_i \rho \right], \tag{9.83}$$

that is

$$\mathbf{F}^K = RT \widetilde{\mu} \nabla \rho - \nabla \widehat{f}, \tag{9.84}$$

where $\widetilde{\mu}$ is the generalized chemical potential (9.46)–(9.47), and therefore the momentum equation becomes

$$\rho \frac{d\mathbf{v}}{dt} + \nabla p' = RT \widetilde{\mu} \nabla \rho, \tag{9.85}$$

where the pressure term has been redefined as,

$$p' = \widetilde{p} + \widehat{f} = \rho \frac{d}{d\rho} (\rho f_{Th}) - RT K \rho \nabla^2 \rho = RT \widetilde{\mu} \rho. \tag{9.86}$$

Alternatively, this equation can also be written as,

$$\rho \frac{d\mathbf{v}}{dt} = \rho \mathbf{F}; \quad \mathbf{F} = -\nabla \psi, \tag{9.87}$$

with

$$\psi = RT \tilde{\mu} \tag{9.88}$$

denoting a sort of potential energy that takes into account all non-local effects.

Comment 9.1 It should be stressed that the Korteweg body force $\mathbf{F}$ is non dissipative, as it arises from the minimum action principle. Its expression in (9.87) is quite intuitive: being proportional to the gradient of the chemical potential (with a minus sign), it pushes the system towards thermodynamic equilibrium and is identically zero at equilibrium. In addition, since this force is reversible, it does not enter explicitly into the energy dissipation term.

Comment 9.2 The Korteweg body force in Eq. (9.87) has the general form (7.27) of any potential force, so that we can apply all the results that we have previously obtained using irreversible thermodynamics. In particular, we can include into ψ also the contributions of any other potential force. For example, in the Boussinesq, quasi-incompressible approximation, the buoyancy force is $\mathbf{F}_g = -\rho g \nabla z$, where g is the gravity acceleration term and z is the vertical coordinate. Accordingly, gravity can be accounted for by simply adding the term gz to ψ in Eq. (9.88).

Comment 9.3 Note that in Eq. (9.87) the pressure term drops out automatically.

9.4.3 Dissipative Terms

When dissipation is taken into account, the equations of motion remains basically the same as in the non-dissipative case, i.e. Eqs. (9.75), (9.77) and (9.78), where a momentum flux (or pressure) tensor, $\mathbf{P}$, and a heat flux vector, $\mathbf{J}^{(q)}$, are added to Eqs. (9.75) and (9.78), respectively. At the end, we obtain:

$$\frac{d\rho}{dt} + \rho \nabla \cdot \mathbf{v} = 0, \tag{9.89}$$

$$\rho \frac{d\mathbf{v}}{dt} + \nabla \cdot \mathbf{P} = \mathbf{F} = -\rho \nabla \psi, \tag{9.90}$$

$$\rho \frac{du}{dt} + \nabla \cdot \mathbf{J}^{(q)} = \dot{q}, \tag{9.91}$$

where $\dot{q} = -\mathbf{P} : \nabla \mathbf{v}$ is the heat source term.

For regular fluids, the heat flux, $\mathbf{J}^{(q)}$, and the momentum flux (i.e. the pressure tensor), $\mathbf{P}$, can be expressed through the following constitutive equations:

$$\mathbf{J}_q = -\kappa \nabla T; \tag{9.92}$$

$$\mathbf{P} = [p - \lambda(\nabla \cdot \mathbf{v})]\mathbf{I} - \eta(\nabla \mathbf{v} + \nabla \mathbf{v}^+), \tag{9.93}$$

where κ and η denote thermal conductivity and shear viscosity, respectively, while $\lambda = \zeta - 2\eta/3$, with ζ indicating the bulk viscosity.

Note again that the thermodynamic pressure drops out automatically from the equation of momentum conservation.

9.5 Multicomponent Systems

For multicomponent systems, Eqs. (9.89), (9.90) and (9.93) are still valid, but η and λ are functions of the mixture composition, in addition to temperature and density, while the Korteweg force $\mathbf{F}$ becomes:

$$\mathbf{F} = -\sum_{k=1}^{n} \phi^{(k)} \nabla \psi^{(k)}, \tag{9.94}$$

where $\phi^{(k)}$ is the mass fraction of component k, while,

$$\psi^{(k)} = RT \tilde{\mu}^{(k)} \tag{9.95}$$

i.e. the potential of the body force acting on component k is proportional to the non-dimensional generalized chemical potential of component k, $\tilde{\mu}^{(k)}$.

The internal energy equation (9.91) is still valid, but the heat flux and heat generation terms become [cf. Eqs. (7.43) and (7.69)]:

$$\dot{q} = -\mathbf{P} : \nabla \mathbf{v} - \sum_{k=1}^{n} \mathbf{J}_d^{(k)} \cdot \nabla \psi^{(k)}, \tag{9.96}$$

and

$$\mathbf{J}^{(q)} = -\kappa \nabla T + \sum_{k=1}^{n} \mathbf{J}_d^{(k)} h^{(k)}, \tag{9.97}$$

where all coupling terms have been neglected. Here, $h^{(k)}$ is the partial enthalpy of component k, while $\mathbf{J}_d^{(k)}$ is the diffusive mass flux of component k, which is determined through the equation of conservation of the chemical species (7.21),

$$\rho \frac{D\phi^{(k)}}{Dt} = -\nabla \cdot \mathbf{J}_d^{(k)} + \sum_{j=1}^{r} \nu^{(kj)} \mathcal{J}^{(j)} \quad (k = 1, 2, \dots, n), \tag{9.98}$$

where the last term expresses the temporal growth of $\rho^{(k)}$ due to chemical reactions.

9.5.1 Incompressible Binary Mixtures

Consider a non reacting binary mixture, composed of two species having the same density, so that ρ is constant and the continuity equation (9.89) reduces to

$$\nabla \cdot \mathbf{v} = 0. \tag{9.99}$$

Accordingly, the pressure appearing in the momentum equation has no real physical meaning, the pressure tensor in (9.90) reduces to:

$$\mathbf{P} = p\mathbf{I} - \eta(\nabla\mathbf{v} + \nabla\mathbf{v}^+), \tag{9.100}$$

and the Korteweg force becomes:

$$\mathbf{F} = \rho\phi\nabla(RT\tilde{\mu}), \tag{9.101}$$

where $\phi = \phi^{(1)}$ is the mass fraction of component 1, while $\tilde{\mu} = \tilde{\mu}^{(1)} - \tilde{\mu}^{(2)}$ is the non-dimensional generalized chemical potential difference.[4]

In the internal energy equation (9.91) the heat generation term (9.96) becomes,

$$\dot{q} = -\mathbf{P}:\nabla\mathbf{v} - \mathbf{J}_\phi \cdot \nabla(RT\tilde{\mu}), \tag{9.102}$$

while the heat flux is expressed through the constitutive relation, [cf. Eqs. (8.33) and (8.35)]

$$\mathbf{J}^{(q)} = -\kappa\nabla T + \mathbf{J}_\phi h^{(12)}, \tag{9.103}$$

where all coupling terms have been neglected. Here, $h^{(12)} = h^{(1)} - h^{(2)}$ is the partial enthalpy difference, while $\mathbf{J}_\phi = \mathbf{J}_d^{(1)}$ is the diffusive mass flux of component 1, which is determined through the equation of conservation of the chemical species,

$$\rho\frac{d\phi}{dt} + \nabla \cdot \mathbf{J}_\phi = 0 \tag{9.104}$$

where $\mathbf{J}_\phi = \phi(\mathbf{v}^{(1)} - \mathbf{v})$ is a diffusive molar flux, with $\mathbf{v}^{(1)}$ denoting the mean velocity of species 1, while the chemical reaction term has been omitted. This term is expressed through the constitutive relation (8.34),

$$\mathbf{J}_\phi = -\rho D^*[\nabla\tilde{\mu}]_T, \tag{9.105}$$

where D^* is an effective diffusivity, $\tilde{\mu} = \tilde{\mu}^{(1)} - \tilde{\mu}^{(2)}$ is the generalized non-dimensional chemical potential difference, while the subscript T indicates that the gradient is taken at constant temperature. Here, it should be stressed that $\tilde{\mu}^{(k)} = \mu^{(k)} + \psi^{(k)}$ [cf. Eq. (7.63)], that is the generalized chemical potential is the sum of the thermodynamic chemical potential and any potential exerted on the k-th chemical species. In our case, that means adding the effect of the non-local part of the free energy.

---

[4]The extra term that is obtained, $\rho\nabla\tilde{\mu}^{(2)}$, being the gradient of a scalar, can be absorbed into the pressure term.

9.5.2 Symmetric Regular Binary Mixtures

Application of this theory is particularly simple in the case of symmetric regular binary mixtures. The theory of regular mixtures was developed by van Laar (a student of van der Waals), who assumed that (a) the two species composing the mixture are of similar size and energies of interaction, and (b) the van der Waals equation of state applies to both the pure fluids and the mixture. Consequently, regular mixtures have negligible excess volume and excess entropy of mixing, i.e. their volume and entropy coincide with those of an ideal gas mixture, with $s_{ex} = 0$ and $v_{ex} = 0$. Therefore, we see that for a regular mixture, since $s_{ex} = -(\partial g_{ex}/\partial T)_{P,x} = 0$, then g_{ex} must be independent of T. In fact, starting from the fundamental expression (9.4) for the Helmholtz free energy and considering that:

$$g_{ex} = f_{ex} + P v_{ex}, \tag{9.106}$$

applying Eqs. (9.8), (9.9) and (9.106) to a system with $v_{ex} = 0$ and constant molar density ρ, we obtain: $g_{ex} = f_{ex} = RT\rho B$, where B is the virial coefficient:

$$B = x_1^2 B^{(11)} + 2x_1 x_2 B^{(12)} + x_2^2 B^{(22)}. \tag{9.107}$$

Here, $B^{(ij)}$ characterizes the repulsive interaction between molecule i and molecule j [see Eq. (9.9)],

$$B^{(ij)} = \frac{1}{2} N_A \int \left[1 - \exp\left(-\frac{\psi^{(ij)}(r)}{kT} \right) \right] d^3 \mathbf{r}, \tag{9.108}$$

where $\psi^{(ij)}$ is the pairwise interaction potential between molecules i and j. In particular, for symmetric solutions, $U^{(11)} = U^{(22)} \neq U^{(12)}$, so that $B^{(11)} = B^{(22)} \neq B^{(12)}$. Accordingly, denoting $x_1 = \phi$, we obtain:

$$g_{ex}(T, P, \phi) = RT\Psi(T, P)\phi(1 - \phi), \tag{9.109}$$

where

$$\Psi(T, P) = 2\rho\left(B^{(12)} - B^{(11)}\right), \tag{9.110}$$

is the so called Margules coefficient. In particular, for an ideal mixture, $B^{(11)} = B^{(12)}$ and therefore $\Psi = 0$. For a mixture composed of van der Waals fluids at constant pressure, substituting the expression (9.10) for B and assuming that the characteristic lengths d and l are the same for the two species, we obtain:

$$\Psi = \frac{2\rho}{RT}\left(c_1^{(11)} - c_1^{(12)}\right) = \frac{4\pi}{3}\frac{\rho N_A^2 l^6}{RT d^3}\left(U_0^{(11)} - U_0^{(12)}\right), \tag{9.111}$$

where $U_0^{(11)}$ and $U_0^{(12)}$ are the strength of the potential between molecules of the same species and that of different species, respectively. From this expression it appears that $\Psi \propto T^{-1}$, thus confirming that g_{ex} is independent of T. In addition, we

Fig. 9.4 Free energy g of a symmetric binary mixture as a function of composition ϕ

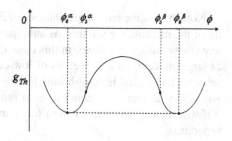

see that when the repulsive forces between unlike molecules are weaker than those between like molecules, i.e. when $U_0^{(12)} < U_0^{(11)}$, then $\Psi > 0$; in the opposite case, $\Psi < 0$. Finally, adding the excess free energy to its ideal part, (which is easily derived generalizing the expression $f_{id} = RT \ln \rho$, valid for single component fluids) we obtain:

$$g_{Th}(T, \phi) = \frac{RT}{M_w} \left[\phi \ln \phi + (1 - \phi) \ln (1 - \phi) + \Psi \phi (1 - \phi) \right]. \quad (9.112)$$

Since chemical stability imposes that $d^2 g / d\phi^2 > 0$, we see that $\Psi = 2$ corresponds to the critical point: when $\Psi < 2$ the two species are always miscible, while when $\Psi > 2$ there is a concentration range where they phase separate. Therefore, we must have: $\Psi = 2T_C / T$, where T_C is the critical temperature. Now consider a typical plot of the free energy as a function of concentration, as in Fig. 9.4: considering that $\mu = dg/d\phi$, we obtain the $\mu - \phi$ plot of Fig. 9.3. In addition, as for $T < T_C$, i.e. $\Psi > 2$, the mixture phase separates, with the equilibrium concentration of the two coexisting phases satisfying $dg/d\phi = 0$, we obtain the $T - \phi$ curve of Fig. 9.3, where the spinodal curve, satisfying $d^2 g / d\phi^2 = 0$, is also indicated. So, we see that the case of incompressible binary mixtures is an exact mirror of the single component case, where concentration ϕ and chemical potential μ replace density ρ and pressure P.

Now, let us determine the constitutive equation for the diffusive mass flux, $\mathbf{J}_\phi$. As we saw, it must be proportional to the gradient of the generalized non-dimensional chemical potential difference, $\tilde{\mu}$. Now, the thermodynamic chemical potential difference, $RT \mu_{Th}$ is:

$$RT \mu_{Th} = \frac{dg_{Th}}{d\phi} = RT \left[\ln \left(\frac{\phi}{1 - \phi} \right) + \Psi (1 - 2\phi) \right]. \quad (9.113)$$

Observing that:

$$[\nabla \mu_{Th}]_T = \frac{d\mu_{Th}}{d\phi} \nabla \phi = \frac{1}{RT} \frac{d^2 g_{Th}}{d\phi^2} \nabla \phi = \left(\frac{1}{\phi(1 - \phi)} - 2\Psi \right) \nabla \phi, \quad (9.114)$$

and imposing that in the dilute limit the constitutive equation (9.105) must reduce to Fick's law, we see that the effective diffusivity must have the following form: $D^* = D \phi (1 - \phi)$ in order to prevent it from diverging as ϕ tends to 0 or to 1.

Note that, as the term in parenthesis of Eq. (9.114) is proportional to the diffusivity of the mixture, we see that, as anticipated, when $\Psi < 2$ diffusivity is positive at any composition. Therefore, in this case, diffusion fluxes are directed from regions of large concentration to regions of low concentration, so that the system tends to relax back to its stable equilibrium, homogeneous configuration. On the other hand, when $\Psi > 2$, there is a concentration range where diffusivity is negative, so that diffusive fluxes are directed from low concentration to large, thus inducing phase separation.

Now, let us add the non-local part of the free energy,

$$g_{NL} = \frac{1}{2}a^2 RT |\nabla \phi|^2, \tag{9.115}$$

where a is a characteristic length, which is approximately equal to the interface thickness. This expression is generally attributed to Cahn and Hilliard [7, 8]. In the case of regular mixtures, i.e. when the two fluids are both van der Waals fluids, proceeding as in Sect. 9.3.1 [see Eqs. (9.33) and (9.34)], we obtain the same expression (9.115), with $a \propto 1/\sqrt{T}$, i.e. $a = \hat{a}\sqrt{\Psi}$, where $\hat{a}$ is a characteristic length. In addition, imposing that at equilibrium the integral of the free energy across the interface region equals the surface tension, proceeding as in Sect. 9.3.3 we see that $\sigma \approx \hat{a}(\rho/M_w)RT$. Considering that the total free energy is the sum of the thermodynamic and non-local parts, i.e. $g = g_{Th} + g_{NL}$, we see that the generalized chemical potential difference becomes, $\tilde{\mu} = \delta g/\delta \phi$ becomes,

$$\tilde{\mu} = \mu_{Th} - RT a^2 \nabla^2 \phi, \tag{9.116}$$

and the constitutive equation for the diffusive material flux can be written as:

$$\mathbf{J}_\phi = -D\phi(1-\phi)[\nabla \tilde{\mu}]_T = -RTD\{[1 - 2\Psi \phi(1-\phi)]\nabla \phi - a^2 \nabla^2 \nabla \phi\}. \tag{9.117}$$

This is the constitutive equation that has been used in all the works on binary mixtures by Mauri and coworkers [14–16, 23, 24, 31]. Note that, apart from the transport coefficients, namely viscosity η, thermal conductivity k and diffusivity D, this model contains two parameters, Ψ and a. The former is known from the equilibrium diagram of the mixture and for regular mixtures $\Psi = 2T_C/T$, where T_C is the critical temperature; on the other hand, a is a characteristic length, which is related to temperature as $a = \hat{a}\sqrt{\Psi}$, where $\hat{a}$ can be evaluated from the surface tension, as $\hat{a} \approx (\sigma\rho)/(RTM_w)$. Consequently, the results of the simulations based on this model can be used as a quantitative predictive tools.

9.6 Problems

Problem 9.1 Determine the equation of state of a single component system near its critical point, using the following variables:

$$\tilde{t} \equiv T_r - 1 = \frac{T - T_C}{T_C}; \qquad \tilde{p} \equiv P_r - 1 = \frac{P - P_C}{P_C}; \qquad \tilde{v} \equiv v_r - 1 = \frac{v - v_C}{v_C},$$

$$(9.118)$$

where $T_r = T/T_C$, $P_r = P/P_C$ and $v_r = v/v_C$ are the reduced variables, while the subscript C indicates critical value. Then in the vicinity of the critical point, i.e. when $\tilde{t} \ll 1$, determine the critical exponents α, β, γ and δ, defined as $c_v \propto \tilde{t}^\alpha$, $\Delta \tilde{v}_e \propto (-\tilde{t})^\beta$, $\kappa_T^{-1} \propto \tilde{t}^\gamma$ and $\tilde{p} \propto \tilde{v}^\delta$, where c_v is the specific volume at constant volume, $\Delta \tilde{v}_e$ is the difference of the specific volumes of the two coexisting phases, κ_T is the isothermal compressibility.

Problem 9.2 Show that the specific heat at constant pressure diverges at the critical point.

Problem 9.3 Consider an incompressible, regular and symmetric binary mixture, with initial composition $\phi_0 = 1/2$, that at time $t = 0$ is instantaneously quenched to a temperature below its critical value. Assuming that the mixture is very viscous, so that convection can be neglected, and using the equation of motion (9.104) and (9.117), show that at leading order the evolution of the concentration field is determined by the equation,

$$\frac{\partial u}{\partial t} = -2\psi \nabla^2 u - \nabla^4 u,$$

where $u = 2\phi - 1 \ll 1$ and $\psi = \Psi - 2$, while the spatial and temporal variables have been made non dimensional in terms of a and $2a^2/D$. Assuming a periodic perturbation, $u = u_0 \exp(i\mathbf{k} \cdot \mathbf{r} + \sigma t)$, determine the wave vector k_{max} that maximizes the exponential growth σ.

References

1. Anderson, D.M., McFadden, G.B., Wheeler, A.A.: Annu. Rev. Fluid Mech. **30**, 139 (1998)
2. Antanovskii, L.K.: Phys. Fluids **7**, 747 (1995)
3. Antanovskii, L.K.: Phys. Rev. E **54**, 6285 (1996)
4. Cahn, J.W.: Acta Metall. **9**, 795 (1961)
5. Cahn, J.W.: J. Chem. Phys. **66**, 3667 (1977)
6. Cahn, J.W., Hilliard, J.E.: J. Chem. Phys. **31**, 688 (1958)
7. Cahn, J.W., Hilliard, J.E.: J. Chem. Phys. **28**, 258 (1958)
8. Cahn, J.W., Hilliard, J.E.: J. Chem. Phys. **31**, 688 (1959)
9. Davis, H.T., Scriven, L.E.: Adv. Chem. Phys. **49**, 357 (1982)
10. de Gennes, P.G.: J. Chem. Phys. **72**, 4756 (1980), and references therein
11. Hohenberg, P.C., Halperin, B.I.: Rev. Mod. Phys. **49**, 435 (1977)
12. Jacqmin, D.: J. Fluid Mech. **402**, 57 (2000)
13. Korteweg, D.J.: Arch. Neerl. Sci. Exactes Nat., Ser. II **6**, 1 (1901)
14. Lamorgese, A.G., Mauri, R.: Phys. Fluids **17**, 034107 (2005)
15. Lamorgese, A.G., Mauri, R.: Phys. Fluids **18**, 044107 (2006)
16. Lamorgese, A.G., Mauri, R.: Int. J. Multiph. Flow **34**, 987 (2008)

17. Landau, L.D., Lifshitz, E.M.: Statistical Physics, Part I. Pergamon, Elmsford (1980), Chap. XIV
18. Landau, L.D., Lifshitz, E.M.: Statistical Physics, Chap. 84
19. Lowengrub, J., Truskinovsky, L.: Proc. R. Soc. Lond., Ser. A **454**, 2617 (1998)
20. Lucretius, T.C.: De Rerum Natura, Book I (50 BCE)
21. Madruga, S., Thiele, U.: Phys. Fluids **21**, 062104 (2009)
22. Mauri, R.: The Diffuse Interface Approach. Springer, Berlin (2010)
23. Molin, D., Mauri, R.: Phys. Fluids **19**, 074102 (2007)
24. Molin, D., Mauri, R.: Chem. Eng. Sci. **63**, 2402 (2008)
25. Onuki, A.: Phys. Rev. E **75**, 036304 (2007)
26. Pismen, L.M.: Phys. Rev. E **64**, 021603 (2001)
27. Pismen, L.M., Pomeau, Y.: Phys. Rev. E **62**, 2480 (2000)
28. Rowlinson, J.S., Widom, B.: Molecular Theory of Capillarity. Oxford University Press, London (1982)
29. Thiele, U., Madruga, S., Frastia, L.: Phys. Fluids **19**, 122106 (2007)
30. van der Waals, J.D.: The thermodynamic theory of capillarity under the hypothesis of a continuous variation of density (1893). Reprinted in J. Stat. Phys. **20**, 200 (1979)
31. Vladimirova, N., Malagoli, A., Mauri, R.: Phys. Rev. E **60**, 2037 (1999), also see p. 6968
32. Widom, B.: J. Phys. Chem. **100**, 13190 (1996)

Chapter 10
Effective Transport Properties

In this chapter we study a particular case of multiphase systems, namely two-phase materials in which one of the phases is randomly dispersed in the other, so that the composite can be viewed on a macroscale as an effective continuum, with well defined properties. In general, the theoretical determination of the parameter for an effective medium requires, as a rule, the solution of a corresponding transport problem at the microscale, which takes into account the morphology of the system and its evolution. As the mathematical problem is well-posed on a microscale, this can be accomplished using, for example, the multiple scale approach shown in Chap. 11; however, the task requires massive computations and is therefore difficult to implement from the practical standpoint. Here, instead, we focus on a deterministic approach to the problem, where the geometry and spatial configuration of the particles comprising the included phase are given and the solution to the microscale problem is therefore sought analytically. As examples, we study the effective thermal conductivity of solid reinforced materials (Sect. 10.1), the effective viscosity of non-colloidal suspensions (Sect. 10.2), the effective permeability of porous materials (10.3) and the effective self- and gradient diffusivities of colloidal suspensions (Sect. 10.4). Then, in Sect. 10.5, an alternative dynamic definition of the transport coefficients is considered, which can also serve as a basis to determine the effective properties of complex systems.

10.1 Heat Conductivity of Composite Solids

Consider a composite solid composed of spherical inclusions dispersed within a continuous phase. As usual, the objective is to determine the transport of heat (and mass as well) on a lengthscale that greatly exceeds that of the microstructure, i.e. both the particle radius and the spacing between the inclusions. Ultimately, effective properties depend on the morphology of the system, as the presence of an inclusion influences the temperature fields of the continuous phase as well as of the surrounding inclusions [2]. In the following, though, we will only consider the dilute case, where particle-particle interactions are neglected, so that we may consider an iso-

R. Mauri, *Non-Equilibrium Thermodynamics in Multiphase Flows*,
Soft and Biological Matter, DOI 10.1007/978-94-007-5461-4_10,
© Springer Science+Business Media Dordrecht 2013

lated sphere of radius a, located in the origin, surrounded by a large volume V of the continuous phase. Then, heat transport is described through the equation,

$$\nabla_i J_i = 0; \quad J_i = -\kappa \nabla_i T, \quad \begin{cases} \kappa = \kappa_p & \text{when } r < a, \\ \kappa = \kappa_0 & \text{when } r > a, \end{cases} \tag{10.1}$$

where κ_p and κ_0 are the heat conductivity of the inclusions and of the continuous phase, respectively. The boundary conditions impose a given temperature gradient at infinity, together with the continuity of temperature and heat flux at the surface of the sphere, i.e.

$$T = G_i r_i \quad \text{at } r \to \infty; \tag{10.2}$$

$$T \text{ and } J_i \text{ continuous at } r = a. \tag{10.3}$$

The general solution of this problem is [cf. Eq. (F.12)]:

$$T(\mathbf{r}) = G_i r_i \left(\lambda' \frac{1}{r^3} + \lambda'' \right). \tag{10.4}$$

Then, imposing that the boundary conditions are satisfied, we find the outer and inner solutions,

$$T(\mathbf{r}) = \left(1 + K \frac{a^3}{r^3} \right) G_i r_i \quad \text{when } r > a; \tag{10.5}$$

and

$$T(\mathbf{r}) = (1 + K) G_i r_i \quad \text{when } r < a, \tag{10.6}$$

where

$$K = \frac{1 - \gamma}{2 + \gamma}; \quad \text{with } \gamma = \frac{\kappa_p}{\kappa_0}. \tag{10.7}$$

At the end, we want to determine the effective Fourier law,

$$\langle J_i \rangle = -\kappa^* \langle \nabla_i T \rangle, \tag{10.8}$$

where the brackets indicate volume average while κ^* denotes the effective heat conductivity. As shown in Problem 10.1, calculating explicitly the volume averages of the general solution, we find,

$$\frac{\kappa^*}{\kappa_0} = 1 - 3K\phi + O(\phi^2). \tag{10.9}$$

In particular, when $\kappa_p = \kappa_0$, i.e. $\gamma = 1$, we find $K = 0$ and then trivially, $\kappa^* = \kappa_0$. In addition, consider the following particular cases, corresponding to perfectly insulating and perfectly conducting inclusions:

$$\frac{\kappa^*}{\kappa_0} = 1 - \frac{3}{2}\phi \quad \text{when } \gamma = 0, \tag{10.10}$$

and

$$\frac{\kappa^*}{\kappa_0} = 1 + 3\phi \quad \text{when } \gamma = \infty. \tag{10.11}$$

This result, obtained by Maxwell in 1873, shows that the leading-order correction to the conductivity is of $O(\phi)$. The extension to less dilute suspensions was developed by Jeffrey,[1] who derived the $O(\phi^2)$ term for a random suspension of spheres by the addition of two-sphere interactions. In fact, Jeffrey used an approach that is both easier and more general (and therefore quite ingenious), considering the average heat flux as

$$\langle \mathbf{J} \rangle = \frac{1}{V} \int_V \mathbf{J} \, d^3\mathbf{r} = -\frac{1}{V} \left[\int_{V_0} \kappa_0 \nabla T \, d^3\mathbf{r} + \sum_{i=1}^{N} \int_{V_i} \kappa_p \nabla T \, d^3\mathbf{r} \right], \tag{10.12}$$

where V_0 and V_i are the total volumes occupied by the continuous phase and the i-th inclusion, respectively, with $V = V_0 + \sum V_i$. This equation can be rewritten as:

$$\langle \mathbf{J} \rangle = -\frac{1}{V} \left[\kappa_0 \int_V \nabla T \, d^3\mathbf{r} + (\kappa_p - \kappa_0) \sum_{i=1}^{N} \int_{V_i} \nabla T \, d^3\mathbf{r} \right],$$

so that, assuming that the inclusions are all identical, with volume V_p, we obtain:

$$\langle \mathbf{J} \rangle = -\kappa_0 \langle \nabla T \rangle + \phi \langle \mathbf{J}^* \rangle, \tag{10.13}$$

where $\phi = NV_p/V$ is the volume fraction occupied by the inclusions, while $\langle \mathbf{J}^* \rangle$ is the average of the additional heat flux $\mathbf{J}^*$ due to the presence of a single inclusion, i.e.,

$$\mathbf{J}^* = -(\kappa_p - \kappa_0) \frac{1}{V_p} \int_{V_p} \nabla T \, d^3\mathbf{r}. \tag{10.14}$$

In the dilute limit and for spherical inclusions, the disturbance of the temperature field induced by a sphere is independent of the presence of all the other spheres, so that applying Eq. (10.6)–(10.7) we see that inside a sphere we have $\nabla T = (1+K)\mathbf{G}$, where $\mathbf{G}$ is the unperturbed temperature gradient (i.e. far from the sphere), which equals at leading order the mean temperature gradient, i.e. $\mathbf{G} = \langle \nabla T \rangle + O(\phi)$. At the end, we obtain at leading order,

$$\mathbf{J}^* = -(\kappa_p - \kappa_0)(1 + K)\phi \langle \nabla T \rangle,$$

so that we obtain the effective Fourier law (10.8), with the following effective conductivity:

$$\frac{\kappa^*}{\kappa_0} = 1 + (\gamma - 1)(1 + K)\phi = 1 - 3\frac{1-\gamma}{2+\gamma}\phi = 1 - 3K\phi + O(\phi^2), \tag{10.15}$$

which coincides with Maxwell's Eq. (10.9).

---

[1]D.J. Jeffrey [8]. The fact that the extension took 100 years to be developed tells us how much more difficult it is.

The Energy Dissipation Approach   Consider the energy dissipated per unit volume, resulting from imposing a given temperature gradient, $\mathbf{G}$, to our system,

$$\dot{E} = -\frac{1}{V}\int_V J_i \nabla_i T\, d^3\mathbf{r} = -\frac{1}{V}\oint_S n_i J_i T\, d^2\mathbf{r}, \tag{10.16}$$

where we have integrated by parts. Here, S is the surface enclosing the fluid volume, and therefore it includes both the outer surface, S_0, at infinity, and the inclusion boundaries, S_i. Now, consider the equality,

$$\dot{E} = -\frac{1}{V}\oint_S n_i J_i T^{(0)}\, d^2\mathbf{r} - \frac{1}{V}\oint_S n_i J_i \left(T - T^{(0)}\right) d^2\mathbf{r}, \tag{10.17}$$

where $T^{(0)} = \mathbf{G}\cdot\mathbf{r}$ define the unperturbed temperature field, i.e. in the absence of inclusions.

Next, let us apply the reciprocal theorem (F.79), where the primed problem corresponds to the unperturbed temperature field, while the double primed problem corresponds to the case under consideration, i.e. with inclusions and with again a constant temperature gradient at infinity,

$$\int_V \mathbf{J}^{(0)}\cdot\nabla T\, d^3\mathbf{r} = \int_V \mathbf{J}\cdot\nabla T^{(0)}\, d^3\mathbf{r}, \tag{10.18}$$

with $\mathbf{J}_0 = -k_0\mathbf{G}$. Therefore, considering that $T = T^{(0)}$ at the outer boundary, Eq. (10.17) becomes:

$$\dot{E} = -\frac{1}{V}\oint_{S_0} n_i J_i^{(0)} T^{(0)}\, d^2\mathbf{r} - \frac{1}{V}\sum_i \oint_{S_i} n_i\left[J_i^{(0)} T + J_i\left(T - T^{(0)}\right)\right] d^2\mathbf{r}. \tag{10.19}$$

The first term on the RHS equals the energy dissipated by the unperturbed temperature field,

$$\dot{E}_0 = k_0\mathbf{G}\cdot\mathbf{G}, \tag{10.20}$$

while the second term is the extra dissipation due to the presence of the inclusions which, assuming that they are all identical, with volume V_p, becomes,

$$\dot{E}^* = -\frac{\phi}{V_p}\oint_{S_p} n_i\left[J_i^{(0)} T + J_i\left(T - T^{(0)}\right)\right] d^2\mathbf{r}. \tag{10.21}$$

Finally, defining the effective conductivity as:

$$\frac{\kappa^*}{\kappa_0} = 1 + \frac{\dot{E}^*}{\dot{E}_0}, \tag{10.22}$$

we obtain the same results that we have seen before.

For example, for perfectly conducting inclusions (i.e. $\kappa_p \to \infty$), in the dilute case, where $T = 0$ at $r = a$, the expression above simplifies as:

$$\dot{E}^* = \frac{\phi}{V_p} \oint_{r=a} n_i J_i T^{(0)} d^2\mathbf{r}, \tag{10.23}$$

and considering that $n_i J_i = 3k_0 G_i r_i / a$ at $r = a$ (the unit vector is directed inward), we obtain: $\dot{E}^* = 3\phi k_0 \mathbf{G} \cdot \mathbf{G}$, so that at the end we find again Maxwell's Eq. (10.11), i.e. $\kappa^*/\kappa_0 = 1 + 3\phi + O(\phi^2)$.

As we have mentioned above, 100 years later, Jeffrey extended Maxwell's result by accounting for two-sphere interactions, obtaining, for $\gamma = \infty$:

$$\frac{\kappa^*}{\kappa_0} = 1 + 3\phi + K_2\phi^2 + O(\phi^3), \tag{10.24}$$

where K_2 depends on the morphology of the composite solid; in particular, $K_2 = 4.51$ for a random dispersion of spheres.

10.2 Effective Viscosity of Suspensions

This problem is quite similar to the evaluation of the effective heat conductivity that was considered in the previous chapter. The following analysis applies to suspensions moving at low Reynolds number, where the fluid is assumed to be Newtonian, with viscosity η_0 and the particles are rigid (i.e. with infinite viscosity), neutrally buoyant spheres. In addition, we restrict our analysis to the dilute case, where particle-particle interactions are neglected, while the more general case is analyzed in [7]. Then, momentum transport is governed by the equation:

$$\nabla \cdot \mathbf{T} = \mathbf{0}; \qquad \nabla \cdot \mathbf{v} = 0, \tag{10.25}$$

where,

$$T_{ij} = -p\delta_{ij} + 2\eta_0 S_{ij}; \quad S_{ij} = \frac{1}{2}(\nabla_i v_j + \nabla_j v_i). \tag{10.26}$$

The boundary conditions are that the flow field is fixed at infinity, while particles are force and torque-free.

Proceeding as in the previous chapter, we obtain:

$$\langle \mathbf{T} \rangle = \frac{1}{V} \int_V \mathbf{T} d^3\mathbf{r} = \frac{1}{V} \left[\int_{V_0} \mathbf{T} d^3\mathbf{r} + \sum_{i=1}^{N} \int_{V_i} \mathbf{T} d^3\mathbf{r} \right],$$

where V_0 and V_i are the total volumes occupied by the continuous phase and the i-th inclusion, respectively, with $V = V_0 + \sum V_i$. Substituting the constitutive relation (10.26) with $\mathbf{S} = \mathbf{0}$ inside the particles, and assuming that the inclusions are all identical, with volume V_p, we obtain an equation very similar to (10.13) and (10.14), i.e.,

$$\langle \mathbf{T} \rangle = -(1 - \phi)\overline{p}\mathbf{I} + 2\eta_0 \langle \mathbf{S} \rangle + \phi \langle \mathbf{T}^* \rangle, \tag{10.27}$$

where $\phi = NV_p/V$ is the volume fraction occupied by the inclusions, while $\langle \mathbf{T}^* \rangle$ is the average of the additional stress tensor $\mathbf{T}^*$ due to the presence of a single inclusion, i.e.,

$$\mathbf{T}^* = \frac{1}{V_p} \int_{V_p} \mathbf{T} d^3\mathbf{r} = \frac{1}{V_p} \oint_{S_p} (\mathbf{n} \cdot \mathbf{T})\mathbf{r} d^2\mathbf{r} = -\frac{1}{V_p} \oint_{S_p} \mathbf{n}(p\mathbf{r}) d^2\mathbf{r}. \tag{10.28}$$

Here we have integrated by parts using the equality $T_{ij} = (\nabla_k T_{kj})r_i$, considering that $\mathbf{T}$ is symmetric and that $\mathbf{S} = 0$ at the surface. In addition, $\overline{p}$ is the average fluid pressure, i.e.

$$\overline{p} = \frac{1}{V_0} \int_{V_0} p d^3\mathbf{r}, \tag{10.29}$$

whose value, though, is uninfluential, as the suspension is incompressible.

In the dilute limit and for spherical inclusions, the disturbance of the flow field induced by a sphere is independent of the presence of all the other spheres. In that case, the pressure induced by an isolated sphere, located in the origin and immersed in a shear flow is given by Eq. (F.46), so that:

$$p(r = a) = -5\eta_0 \frac{r_i r_k}{a^2} G_{ik}, \tag{10.30}$$

where $\mathbf{G}$ is the imposed shear rate which, within an $O(\phi)$ correction, coincides with the mean shear rate $\langle \mathbf{S} \rangle$, i.e. $\mathbf{G} = \langle \mathbf{S} \rangle + O(\phi)$. Finally, substituting (10.30) into (10.28), and considering that $\mathbf{I}:\mathbf{G} = 0$, we obtain:

$$\mathbf{T}^* = 5\eta_0 \langle \mathbf{S} \rangle. \tag{10.31}$$

Now, defining the effective suspension viscosity, η^*, as:

$$\langle \mathbf{T} \rangle^d = 2\eta^* \langle \mathbf{S} \rangle, \tag{10.32}$$

where the superscript d indicates the deviatoric part, substituting (10.31) into (10.27) we obtain:

$$\frac{\eta^*}{\eta_0} = 1 + \frac{5}{2}\phi + O(\phi^2). \tag{10.33}$$

This is the celebrated formula obtained by Einstein [6].

Correction to Einstein's formula was obtained by Batchelor and Green [4] as,

$$\frac{\eta^*}{\eta_0} = 1 + \frac{5}{2}\phi + K_2\phi^2 + O(\phi^3). \tag{10.34}$$

where the second-order coefficient K_2 depends on the pair particle distribution. In particular, assuming a uniform distribution, we find $K_2 = 5.2$, which is in excellent agreement with experiments [10].

Identical results can be obtained using the energy dissipation approach. In fact, proceeding as in the previous chapter, we obtain:

$$\frac{\eta^*}{\eta_0} = 1 + \frac{\dot{E}^*}{\dot{E}_0}, \tag{10.35}$$

where

$$\dot{E}_0 = 2\eta_0 \mathbf{G} : \mathbf{G} \tag{10.36}$$

is the energy dissipated by the unperturbed velocity field, with $\mathbf{G}$ (with $\mathbf{G} = \mathbf{G}^+$ and $\mathbf{I}:\mathbf{G} = 0$) denoting the imposed shear rate, while $\dot{E}^*$ is the extra dissipation due to the presence the inclusions,

$$\dot{E}^* = -\frac{\phi}{V_p} \oint_{S_p} n_i \left[T_{ij}^{(0)} v_j + T_{ij} (v_j - v_j^{(0)}) \right] d^2\mathbf{r}, \tag{10.37}$$

where we have assumed that inclusions are all identical, with volume V_p. Here, $\mathbf{v}^{(0)} = \mathbf{G} \cdot \mathbf{r}$ is the unperturbed velocity field, while $\mathbf{v}$ is the velocity field of the suspension, with $\mathbf{v} = \mathbf{0}$ and $\mathbf{S} = \mathbf{0}$ at the particle surface. Therefore, we obtain:

$$\dot{E}^* = -\frac{\phi}{V_p} G_{ij} \oint_{r=a} n_i \, p \, r_j \, d^2\mathbf{r}, \tag{10.38}$$

which reproduces Eq. (10.28), thus yielding $\dot{E}^* = 5\eta_0 \phi \mathbf{G}:\mathbf{G}$, so that at the end we obtain again Einstein's equation (10.33).

10.3 Permeability of Porous Media

The homogeneous flow of a Newtonian liquid through a porous medium at low Reynolds number is governed by the Darcy phenomenological law,

$$\langle \nabla p \rangle = \eta \mathbf{k}^{-1} \cdot \mathbf{V}, \tag{10.39}$$

where the brackets indicate averaging, ∇p is the pressure gradient, η is the fluid viscosity, $\mathbf{V}$ its mean velocity, while the coefficient $\mathbf{k}$ depends only on the morphology of the material, has the units of an area (i.e, typically, its magnitude is that of the square of the pore size) and is referred to as the permeability of the porous material.[2] Clearly, $\langle \nabla p \rangle$ is the mean body force, i.e. the force per unit volume, exerted

---

[2]To differentiate it from the "permeability" used in Biophysics, where it denotes the material transfer coefficient (having the units of a velocity), sometimes k is referred to as Darcy's permeability.

by the fluid on the solid particles that constitute the matrix of the porous material. Now, at low Reynolds number the drag force on each particle, $\mathbf{F}_1$, is proportional to the unperturbed velocity, $\mathbf{V}$ through a symmetric translation resistance tensor, $\boldsymbol{\zeta}$ (see Problem 2.1). Therefore, we obtain Eq. (10.39) with,

$$\mathbf{k}^{-1} = \frac{\phi}{\eta V_p} \langle \boldsymbol{\zeta} \rangle, \tag{10.40}$$

where V_p is the mean volume occupied by a single particle, while the averaging is taken over all possible morphologies, i.e. shape and orientation of the inclusions, plus the influence of the surrounding particles. In particular, for isotropic media, we have: $\langle \boldsymbol{\zeta} \rangle = \zeta \mathbf{I}$, where $\zeta = \frac{1}{3} \zeta_{ij} \delta_{ij}$ is the drag coefficient which, for uniform spherical inclusions of radius a, in the dilute limit, reduces to $\zeta_0 = 6\pi \eta a$, so that,

$$\mathbf{k} = k^* \mathbf{I}, \quad \text{where } k^* = \frac{2a^2}{9\phi}. \tag{10.41}$$

The constitutive relation (10.39) can also be interpreted in an alternative way, if we refer it to a reference frame moving at constant speed $\mathbf{V}$. Then, the permeability $\mathbf{k}$ can be seen as the mean instantaneous velocity of a suspended particle under the action of a normalized pressure gradient.[3] Now, when particle-particle interactions are taken into account, the drag coefficient $\zeta(\phi)$ of a monodisperse macroscopically uniform and isotropic suspension of spheres can be written as:

$$\zeta(\phi) = \zeta_0 [1 + K(\phi)], \tag{10.42}$$

where $K(\phi)$ depends on the particle distribution. As shown by Batchelor [1, 3], in a dilute suspension of rigid spheres with uniform probability of all accessible configurations, we have:

$$K(\phi) = 6.55\,\phi + O(\phi^2), \tag{10.43}$$

showing that the sedimentation velocity decreases as the particle concentration increases. This is due to the fact that, although particles tend to help each other in sedimenting, this effect is more than compensated by the return flow of the fluid, that pulls the particles back up. The derivation of this result is not straightforward, as a simple summation of the velocity changes due to the hydrodynamic pairwise interaction between a tagged particle and all the other surrounding particles yields a non-converging integral. The renormalization procedure was developed by G.K. Batchelor; in the Appendix F.4, we present a slightly different derivation.

At the end, we obtain,

$$\mathbf{k} = k(\phi)\mathbf{I}, \quad \text{where } k(\phi) = k_0 [1 - 6.55\phi + O(\phi^2)], \tag{10.44}$$

---

[3]This statement is formally proven in Problem 11.3 using the method of homogenization.

where $k_0 = k^*$ is the permeability (10.41) at infinite dilution, showing that the hydrodynamic hindrance causes permeability to decrease as volume fraction increases, at least for dilute suspensions.[4]

10.4 Diffusion in Colloidal Suspensions

Consider a suspension of neutrally buoyant Brownian particles.[5] As we have mentioned in previous chapters, there are two types of diffusivities that one can define, namely self-diffusivity, $D^{(s)}$ and gradient diffusivity, D. In the dilute limit, when particle-particle interactions can be neglected, the Stokes-Einstein relation reveals that these two diffusivities are equal to each other, in agreement with the fluctuation-dissipation theorem. However, when particle-particle interactions are taken into account, the relation between fluxes and forces ceases to be linear, as diffusivities are themselves functions of the concentration, and therefore most of the results of non-equilibrium thermodynamics, such as the fluctuation-dissipation theorem, cannot be applied. In fact, we will see below that self-diffusivity, D^s, and gradient diffusivity, D, are not equal to each other, as at leading order we have: $D^s(\phi) = D_0(1 - 6.55\phi)$ and $D(\phi) = D_0(1 + 1.45\phi)$, where $D_0 = kT/\zeta_0$ is the diffusivity at infinite dilution. Just as in the case examined in Sect. 4.5, we cannot really speak of a violation of the FD theorem, but simply of a case where it cannot be applied.

10.4.1 Self Diffusion

Self diffusivity describes the temporal growth of the mean square displacement undergone by the Brownian particles moving randomly in a uniform suspension.[6] Accordingly, we may define the self diffusion tensor using (2.29) as:

$$D^s_{ij} = \frac{1}{2} \lim_{t \to \infty} \frac{d}{dt} \langle X_i(t) X_j(t) \rangle, \tag{10.45}$$

where $\mathbf{X}(t)$ is the position of the Brownian particle at time t, referred to a fixed laboratory reference frame (i.e. the walls of the vessel containing the suspension), assuming that $\mathbf{X}(0) = \mathbf{0}$. Denoting by $\mathbf{V}^{(p)}(t)$ the velocity of the Brownian particle referred to this frame of reference, we have:

$$\mathbf{X}(t) = \int_0^t \mathbf{V}^{(p)}(\tau) \, d\tau, \tag{10.46}$$

---

[4]Both experiments and numerical simulations show that permeability continues to decrease almost linearly with volume fraction, until it becomes zero at the percolation limit.

[5]These, so called, colloidal particles are small enough to undergo thermal fluctuations and yet large enough to present clear phase boundaries; they have diameter from 5 to 200 nanometers.

[6]This process is sometimes called Knudsen *effusion*.

where the superscript "p" stands for "particle". Clearly, as in the laboratory reference frame the total volumetric suspension flux is zero, the volume particle flux must be counterbalanced by the volume fluid flux (trivially, if the particles sediment down, the fluid must move up, in the opposite direction), so that $\mathbf{V}^{(p)}$ is the particle velocity relative to the volume-averaged suspension velocity, which we assume to be zero. As we are dealing with neutrally buoyant suspensions, though, the volume mean velocity coincides with the mass averaged velocity (7.20), so that $\mathbf{V}^{(p)}(t)$ is really a diffusive-only velocity. Finally, assuming that $\mathbf{V}^{(p)}(t)$ is a stationary random function, so that $\langle \mathbf{V}^{(p)}(t)\mathbf{V}^{(p)}(t+\tau)\rangle$ depends only on τ, substituting (10.46) into (10.45) we obtain:

$$D_{ij}^s = \frac{1}{2}\int_0^\infty \langle V_i^{(p)}(0)V_j^{(p)}(\tau)\rangle d\tau, \qquad (10.47)$$

which is the Green-Kubo relation (2.28).

Now, referring to the results of Sect. 3.3, when the general stationary random variable $\mathbf{x}$ is identified with $\mathbf{V}^{(p)}$, for times $t > m/\|\boldsymbol{\zeta}\|$, the Langevin equation becomes:

$$\boldsymbol{\zeta} \cdot \mathbf{V}^{(p)} = \mathbf{f}, \qquad (10.48)$$

where $\boldsymbol{\zeta}$ denotes the translation tensor, $\|\boldsymbol{\zeta}\|$ is its norm, while $\mathbf{f}$ is a random force, with $\langle f_i\rangle = 0$ and $\langle f_i(t)f_j(t+\tau)\rangle = 2kT\zeta_{ij}\delta(\tau)$. Therefore, substituting (10.48) into (10.47), we obtain:

$$\mathbf{D}^s = kT\boldsymbol{\zeta}^{-1}, \qquad (10.49)$$

which is a simple generalization of the Stokes-Einstein relation. This is the self-diffusivity tensor at a given time, corresponding to a given morphology of the suspension; accordingly, the inverse translation tensor in (10.49) must be averaged out over all the system configurations, namely the particle orientation and the position of all its surrounding particles.

As we saw in the previous section, the translation resistance tensor of a very dilute suspension of randomly distributed spherical particles of radius a is $\zeta_{ij} = \zeta(\phi)\delta_{ij}$, where $\zeta(\phi)$ is given by Batchelor's result, Eq. (10.42)–(10.43). So, at the end, we obtain:

$$D^s(\phi) = D_0\left[1 - 6.55\phi + O\left(\phi^2\right)\right], \qquad (10.50)$$

where $D_0 = kT/\zeta_0$ is the diffusivity at infinite dilution, showing that the hydrodynamic hindrance to particle movement causes self-diffusivity to decrease as volume fraction increases, at least for dilute suspensions.[7]

---

[7] As for permeability, both experiments and numerical simulations show that self-diffusivity continues to decrease almost linearly with volume fraction, until it becomes zero at the percolation limit.

10.4.2 Gradient Diffusion

Now consider the isothermal and isobaric diffusion of identical spherical particles down a concentration gradient. The diffusive particle flux is defined through (7.22) as,

$$\mathbf{J}_d^{(p)} = \rho\phi\big(\mathbf{v}^{(p)} - \bar{\mathbf{v}}\big), \tag{10.51}$$

where ρ is the constant suspension mass density, ϕ is the particle mass (and volume, in this case) fraction, $\mathbf{v}^{(p)}$ is the mean particle velocity, while $\bar{\mathbf{v}}$ is the mass (and volume as well) averaged suspension velocity, which here is assumed to be zero.[8] Now, considering that, at uniform temperature and pressure, the mean relative particle velocity is equal to the mean force $\mathbf{F}_1$ exerted on a single particle divided by the drag coefficient ζ, and that, in turn, $\mathbf{F}_1$ equals the thermodynamic force $-\nabla\mu^{(p)}$ exerted on one mole of particles divided by the Avogadro number, N_A, we obtain:

$$\mathbf{J}_d^{(p)} = -\frac{\rho\phi}{N_A\zeta}\nabla\mu^{(p)}; \quad \nabla\mu^{(p)} = \frac{\partial\mu^{(p)}}{\partial\phi}\nabla\phi. \tag{10.52}$$

Therefore, we see that gradient diffusion is described through the constitutive relation (8.43), i.e.,

$$\mathbf{J}_d^{(p)} = -\rho D\nabla\phi, \tag{10.53}$$

where D is the gradient diffusivity,

$$D = \frac{\phi}{N_A\zeta}\frac{\partial\mu^{(p)}}{\partial\phi}. \tag{10.54}$$

This result is equivalent to (8.63), i.e.,

$$\mathbf{J}_d^{(p)} = -\frac{L}{T}\nabla\mu^{(pf)}, \tag{10.55}$$

where L is the Onsager coefficient, while $\mu^{(pf)} = \mu^{(p)} - \mu^{(f)}$ is the chemical potential difference, with the superscript "f" standing for "fluid". In fact, applying the Gibbs-Duhem relation, i.e.,

$$\nabla\tilde{\mu}^{(pf)} = \frac{1}{1-\phi}\nabla\mu^{(p)}, \tag{10.56}$$

and equating (10.53) and (10.55), we see that,

$$\zeta = \frac{\rho T\phi(1-\phi)}{N_A L}. \tag{10.57}$$

---

[8]Trivially, $\bar{\mathbf{v}} = \phi\mathbf{v}^{(p)} + (1-\phi)\mathbf{v}^{(f)}$, where $\mathbf{v}^{(f)}$ is the mean fluid velocity.

Now, we turn to evaluating the chemical potential of a suspension of hard spheres. Although the complete theory is available elsewhere,[9] here we note that the particle free energy can be evaluated following the procedure of Sect. 9.2.1, replacing ρ with the particle molar concentration $c^{(p)}$ and assuming that the interaction potential $\psi(r)$ between the suspended particles can be expressed through Eq. (9.6), with $U_0 = 0$ and $d = 2r_p$, i.e.

$$\psi(r) = \begin{cases} 0 & (r > 2r_p), \\ \infty & (r < 2r_p) \end{cases} \qquad (10.58)$$

where r_p is the hard-core particle radius. At the end, we obtain the thermodynamic free energy [see Eq. (9.5)–(9.11) with $c_1 = 0$],

$$f(T, c^{(p)}) = f_0 + RT[\log c^{(p)} + c^{(p)} B], \qquad (10.59)$$

where $B = 4V_p$ is the virial coefficient, with V_p indicating the volume occupied by 1 mole of the suspended particles. Accordingly, considering that $c^{(p)} V_p = \phi^{(p)} = \phi$ (we remind that here volume fractions are identically equal to mass fractions), and applying Eq. (9.48) we obtain:

$$\mu^{(p)} = \frac{d(c^{(p)} f)}{dc^{(p)}} = RT[1 + \log c^{(p)} + 8\phi], \qquad (10.60)$$

and therefore,

$$\phi \frac{d\mu^{(p)}}{d\phi} = RT(1 + 8\phi). \qquad (10.61)$$

Predictably, the 8ϕ correction appearing in Eq. (10.61) is the excluded volume. Finally, substituting (10.61) and (10.42)–(10.43) into (10.54) yields:

$$D(\phi) = D_0(1 + 1.45\phi + O(\phi^2)), \qquad (10.62)$$

showing that the excluded volume effect, which tends to increase diffusion, overweighs the hydrodynamic retardation, so that in dilute suspensions gradient diffusivity slightly increases with volume fraction.

Comment Both self-diffusion and gradient diffusion can be interpreted in terms of an average velocity. However, gradient diffusion describes the collective migration of particles subjected to a concentration gradient, and so the related mean particle velocity, $v^{(p)}$, is a "thermodynamic" quantity. On the other hand, self diffusion depends on the mean instantaneous velocity, V, of a single particle in a uniform suspension as we follow it during its motion, and therefore coincides with the mean velocity of the Darcy law (10.39). To see that, consider that the force acting on a single particle is $F_1 = -\nabla \mu_1 = -kT \nabla \phi / \phi$, where $\mu_1 = \mu_0 + kT \ln \phi$ is the single

---

[9]See [11].

particle chemical potential.[10] So, at the end, considering that $\mathbf{F}_1 = \boldsymbol{\zeta} \cdot \mathbf{V}$, we find the material flux $\mathbf{J}_\phi = \mathbf{V}\phi = -\mathbf{D}^s \cdot \nabla\phi$, where, as expected, $\mathbf{D}^s = kT\boldsymbol{\zeta}^{-1}$ is the self diffusivity (10.49), while $\mathbf{V}$ satisfies Eq. (10.39).

10.5   Dynamic Definition of Transport Coefficients

10.5.1   Heat and Mass Diffusion

We have seen that properties such as mass and heat diffusivities are related to the dispersion of mass and energy through the fluctuation-dissipation theorem. The premier example of this class of relations is Einstein's classical formula (3.23), showing that the coefficient of molecular diffusion of colloidal particles, D, is proportional to the time derivative of the mean square displacement of one of those particles as it diffuses through the system. This result was then generalized through the Kubo relation (2.29), allowing to define in the same way any Onsager phenomenological coefficient $\mathbf{L}$.

Let us see this case once again. Assume that a unit mass (or heat) disturbance is introduced at the origin $\mathbf{r} = \mathbf{0}$ and at time $t = 0$ in an otherwise homogeneous medium, where it diffuses with diffusivity D. The governing equation reads,

$$\mathcal{L}_c(\Pi) \equiv \partial_t \Pi - D\nabla^2 \Pi = \delta(\mathbf{r})\delta(t), \tag{10.63}$$

with

$$\Pi(\mathbf{r}, t) = 0 \quad \forall t < 0, \tag{10.64}$$

stating that the system is initially unperturbed. Here, $\Pi(\mathbf{r}, t)$ is the propagator of the diffusion equation, i.e. a Wiener process (see Sects. 4.6 and 6.1), denoting, physically, the probability that a tracer introduced in the origin at time $t = 0$ is found at location $\mathbf{r}$ at a later time t.

Taking the Fourier transforms $\widehat{\Pi}(\mathbf{k}, \omega)$ of $\Pi(\mathbf{r}, t)$,

$$\widehat{\Pi}(\mathbf{k}, \omega) = \int \Pi(\mathbf{r}, t)e^{i(\mathbf{k}\cdot\mathbf{r}+\omega t)} d^3r\, dt, \tag{10.65}$$

with,

$$\Pi(\mathbf{r}, t) = \int \widehat{\Pi}(\mathbf{k}, \omega)e^{-i(\mathbf{k}\cdot\mathbf{r}+\omega t)} \frac{d^3k}{(2\pi)^3} \frac{d\omega}{2\pi}, \tag{10.66}$$

Eq. (10.63) yields:

$$\widehat{\Pi}(\mathbf{k}, \omega) = \left(i\omega D^{-1} + k^2\right)^{-1}, \tag{10.67}$$

---

[10]Note that here no excluded volume effect has been accounted for; it is as if the tagged particle belongs to a different chemical species than the other suspended particles.

with $k = |\mathbf{k}|$. Finally, anti-transforming (10.67) we find:

$$\Pi(\mathbf{r}, t) = (4\pi Dt)^{-3/2} \exp\left(-\frac{r^2}{4Dt}\right) H(t), \qquad (10.68)$$

with $H(t)$ denoting the Heaviside step function. It is easy to see that, since the propagator decays exponentially fast, its integral converges, with,

$$\int_{-\infty}^{\infty} \Pi(\mathbf{r}, t)\, d^3\mathbf{r} = 1, \qquad (10.69)$$

as one would expect, as a consequence of mass (or energy) conservation.

Dynamic Definition of Diffusivity   Now, let us define the central second moment of the propagator [see Eq. (A.26)],

$$\mu_{jk}^{(2)}(t) = \int_{-\infty}^{\infty} r_j \Pi(\mathbf{r}, t) r_k\, d^3\mathbf{r}. \qquad (10.70)$$

Since Π is a Gaussian, we easily find: $\mu_{jk}^{(2)}(t) = 2D_{jk}t$, with $D_{jk} = D\delta_{jk}$, so that,

$$D_{jk} = \frac{1}{2}\frac{d}{dt}\mu_{jk}^{(2)}(t). \qquad (10.71)$$

This indicates that the diffusivity measures the temporal growth of the second moment of the probability distribution. Alternately, we can rewrite this equation as,

$$D_{jk} = \frac{1}{2}\int r_j \dot{\Pi}(\mathbf{r}, t) r_k\, d^3\mathbf{r}. \qquad (10.72)$$

So, after introducing a Brownian particle in an otherwise homogeneous medium and subsequently monitoring its temporal stochastic spread, we may define diffusivity as the temporal growth of the central second moment of the concentration field. This dynamic definition could be applied as well to evaluate the effective diffusivity within a complex material, assuming that the timescale τ of the measuring process is much larger than the typical time that is necessary for the tracer to sample all positions within the microstructure.[11] Then, since the molecular diffusivity is a strong function of position,[12] i.e. $D = D(\mathbf{r})$, Eq. (10.72) indicates that the effective diffusivity measures a sort of mean diffusion of a solute particle as it spreads within the medium. This nontraditional definition of diffusivity has been applied by Brenner and coworkers [5] to determine the effective diffusivity in spatially periodic complex materials, obtaining results that are identical to those derived using the traditional definition, based on coarse-graining averaging procedures (see Sect. 11.3).

---

[11] For example, if we are considering the diffusive spread of a contaminant in a macroscopically homogeneous complex material, we must assume that $\tau \gg L^2/D$, where L is the typical dimension of the microstructure (e.g. the pore size), while D is the molecular diffusivity.

[12] In particular, in porous material, $D = 0$ within the inclusions.

10.5.2 Momentum Diffusion

Now, we intend to see that, in analogy with heat and mass diffusion, the kinematic viscosity describing the diffusion of momentum can be defined as the temporal growth of the central second moment of the velocity field generated by an impulsive perturbation initially introduced in the system [9].

The analogy between energy (and mass) diffusion, which is governed by the heat equation, and momentum diffusion, which obeys the unsteady Stokes equation, is not obvious, as there is a fundamental difference between the two processes: the effect of an impulsive perturbation decays exponentially fast in heat diffusion, and as r^{-3} in momentum diffusion. Accordingly, while the mass (or energy) initially introduced in the system is conserved (in the absence of "source" terms) as it spreads through the system, this seems not to be the case in momentum transport. To see that, consider the Stokes equations describing the flow field induced by a disturbance $\mathbf{F}$, that is introduced at the origin $\mathbf{r} = \mathbf{0}$ and at time $t = 0$ in an otherwise homogeneous quiescent fluid with unit density and viscosity η:

$$\mathcal{L}_{\mathbf{u}}(\mathbf{u}, p) \equiv \partial_t \mathbf{u} + \nabla p - \eta \nabla^2 \mathbf{u} = \mathbf{F}\delta(\mathbf{r})\delta(t), \tag{10.73}$$

$$\nabla \cdot \mathbf{u} = 0, \tag{10.74}$$

with $\mathbf{u}(\mathbf{r}, t)$ and $p(\mathbf{r}, t)$ denoting the fluid velocity and the pressure, respectively. Since (10.73) and (10.74) are linear equations, their solution can be expressed in terms of the propagators $\mathbf{\Pi}(\mathbf{r}, t)$ and $\boldsymbol{\pi}(\mathbf{r}, t)$ as

$$\mathbf{u}(\mathbf{r}, t) = \mathbf{\Pi}(\mathbf{r}, t) \cdot \mathbf{F}, \qquad p(\mathbf{r}, t) = \boldsymbol{\pi}(\mathbf{r}, t) \cdot \mathbf{F}, \tag{10.75}$$

where $\mathbf{\Pi}$ and $\boldsymbol{\pi}$ satisfy the following equations:

$$\mathcal{L}_{\mathbf{u}}(\mathbf{\Pi}, \boldsymbol{\pi}) = \mathbf{I}\delta(\mathbf{r})\delta(t), \qquad \nabla \cdot \mathbf{\Pi} = \mathbf{0}, \tag{10.76}$$

subjected to the condition,

$$\mathbf{\Pi}(\mathbf{r}, t) = \mathbf{0} \quad \forall t < 0, \tag{10.77}$$

stating that the fluid is initially at rest. Taking the Fourier transforms $\widehat{\mathbf{\Pi}}(\mathbf{k}, \omega)$ and $\widehat{\boldsymbol{\pi}}(\mathbf{k}, \omega)$ [see (10.65) and (10.66)] of $\mathbf{\Pi}(\mathbf{r}, t)$ and $\boldsymbol{\pi}(\mathbf{r}, t)$, respectively, (10.76) yields:

$$\widehat{\boldsymbol{\pi}}(\mathbf{k}, \omega) = -i\frac{\mathbf{k}}{k^2}, \tag{10.78}$$

$$\widehat{\mathbf{\Pi}}(\mathbf{k}, \omega) = \left(\mathbf{I} - \frac{\mathbf{k}\mathbf{k}}{k^2}\right)\left(i\omega\eta^{-1} + k^2\right)^{-1}, \tag{10.79}$$

with $k = |\mathbf{k}|$. Finally, antitransforming (10.78) and (10.79) we find:

$$\boldsymbol{\pi}(\mathbf{r}, t) = \nabla\left(\frac{1}{r}\right)\delta(t), \tag{10.80}$$

$$\boldsymbol{\Pi}(\mathbf{r}, t) = \boldsymbol{\Pi}_1(\mathbf{r}, t) + \boldsymbol{\Pi}_2(\mathbf{r}, t) + \boldsymbol{\Pi}_3(\mathbf{r}), \qquad (10.81)$$

with $r = |\mathbf{r}|$, where:

$$\boldsymbol{\Pi}_1(\mathbf{r}, t) = (4\pi \eta t)^{-3/2} \exp\left(-\frac{r^2}{4\eta t}\right) H(t)\mathbf{I}, \qquad (10.82)$$

$$\boldsymbol{\Pi}_2(\mathbf{r}, t) = -\frac{1}{4\pi} \nabla\nabla\left[\frac{1}{r}\, \mathrm{erfc}\left(\frac{r}{2\sqrt{\eta t}}\right)\right] H(t), \qquad (10.83)$$

$$\boldsymbol{\Pi}_3(\mathbf{r}) = \frac{1}{4\pi} \nabla\nabla\left(\frac{1}{r}\right) H(t), \qquad (10.84)$$

with $H(t)$ denoting the Heaviside step function.

The function $\boldsymbol{\Pi}_1$ defined in (10.82) is just the fundamental solution of the diffusion equation, i.e.,

$$\mathcal{L}_\mathbf{u}(\boldsymbol{\Pi}_1, \mathbf{0}) = \partial_t \boldsymbol{\Pi}_1 - \eta \nabla^2 \boldsymbol{\Pi}_1 = \mathbf{I}\delta(\mathbf{r})\,\delta(t). \qquad (10.85)$$

Since $\boldsymbol{\Pi}_1$ is not divergence-free, introducing an impulsive unit source of momentum results also in the appearance of a steady velocity field, $\boldsymbol{\Pi}_3$, which is created through the action of the delta-like pressure field, π. Thus $\boldsymbol{\Pi}_3$ and π are determined by mass conservation (indeed they are independent of the viscosity η) and solve the following equation,

$$\mathcal{L}_\mathbf{u}(\boldsymbol{\Pi}_3, \pi) = \eta\, \nabla\nabla\delta(\mathbf{r})\, H(t). \qquad (10.86)$$

Finally $\boldsymbol{\Pi}_2$ describes the decay in time of $\boldsymbol{\Pi}_3$, and satisfies the equation:

$$\mathcal{L}_\mathbf{u}(\mathbf{P}_2, \mathbf{0}) = -\eta\, \nabla\nabla\delta(\mathbf{r})\, H(t). \qquad (10.87)$$

From this analysis we see that the time-dependent part of the propagator,

$$\boldsymbol{\Pi}^* = \boldsymbol{\Pi}_1 + \boldsymbol{\Pi}_2 = \boldsymbol{\Pi} - \boldsymbol{\Pi}_3, \qquad (10.88)$$

satisfies the diffusion-like equation,

$$\mathcal{L}_\mathbf{u}(\boldsymbol{\Pi}^*, \mathbf{0}) = \mathbf{I}\delta(\mathbf{r})\,\delta(t) - \eta\, \nabla\nabla\delta(\mathbf{r})\, H(t). \qquad (10.89)$$

Unlike $\boldsymbol{\Pi}$, which has a slow rate of spatial decay, namely r^{-3}, $\boldsymbol{\Pi}^*$ decays exponentially fast. In addition, integrating (10.89) we obtain the normalization condition,

$$\int \boldsymbol{\Pi}^* d^3\mathbf{r} = \mathbf{I}. \qquad (10.90)$$

stating that the momentum originally introduced is conserved as it spreads through the system.

Equations (10.89) and (10.90) indicate that the time-dependent propagator $\boldsymbol{\Pi}^*$ is conserved and satisfies a diffusion-like equation; therefore it is $\boldsymbol{\Pi}^*$, not $\boldsymbol{\Pi}$, that is the analogous of the propagator of the heat equation. Now, the propagator of the heat equation can also be interpreted as the conditional probability describing the random motion of a passive heat tracer as it diffuses through the system. This simple physical interpretation cannot be easily applied to momentum diffusion, as some of the components of Π^* might be negative, so that Π^* does not have any obvious probabilistic meaning. However, it is tempting to interpret Π^* as the result of a stochastic process describing the transport of momentum, just as for the heat equation. The first symptom of this analogy is studied below, where we will develop a dynamical definition of viscosity based upon Π^*.

Dynamic Definition of Viscosity   Viscosity is in general a forth-order tensor, η_{ijkl}, defined as the proportionality term between the deviatoric stress tensor T_{ij} and the rate of strain dyadic $S_{ij} = (u_{i,j} + u_{j,i})/2$,

$$T_{ij} = 2\,\eta_{ijkl}\,S_{kl}, \tag{10.91}$$

where $u_{i,j} = \partial u_i/\partial x_j$. Due to the symmetry of both $\mathbf{T}$ and $\mathbf{S}$, η_{ijkl} must satisfy the following symmetry relations:

$$\eta_{ijkl} = \eta_{jikl} = \eta_{ijlk}. \tag{10.92}$$

In addition, due to Onsager's reciprocity relation, η_{ijkl} satisfies the additional condition,

$$\eta_{ijkl} = \eta_{klij}. \tag{10.93}$$

Considering that both $\mathbf{T}$ and $\mathbf{S}$ are traceless, η_{ijkl} is defined only up to an arbitrary tetradic α_{ijkl} such that

$$\alpha_{ijkl} = b\,\delta_{ij}\,\delta_{kl}, \tag{10.94}$$

for any arbitrary constant b. In the sequel, whenever we identify a certain tetradic as the viscosity η_{ijkl}, it is understood that any tensor α_{ijkl} satisfying (10.94) can be arbitrarily added.

In the case of isotropic fluids we have:

$$\eta_{ijkl} = \frac{1}{2}\,\eta(\delta_{ik}\delta_{jl} + \delta_{il}\delta_{jk}). \tag{10.95}$$

Now consider the following relations, which are easily obtained from (10.82) and (10.83):

$$\int x_i\,\Pi_{jk}^{(1)}\,x_l\,d^3\mathbf{r} = 4\,\eta\,t\,\delta_{il}\delta_{jk}, \tag{10.96}$$

$$\int x_i\,\Pi_{jk}^{(2)}\,x_l\,d^3\mathbf{r} = -2\,\eta\,t\,(\delta_{ik}\delta_{jl} + \delta_{ij}\delta_{kl}). \tag{10.97}$$

Comparing these results with (10.95) we obtain:

$$\eta_{ijkl} = \frac{1}{2}\frac{d}{dt}\left(\mu_{ijkl}^{(2)*}\right)^{sym},$$ (10.98)

where

$$\mu_{ijkl}^{(2)*} = \int x_i\,\Pi_{jk}^*\,x_l\,d^3\mathbf{r},$$ (10.99)

and

$$(A_{ijkl})^{sym} = \frac{1}{4}\left(A_{ijkl} + A_{jikl} + A_{ijlk} + A_{jilk}\right).$$ (10.100)

An alternative definition, and a more general one, in fact, is:

$$\eta_{ijkl} = \frac{1}{2}\left(\int x_i\,\dot{\Pi}_{jk}\,x_l\,d^3\mathbf{r}\right)^{sym}.$$ (10.101)

The expression (10.101) shows that viscosity measures the temporal growth of the second moment of the unsteady Stokes propagator. This equation generalizes the analogous definition (10.72) of diffusivity in heat (and mass) transport[13] and, in fact, can also be applied to measure the configuration-specific viscosity of complex fluids, such as suspensions, where viscosity is a strong function of the position, i.e. $\eta = \eta(\mathbf{r})$. In particular, the nontraditional definition (10.101) of viscosity has been applied by Mauri and Brenner[14] to determine the effective viscosity of periodic suspensions, obtaining results that are identical to those derived using the traditional definition (10.91).

10.6 Problems

Problem 10.1 Prove Eq. (10.9).

Problem 10.2 Prove Eq. (10.33) by averaging **T** and **S** by "brute force".

Problem 10.3 Derive the dynamical definition of viscosity (10.101) by substituting the equations of motion (10.76) into the RHS of Eq. (10.101).

---

[13]It should be added that in heat and mass diffusion, due to its probabilistic interpretation, the second moment of the unsteady heat equation coincides with the mean square displacement of a passive tracer as it diffuses through the system. As noted here, this simple interpretation cannot be extended to momentum transfer.

[14]A summary of this work and its most relevant results can be found in [5, Chap. 11].

References

1. Batchelor, G.K.: J. Fluid Mech. **52**, 245 (1972)
2. Batchelor, G.K.: Transport properties of two-phase materials with random structure. Ann. Rev. Fluid Mech. **6**, 227–255 (1974)
3. Batchelor, G.K.: J. Fluid Mech. **74**, 1 (1976)
4. Batchelor, G.K., Green, J.T.: J. Fluid Mech. **56**, 401–427 (1972)
5. Brenner, H., Edwards, D.A.: Macrotransport Processes. Butterworth, Boston (1991)
6. Einstein, A.: Investigation on the Theory of the Brownian Movement. Dover, New York (1956), Chap. III.2
7. Guazzelli, E., Morris, J.F.: Physical Introduction to Suspension Dynamics. Cambridge University Press, Cambridge (2011). Chaps. 6–7
8. Jeffrey, D.J.: Proc. R. Soc. Lond. A **335**, 355 (1973)
9. Mauri, R., Rubinstein, J.: Chem. Eng. Commun. **148**, 385 (1996)
10. Pasquino, R., Grizzuti, N., Maffettone, P.L., Greco, F.F.: J. Rheol. **52**, 1369–1384 (2008)
11. Russel, W.B., Saville, D.A., Schowalter, W.R.: Colloidal Dispersions. Cambridge University Press, Cambridge (1989), Chaps. 10–12

Chapter 11
Multiple Scale Analysis

In Chap. 10, we have assumed to know both the structure of the effective equations
and the micro-scale morphology of the multiphase systems, and thus we focussed
on how to determine the effective properties appearing in the associated constitutive
equations. On the other hand, in this chapter we only assume to know the governing
equations, describing the transport of momentum, energy and mass at the microscale level, and then we intend to average them out, to find the effective equations
at the macro- (or meso-) scale. First, in Sect. 11.1, we show how to perform a direct volume averaging, using a multi-pole expansion technique. Clearly, though,
any averaging procedure must assume a clear separation of scales, that is the typical
length- and time-scales at the micro-level must be much smaller than their macroscopic (or mesoscopic) counterparts. Accordingly, the most natural way to move up
from one scale to the other, and thus determine the effective equations of a multiphase system, is by using multiple scale analysis. In Sect. 11.2, first we explain the
idea underlying this approach and then show two examples of application to derive
the Smoluchowsky equation and study Taylor dispersion. Finally, in Sect. 11.3, this
approach is generalized, describing the coarse-graining homogenization procedure
and thus show how some results on deterministic chaos can be found. In particular,
we see that the transport of colloidal particles in non-homogeneous random velocity
fields is described through a convection-diffusion equation that can also be derived
from the Stratonovich stochastic process seen in Chap. 5.

11.1 Volume Averaging

In this section we determine the effective equations describing the flow through
porous media by direct volume averaging of the governing equations at the
microscale. Although other approaches have been proposed, such as moment-
matching [3], ensemble averaging [11], and multiple scale perturbation expansion
(see next sections), this is perhaps the simplest of all methods.

Let us consider the steady, slow motion of a viscous fluid of viscosity η_0 and
unit density through a bed of rigid spheres of radius a, either fixed or neutrally

R. Mauri, *Non-Equilibrium Thermodynamics in Multiphase Flows*,
Soft and Biological Matter, DOI 10.1007/978-94-007-5461-4_11,
© Springer Science+Business Media Dordrecht 2013

buoyant, located at positions $\mathbf{r}_N$. For low solid volume fractions the influence of the spheres can be modeled through singular multipole force distributions centered at $\mathbf{r}_N$, leading to the steady state Stokes equations of motion (F.52),

$$\nabla p - \eta_0 \nabla^2 \mathbf{v} = \mathbf{F} = \sum_N \sum_{n=0}^{\infty} \mathbf{F}_N^{(n+1)} (\cdot)^n \underbrace{\nabla\nabla \ldots \nabla}_{n \; times} \delta(\mathbf{r} - \mathbf{r}_N); \qquad \nabla \cdot \mathbf{v} = 0,$$

(11.1)

where $\nabla^n = \nabla\nabla \cdots \nabla$ (n times), while $\mathbf{F}_N^{(n)}$ is an n-th order constant tensor expressing the strength of the n-th multipole of the N-th sphere. Thus, $\mathbf{F}^{(1)}$ is the force exerted by the sphere on the fluid, $\mathbf{F}^{(2)}$ is the corresponding moment of dipole, $\mathbf{F}^{(3)}$ the moment of quadrupole, etc. In turn, due to the linearity of the Stokes equation, these multipole strengths are proportional to the gradients of the unperturbed velocity at the center of the spheres, that is the velocity field in the absence of the N-th particle, i.e.,

$$\mathbf{F}_N^{(m)} = -\sum_n \mathbf{R}_N^{(mn)} (\cdot)^n \nabla^{n-1} \mathbf{U}|_{\mathbf{r}=\mathbf{r}_N}, \tag{11.2}$$

where $\mathbf{R}_N^{(mn)}$ is a grand resistance matrix of the N-th sphere. In the following we shall assume that the averaging volume is sufficiently far from the boundaries of the medium that the grand resistance matrix is the same for each particle, so that the subscript N can be dropped from $\mathbf{R}_N^{(mn)}$. Now we shall proceed to take the volume average of the Stokes equations over a domain V, comprising many particles and in which, at the same time, the unperturbed velocity and pressure fields do not vary appreciably. This, in essence, requires that a separation of scales exists, that is, the macroscale over which velocity and pressure gradient vary greatly exceeds the microscale, e.g. the typical particle-particle distance. The main point is that, if a large number of particles is located within the averaging volume V, the average, or macroscale, velocity and pressure fields, $\langle \mathbf{u} \rangle$ and $\langle p \rangle$, are given by:

$$\langle \mathbf{u} \rangle = \langle \mathbf{u}_N \rangle; \qquad \langle p \rangle = \langle p_N \rangle, \tag{11.3}$$

where the bracket denotes volume average over V. Using these definitions, we can easily show that

$$\sum_N \langle \nabla^m \delta(\mathbf{r} - \mathbf{r}_N) \mathbf{u}_N \rangle = (-1)^m \frac{3\phi}{4\pi a^3} \nabla^m \langle \mathbf{u} \rangle. \tag{11.4}$$

Finally, taking the average of the Stokes equations (11.1) and (11.2) and using the above relations (11.3) and (11.4), we find the following generalized Brinkman equation:

$$\nabla_i \langle p \rangle = -\eta_0 k_{ij}^{-1} \langle u_j \rangle - 2\xi_{ijk} \nabla_j \langle u_k \rangle + \eta_{ijk\ell} \nabla_j \nabla_k \langle u_\ell \rangle; \qquad \nabla_i \langle u_i \rangle = 0, \quad (11.5)$$

where k_{ij}, ξ_{ijk} and $\eta_{ijk\ell}$ are the permeability, coupling and viscosity tensors, respectively, with:

$$k_{ij}^{-1} = \frac{3\phi}{4\pi a^3 \eta_0} R_{ij}^{(11)};$$

(11.6)

$$\xi_{ijk} = \frac{3\phi}{8\pi a^3} (R_{ijk}^{(12)} - R_{jik}^{(21)});$$

(11.7)

$$\eta_{ijk\ell} = \eta_0(\delta_{ik}\delta_{j\ell} + \delta_{i\ell}\delta_{jk}) + \frac{3\phi}{4\pi a^3} (R_{ijk\ell}^{(22)} - R_{jik\ell}^{(13)} - R_{ikj\ell}^{(31)}).$$

(11.8)

Note that, by construction,

$$R_{ijk\ell}^{(13)} = R_{ikj\ell}^{(13)}; \qquad R_{ijk\ell}^{(31)} = R_{jik\ell}^{(31)}.$$

(11.9)

Since our averaging procedure implies a separation of scales, we have tacitly assumed that $|\mathbf{u}| \gg \ell|\nabla \mathbf{u}|$, where $\ell = V^{1/3}$ is the linear dimension of the averaging volume. Therefore, the three terms at the righthand side of Eq. (11.5) are of decreasing magnitude, thus justifying why we have neglected higher-order velocity gradient terms.

11.1.1 Onsager's Reciprocity Relations

The average energy dissipated per unit time and volume, $\dot{E}$, is:

$$\dot{E} = 2\eta_0 \langle \nabla \mathbf{u} : \nabla \mathbf{u} \rangle + \langle \mathbf{F} \cdot \mathbf{u} \rangle.$$

(11.10)

Substituting Eq. (11.1) for the generalized force $\mathbf{F}$ into (11.10), and applying the averaging rule (11.4) we obtain:

$$\dot{E} - 2\eta_0 \langle \nabla \mathbf{u} : \nabla \mathbf{u} \rangle = \frac{3\phi}{4\pi a^3} \sum_m (-1)^m \mathbf{F}^{(m+1)} (\cdot)^{m+1} \nabla^m \langle \mathbf{u} \rangle,$$

(11.11)

showing that the $(m+1)$-th pole strength $\mathbf{F}^{(m+1)}$ is conjugated with the m-th mean velocity gradient. Therefore, since these two quantities are linearly related through Eq. (11.2) the Onsager relations state that the proportionality term, i.e. $\mathbf{R}^{(mn)}$, is a symmetric matrix. To exemplify what that means, let us rewrite Eq. (11.2) as

$$-F_i^{(1)} = R_{ij}^{(11)} u_j + R_{ijk}^{(12)} \nabla_j u_k + R_{ijk\ell}^{(13)} \nabla_j \nabla_k u_\ell;$$

(11.12)

$$-F_{ij}^{(2)} = R_{ijk}^{(21)} u_k + R_{ijk\ell}^{(22)} \nabla_k u_\ell;$$

(11.13)

$$-F_{ijk}^{(3)} = R_{ijk\ell}^{(31)} u_\ell,$$

(11.14)

where the subscripts N have been dropped from $\mathbf{F}^{(m)}$ and $\mathbf{u}$ for simplicity. In these equations, we have not considered higher velocity gradient terms, since we saw that they are not required to determine Brinkman's effective equations. Now, due to the symmetry of the grand resistance matrix, we find:

$$R_{ij}^{(11)} = R_{ji}^{(11)}; \qquad R_{ijk}^{(12)} = R_{jki}^{(21)}; \qquad R_{ijk\ell}^{(13)} = R_{jk\ell i}^{(31)}; \qquad R_{ijk\ell}^{(22)} = R_{k\ell ji}^{(22)}. \tag{11.15}$$

Applying these symmetry relations to Eq. (11.6)–(11.8), we obtain:

$$k_{ij} = k_{ji} \tag{11.16}$$

while (11.7) and (11.8) can be rewritten as:

$$\xi_{ijk} = \frac{3\phi}{8\pi a^3}\left(R_{ijk}^{(12)} - R_{kji}^{(12)}\right) = -\xi_{kji} \tag{11.17}$$

and

$$\eta_{ijk\ell} = \eta_0(\delta_{ik}\delta_{j\ell} + \delta_{i\ell}\delta_{jk}) + \frac{3\phi}{4\pi a^3}\left(R_{ijk\ell}^{(22)} - R_{jik\ell}^{(13)} - R_{\ell ikj}^{(13)}\right) = \eta_{k\ell ij}, \tag{11.18}$$

where we have considered that, by construction $\eta_{ijk\ell} = \eta_{ikj\ell}$. These relations show that the permeability and the effective viscosity are identically symmetric tensors, while the coupling term is antisymmetric.

11.1.2 Brinkman's Equation

At leading order, we neglect the influence of particle-particle interactions, so that we may determine the grand resistance matrix by studying the flow field around an isolated sphere.

First, consider the case of neutrally buoyant suspended particles. Then, $\mathbf{F}^{(1)} = \mathbf{0}$, and, as shown in Appendix F [cf. Eqs. (F.61)], the antisymmetric component of $\mathbf{F}^{(2)}$ (i.e. the rotlet) is zero, so that:

$$R_{ij}^{(11)} = R_{ijk}^{(12)} = R_{ijk\ell}^{(13)} = 0; \qquad R_{ijk\ell}^{(22)} = \frac{10}{3}\pi\eta a^3(\delta_{ik}\delta_{j\ell} + \delta_{i\ell}\delta_{jk}). \tag{11.19}$$

Consequently, we find:

$$k_{ij}^{-1} = \xi_{ijk} = 0; \qquad \eta_{ijk\ell} = \eta^*(\delta_{ik}\delta_{j\ell} + \delta_{i\ell}\delta_{jk}), \tag{11.20}$$

where

$$\eta^* = \eta_0\left(1 + \frac{5}{2}\phi\right) \tag{11.21}$$

is Einstein's effective viscosity (10.33), and the effective equation (11.5) becomes:

$$\nabla_i\langle p\rangle = \eta^*\nabla^2\langle u_i\rangle; \qquad \nabla_i\langle u_i\rangle = 0. \tag{11.22}$$

Now, let us consider the more interesting case of a fixed bed of spherical particles. In this case, in Appendix F we see that [cf. Eqs. (F.55) and (F.60)]:

$$R_{ij}^{(11)} = 6\pi \eta a \delta_{ij}; \qquad R_{ijk}^{(12)} = 0; \qquad R_{ijk\ell}^{(13)} = \pi \eta a^3 \delta_{i\ell} \delta_{jk}; \qquad (11.23)$$

and

$$R_{ijk\ell}^{(22)} = \pi \eta a^3 \left(\frac{16}{3} \delta_{ik} \delta_{j\ell} + \frac{4}{3} \delta_{i\ell} \delta_{jk} \right). \qquad (11.24)$$

Substituting these results into (11.16)–(11.18) we obtain:

$$k_{ij}^{-1} = \frac{9\phi}{2a^2} \delta_{ij}; \qquad \xi_{ijk} = 0, \qquad (11.25)$$

and

$$\eta_{ijk\ell} = \eta_0 \left(1 + \frac{5}{2}\phi \right) \delta_{ik} \delta_{j\ell} + \eta_0 (1 + \phi) \delta_{i\ell} \delta_{jk}. \qquad (11.26)$$

This shows that the flow of a Newtonian fluid through a dilute bed of solid spheres is described by the following, so called, *Brinkman* equation:

$$\nabla_i \langle p \rangle + \frac{\eta_0}{k^*} \langle u_j \rangle = \eta^* \nabla^2 \langle u_i \rangle; \qquad \nabla_i \langle u_i \rangle = 0, \qquad (11.27)$$

with permeability $k^* = (2a^2)/(9\phi)$ and effective viscosity $\eta^* = \eta_0 (1 + \frac{5}{2}\phi)$. Equation (11.27) shows that the viscous term in Brinkman's equation is expressed via the Einstein effective viscosity (10.33) [6, 13].

Brinkman's equations can also be written as a momentum conservation equation, in terms of the mean force density **f** (i.e. the force per unit volume exerted by the spheres onto the fluid), the mean body couple density **g** and the mean stress tensor **T** as:

$$\mathbf{f} + \frac{1}{2}\nabla \times \mathbf{g} + \nabla \cdot \mathbf{T} = \mathbf{0}; \qquad \nabla \cdot \langle \mathbf{u} \rangle = 0, \qquad (11.28)$$

with the constitutive relations,

$$\mathbf{f} = -\frac{\eta_0}{k^*} \langle \mathbf{u} \rangle; \qquad k^* = \frac{2a^2}{9\phi}; \qquad (11.29)$$

$$\mathbf{g} = -\chi^* \langle \boldsymbol{\omega} \rangle; \qquad \chi^* = 6\eta_0 \phi; \qquad (11.30)$$

$$\mathbf{T} = -\langle p \rangle \mathbf{I} + 2\tilde{\eta} \langle \mathbf{S} \rangle; \qquad \tilde{\eta} = \eta_0 \left(1 + \frac{7}{4}\phi \right) \qquad (11.31)$$

where $\langle \mathbf{S} \rangle$ is the mean rate of strain, $\langle \boldsymbol{\omega} \rangle = \frac{1}{2}\nabla \times \langle \mathbf{u} \rangle$ is the mean angular velocity of the fluid, while χ^* and $\tilde{\eta}$ denote the effective spin, or rotational, viscosity and the effective stress viscosity, respectively.

It is important to note that the effective shear viscosity η^* appearing in the Brinkman equation (11.27) is different from the stress viscosity $\tilde{\eta}$, the latter being defined as the ratio between the symmetric part of the mean deviatoric stress tensor and the mean rate of shear. This can be seen rewriting (11.26) as:

$$\eta_{ijk\ell} = \tilde{\eta}(\delta_{ik}\delta_{j\ell} + \delta_{i\ell}\delta_{jk}) + \frac{1}{8}\chi^*(\delta_{ik}\delta_{j\ell} - \delta_{i\ell}\delta_{jk}), \qquad (11.32)$$

showing that $\chi^*/8 = 3\phi/4$ contributes an extra term to the shear viscosity [9], i.e. $\eta^* = \tilde{\eta} + \chi^*/8$.

The approach described in this section can be generalized to the case of dilute beds of particles of any shape and even to the non dilute case. In particular, it is important to note that, whenever the system morphology lacks mirror symmetry (think of the screwlike particles mentioned in Problem 2.1 of Chap. 2), fluid translational and angular velocities are coupled to each other, i.e. $\mathbf{R}_N^{(12)}$ is not identically zero, so that the coupling term $2\boldsymbol{\xi} : \nabla\mathbf{u}$ in the generalized Brinkman equation (11.5) will not vanish.

11.2 Multiple Scale Analysis

Perhaps the most important among the approximation methods in multiphase rheology is the multiple scale analysis, a perturbation technique where the effective governing equations are represented through an asymptotic expansion in terms of a small parameter that appears naturally in the problem. In particular, when we want to determine the effective equations of slow, or macro, variables, smearing out the effects of all the other faster, or micro, variables, the small parameter is the ratio between the two time- or length-scales. Accordingly, after expanding the governing equations as regular perturbations of the small parameter ϵ, coefficients of like powers of ϵ are equated, producing a set of boundary value problems to determine the effective coefficients appearing in the final, effective equations.

In this section, this method is described by solving two problems. In the first we see how the Kramers equations, that is the Fokker-Planck equation in phase space, reduces naturally to the Smoluchowsky equation in the configuration space only, when the effect of the faster momentum variable is averaged out. The second example analyzes Taylor dispersion, occurring when two miscible liquids are pumped in a capillary, resulting in an enhanced effective diffusivity, due to the non uniform velocity field. Finally, this technique is applied to study heat and mass transport through porous media.

11.2.1 Smoluchowsky's Equation

As we saw in Chap. 3, when we describe the motion of a free Brownian particle as a function of its position $\mathbf{r}$ and velocity $\mathbf{v}$ using the Langevin equation, we see that thermal fluctuations directly affect the velocity field, which in turn, at longer timescales, causes diffusion in the physical space. This can be seen even more clearly using the Fokker-Planck equation in the $(\mathbf{r}, \mathbf{v})$ phase space, which is often referred to as Kramers' equation [cf. Eqs. (4.90) and (4.91)], i.e.,

$$\frac{\partial \Pi}{\partial t} + \frac{\partial}{\partial \mathbf{r}} \cdot \mathbf{J}_r + \frac{\partial}{\partial \mathbf{v}} \cdot \mathbf{J}_v = 0, \tag{11.33}$$

where,

$$\Pi = \Pi(\mathbf{r}, \mathbf{v}, t); \qquad \mathbf{J}_r = \mathbf{v}\Pi; \qquad \mathbf{J}_v = -\frac{\zeta}{m}\mathbf{v}\Pi - \frac{kT\zeta}{m^2}\frac{\partial \Pi}{\partial \mathbf{v}}. \tag{11.34}$$

This equation can be easily obtained by applying the rules of Sect. 4.4, considering the phenomenological equations $\dot{\mathbf{r}} = \mathbf{v}$ and $\dot{\mathbf{v}} = -\zeta \mathbf{v}/m$, together with the Maxwellian distribution, leading to $g_{vv} = m/kT$ (no such equilibrium distribution exists in the configuration space, so that $g_{rv} = g_{rr} = 0$).

Now define a slow and a fast non-dimensional variable, $\tilde{\mathbf{x}} = \mathbf{r}/L$ and $\tilde{\mathbf{y}} = \mathbf{v}/V$, respectively, where L is a characteristic linear dimension, while $V = \sqrt{kT/m}$ is a typical thermal fluctuation speed. The characteristic lengthscales associated with these variables are $\tau_x = L/V$ and $\tau_y = m/\zeta$, with

$$\epsilon = \frac{\tau_y}{\tau_x} = \frac{mV}{L\zeta} \ll 1. \tag{11.35}$$

Scaling time in terms of a longer, τ_x/ϵ, timescale, Kramers' equation in non-dimensional form becomes:

$$\frac{\partial \Pi}{\partial \tilde{t}} + \frac{1}{\epsilon}\tilde{\mathbf{y}} \cdot \nabla_{\tilde{\mathbf{x}}}\Pi = \frac{1}{\epsilon^2}\nabla_{\tilde{\mathbf{y}}} \cdot (\tilde{\mathbf{y}}\Pi + \nabla_{\tilde{\mathbf{y}}}\Pi) = 0, \tag{11.36}$$

with $\tilde{t} = tV\epsilon/L$, $\nabla_{\tilde{\mathbf{x}}} = \partial/\partial\tilde{\mathbf{x}}$ and $\nabla_{\tilde{\mathbf{y}}} = \partial/\partial\tilde{\mathbf{y}}$, to be solved assuming that $\Pi \to 0$ exponentially as $|\tilde{\mathbf{y}}| \to \infty$.

Now we expand the probability distribution as a power series of ϵ,

$$\Pi(\tilde{\mathbf{x}}, \tilde{\mathbf{y}}, \tilde{t}) = \sum_{n=0}^{\infty} \epsilon^n \Pi_n(\tilde{\mathbf{x}}, \tilde{\mathbf{y}}, \tilde{t}). \tag{11.37}$$

At leading order we obtain:

$$O(\epsilon^{-2}): \quad \nabla_{\tilde{\mathbf{y}}} \cdot (\tilde{\mathbf{y}}\Pi_0 + \nabla_{\tilde{\mathbf{y}}}\Pi_0) = 0; \qquad \lim_{|\tilde{\mathbf{y}}|\to\infty} \Pi_0 \approx e^{-y^2}. \tag{11.38}$$

This problem can be easily solved obtaining,

$$\Pi_0(\tilde{\mathbf{x}}, \tilde{\mathbf{y}}, \tilde{t}) = \overline{\Pi}(\tilde{\mathbf{x}}, \tilde{t})\Pi_v(\tilde{\mathbf{y}}), \tag{11.39}$$

where

$$\Pi_v(\tilde{\mathbf{y}}) = Ce^{-\tilde{y}^2/2}, \tag{11.40}$$

is the Maxwellian distribution, with C denoting a normalization constant. Since at the end we want to be able to describe the motion of the Brownian particle in configurational space in terms of the leading-order term, $\Pi_x(\tilde{\mathbf{x}}, \tilde{t})$, by taking advantage of the fact that the higher-order terms in (11.37) are defined within an arbitrary constant, without lack of generality we can impose the following condition,

$$\langle \Pi(\tilde{\mathbf{x}}, \tilde{\mathbf{y}}, \tilde{t}) \rangle = \overline{\Pi}(\tilde{\mathbf{x}}, \tilde{t}) \quad \text{where} \quad \langle A \rangle = \int_{-\infty}^{\infty} A(\tilde{\mathbf{y}}) \, d\tilde{\mathbf{y}}. \tag{11.41}$$

Consequently, we find $C = \pi^{-1/2}$, so that $\langle \Pi_v \rangle = 1$, while $\langle \Pi_n \rangle = \overline{\Pi} \delta_{n,0}$.

At the next order, we find:

$$O(\epsilon^{-1}): \quad \tilde{\mathbf{y}} \cdot \nabla_{\tilde{\mathbf{x}}} \Pi_0 = \nabla_{\tilde{\mathbf{y}}} \cdot (\tilde{\mathbf{y}} \Pi_1 + \nabla_{\tilde{\mathbf{y}}} \Pi_1); \qquad \lim_{|\tilde{\mathbf{y}}| \to \infty} \Pi_1 \approx e^{-\tilde{y}^2}. \tag{11.42}$$

This problem is well posed; in fact, its solvability condition, imposing that the integral of the LHS and the RHS are equal to each other, is here satisfied identically. At the end, Eq. (11.42) can be solved by imposing:

$$\Pi_1(\tilde{\mathbf{x}}, \tilde{\mathbf{y}}, \tilde{t}) = -\tilde{\mathbf{B}}(\tilde{\mathbf{y}}) \cdot \nabla_{\tilde{\mathbf{x}}} \Pi_0, \tag{11.43}$$

where the so-called $\tilde{\mathbf{B}}$-field satisfies the following problem:

$$\tilde{\mathbf{y}} \cdot \nabla_{\tilde{\mathbf{y}}} \tilde{\mathbf{B}} - \nabla_{\tilde{\mathbf{y}}}^2 \tilde{\mathbf{B}} = \tilde{\mathbf{y}}, \tag{11.44}$$

with the additional constraint that $\langle \Pi_1 \rangle = 0$, i.e.:

$$\langle \tilde{\mathbf{B}}(\tilde{\mathbf{y}}) \Pi_v(\tilde{\mathbf{y}}) \rangle = C \int_{-\infty}^{\infty} \tilde{\mathbf{B}}(\tilde{\mathbf{y}}) e^{-\tilde{y}^2/2} \, d\tilde{\mathbf{y}} = 0. \tag{11.45}$$

So, at the end, we obtain:

$$\tilde{\mathbf{B}}(\tilde{\mathbf{y}}) = \tilde{\mathbf{y}}. \tag{11.46}$$

At the next order, we obtain:

$$O(1): \quad \frac{\partial \Pi_0}{\partial \tilde{t}} + \tilde{\mathbf{y}} \cdot \nabla_{\tilde{\mathbf{x}}} \Pi_1 = \nabla_{\tilde{\mathbf{y}}} \cdot (\tilde{\mathbf{y}} \Pi_2 + \nabla_{\tilde{\mathbf{y}}} \Pi_2); \qquad \lim_{|\tilde{\mathbf{y}}| \to \infty} \Pi_2 \approx e^{-\tilde{y}^2}. \tag{11.47}$$

Now we impose that this problem is well posed, that is the solvability condition is satisfied. Accordingly, integrating the above equation over $\tilde{\mathbf{y}}$ and considering that $\langle \Pi_0 \rangle = \overline{\Pi}$, we find:

$$\frac{\partial \overline{\Pi}}{\partial \tilde{t}} = \nabla_{\tilde{\mathbf{x}}} \cdot \tilde{\mathbf{D}} \cdot \nabla_{\tilde{\mathbf{x}}} \overline{\Pi}, \tag{11.48}$$

where

$$\tilde{\mathbf{D}} = \frac{1}{\sqrt{\pi}} \int_{-\infty}^{\infty} \tilde{\mathbf{y}} \tilde{\mathbf{y}} e^{-y^2/2} \, d\tilde{\mathbf{y}} = \mathbf{I}. \tag{11.49}$$

Finally, going back to dimensional variables, we obtain the well-known Smoluchowsky equation,

$$\frac{\partial \overline{\Pi}}{\partial t} = D \nabla_r^2 \overline{\Pi}; \quad D = \tilde{D} \epsilon V L = \frac{kT}{\zeta}, \tag{11.50}$$

where D is the usual molecular diffusivity.

11.2.2 Taylor Dispersion

Consider the transport of a colloidal solute in a channel of half-width Y, convected by a solvent, flowing with velocity $v(y)$ in the longitudinal z-direction, where $-Y \leq y \leq Y$ denotes the transversal coordinate. As the solute particles diffuse in the y-direction, they will experience different velocities, so that what initially is a concentrated spot of solute particles will subsequently spread longitudinally, exhibiting a so-called Taylor dispersion, named after G.I. Taylor [21–23]. From an elementary dimensional analysis, we see that the Taylor dispersion coefficient is $D^* \propto V^2 \tau$, where V is the mean longitudinal velocity, while τ is the characteristic time that is necessary for the solute to sample all positions in the cross section, i.e. in the y-direction. Therefore, since $\tau = Y^2/D$, where D is the molecular diffusivity, we easily obtain for the effective diffusivity:

$$D^* \propto D N_{Pe}^2, \tag{11.51}$$

where $N_{Pe} = VY/D$ is the Peclet number.

As we want to describe the dispersion process at the macroscale $L \gg Y$, it is natural to choose $\epsilon = Y/L$ as our small parameter. In fact, the timescale in the slow z-variable is $\tau_z = L/V$, as longitudinal transport is convection-driven, while for the y-variable we have $\tau_y = Y^2/D$, as the solute particles move transversally by diffusion only. So at the end we have: $\tau_y/\tau_z = \epsilon N_{Pe}$; accordingly, imposing that $\tau_y \ll \tau_z$, that means that $\epsilon \ll N_{Pe}^{-1}$.

The transport of a dilute solute having concentration $c = c(y, z, t)$ is described through the convection-diffusion equation,

$$\frac{\partial c}{\partial t} + V(y) \frac{\partial c}{\partial z} = D \left(\frac{\partial^2 c}{\partial z^2} + \nabla_y^2 c \right), \tag{11.52}$$

where $\nabla_y^2 c = \partial^2 c/\partial y^2$, while $V(y)$ is the velocity field of the solvent, with V denoting its mean value. This governing equation is to be solved with given initial conditions and no-flux boundary conditions at the walls,

$$\mathbf{n} \cdot \nabla_y c = \frac{\partial c}{\partial y} = 0 \quad \text{at } y = \pm Y. \tag{11.53}$$

Scaling time, as in the previous subsection, in terms of a longer timescale τ_z/ϵ, and referring the longitudinal coordinate z to a moving reference frame, define the following scaling,

$$\tilde{x} = \frac{z - V_c t}{L}; \qquad \tilde{y} = \frac{y}{Y}; \qquad \tilde{t} = \frac{t}{L^2/D}, \qquad \tilde{v}(\tilde{y}) = \frac{V(y) - V_c}{V}, \qquad (11.54)$$

where V_c is a characteristic velocity to be determined using the solvability condition. Then, the governing equation becomes:

$$\frac{\partial c}{\partial \tilde{t}} + \frac{1}{\epsilon} N_{Pe} \tilde{v} \frac{\partial c}{\partial \tilde{x}} = \frac{\partial^2 c}{\partial \tilde{x}^2} + \frac{1}{\epsilon^2} \nabla_{\tilde{y}}^2 c. \qquad (11.55)$$

Now, let us expand the concentration field as a power series,

$$c(\epsilon, \tilde{x}, \tilde{y}, \tilde{t}) = \sum_{n=0}^{\infty} \epsilon^n c_n(\tilde{x}, \tilde{y}, \tilde{t}). \qquad (11.56)$$

At leading order, we find:

$$\nabla_{\tilde{y}}^2 c_0 = 0; \quad \text{with B.C.:} \quad \frac{\partial c_0}{\partial y} = 0 \quad \text{at } \tilde{y} = \pm 1, \qquad (11.57)$$

showing that $c_0 = c_0(\tilde{x}, \tilde{t})$. Note that here the solvability condition is satisfied identically.

Since at the end we want to be able to describe the transport of the solute in terms of the leading-order term, by taking advantage of the fact that the higher-order terms in (11.56) are defined within an arbitrary constant, without lack of generality we can impose the following condition,

$$\langle c(\tilde{x}, \tilde{y}, \tilde{t}) \rangle = c_0(\tilde{x}, \tilde{t}) \quad \text{where } \langle A \rangle = \frac{1}{2} \int_{-1}^{1} A(\tilde{y}) d\tilde{y}, \qquad (11.58)$$

so that $\langle c_n \rangle = c_0 \delta_{n,0}$.

At the next order we find:

$$N_{Pe} \tilde{v} \frac{\partial c_0}{\partial \tilde{x}} = \nabla_{\tilde{y}}^2 c_1; \quad \text{with B.C.:} \quad \frac{\partial c_1}{\partial \tilde{y}} = 0 \quad \text{at } \tilde{y} = \pm 1. \qquad (11.59)$$

The solvability condition (i.e. the integral of the LHS over the y-variable must equal that of the RHS) here imposes that $\langle \tilde{v} \rangle = 0$, which shows that $V_c = V$, i.e. the characteristic velocity equals the mean solvent velocity.

Now, defining Brenner's function,[1] $\tilde{B}(\tilde{y})$, as,

$$c_1(\tilde{x}, \tilde{y}, \tilde{t}) = N_{Pe} \tilde{B}(\tilde{y}) \frac{\partial c_0}{\partial \tilde{x}}(\tilde{x}, \tilde{t}), \qquad (11.60)$$

---

[1]The $\tilde{B}$-function is named after H. Brenner, who wrote many fundamental papers on generalized Taylor dispersion. See [2, 4].

substituting (11.60) into (11.59), we see that it satisfies the following problem:

$$\nabla_{\tilde{y}}^2 \tilde{B} = \tilde{v}; \quad \text{with B.C.:} \quad \frac{d\tilde{B}}{d\tilde{y}} = 0 \quad \text{at } \tilde{y} = \pm 1. \tag{11.61}$$

Clearly, Brenner's function is determined within an arbitrary (and, it turns out, irrelevant) constant, which nevertheless can be determined imposing $\langle c_1 \rangle = 0$, i.e., $\langle \tilde{B} \rangle = 0$.

Finally, proceeding to the next order we find:

$$\frac{\partial c_0}{\partial \tilde{t}} + N_{Pe}\tilde{v}\frac{\partial c_1}{\partial \tilde{x}} = \frac{\partial^2 c_0}{\partial \tilde{x}^2} + \nabla_{\tilde{y}}^2 c_2; \quad \text{with B.C.:} \quad \frac{\partial c_2}{\partial \tilde{y}} = 0 \quad \text{at } \tilde{y} = \pm 1. \tag{11.62}$$

Applying the solvability condition and substituting (11.60) we obtain:

$$\frac{\partial c_0}{\partial \tilde{t}} = \tilde{D}\frac{\partial^2 c_0}{\partial \tilde{x}^2}, \tag{11.63}$$

where $\tilde{D}$ is a non-dimensional effective diffusivity,

$$\tilde{D} = 1 + \alpha N_{Pe}^2; \quad \alpha = -\langle \tilde{v}\tilde{B} \rangle. \tag{11.64}$$

Substituting (11.61) into (11.64), integrating by parts and considering the no-flux boundary conditions, we see that α can be written in the following equivalent form:

$$\alpha = \left\langle \left[\frac{dB}{d\tilde{y}} \right]^2 \right\rangle. \tag{11.65}$$

Therefore, going back to dimensional variables, we see that the effective equation reads:

$$\frac{\partial \bar{c}}{\partial t} + V\frac{\partial \bar{c}}{\partial z} = D^*\frac{\partial^2 \bar{c}}{\partial z^2}, \tag{11.66}$$

where the effective, Taylor diffusivity D^* is:

$$D^* = D\left(1 + \alpha N_{Pe}^2\right). \tag{11.67}$$

(A) Poiseuille Flow in a Channel In this case, $V(y) = \frac{3}{2}V(1 - y^2/Y^2)$. Therefore, substituting $\tilde{v} = \frac{1}{2} - \frac{3}{2}\tilde{y}^2$ into (11.61), we easily find:

$$\frac{dB}{dy} = \frac{1}{2}(\tilde{y} - \tilde{y}^3) \quad \text{and} \quad \tilde{B} = -\frac{7}{120} + \frac{1}{4}\tilde{y}^2 - \frac{1}{8}\tilde{y}^4, \tag{11.68}$$

where the constant in the expression of $\tilde{B}(y)$ has been determined imposing that $\langle \tilde{B} \rangle = 0$. Substituting (11.68) into (11.64) or (11.65) we easily find: $\alpha = 2/105$. Therefore, we obtain the effective equation (11.66), with:

$$D^* = D + \frac{2}{105}\frac{V^2 Y^2}{D}. \tag{11.69}$$

(B) Poiseuille Flow in a Pipe   Since all the results that we have seen so far can be applied to any one-directional flow field, consider a Poiseuille flow in a circular conduit of radius R, $V(r) = 2V(1 - \tilde{r}^2)$, with $\tilde{r} = r/R$, where V denotes the mean solvent velocity. Then, Eq. (11.61) becomes, in radial coordinates:

$$\nabla_{\tilde{r}}^2 \tilde{B} = \frac{1}{\tilde{r}} \frac{d}{d\tilde{r}} \left(\tilde{r} \frac{d\tilde{B}}{d\tilde{r}} \right) = \tilde{v}; \quad \text{with B.C.:} \quad \frac{d\tilde{B}}{d\tilde{r}} = 0 \quad \text{at } \tilde{r} = 1, \tag{11.70}$$

which is easily solved obtaining:

$$\frac{d\tilde{B}}{d\tilde{r}} = \frac{1}{2}(\tilde{r} - \tilde{r}^3) \quad \text{and} \quad \tilde{B} = -\frac{1}{12} + \frac{1}{2}\tilde{r}^2 - \frac{1}{8}\tilde{r}^4. \tag{11.71}$$

Then, we find:

$$\alpha = \left\langle \left[\frac{d\tilde{B}}{d\tilde{r}} \right]^2 \right\rangle = 2 \int_0^1 \left(\frac{d\tilde{B}}{d\tilde{r}} \right)^2 \tilde{r} d\tilde{r} = \frac{1}{48}, \tag{11.72}$$

so that the effective diffusivity D^* becomes:

$$D^* = D + \frac{1}{48} \frac{V^2 R^2}{D}. \tag{11.73}$$

11.2.3 More on Taylor Dispersion

I. Velocity Covariance

The solution of Eq. (11.61) (in its dimensional form) can also be expressed as:

$$B(y) = -\int_0^\infty dt \int_{-Y}^Y dy_0 \left[\Pi(y, t|y_0) v(y_0) \right], \tag{11.74}$$

where $v(y) = V(y) - V$, while $\Pi(y, t|y_0)$ is the propagator (or Green function) of the diffusion equation along the y direction. $\Pi(y, t|y_0)$ is defined through the equation,

$$\frac{\partial \Pi}{\partial t} - D\nabla_y^2 \Pi = \delta(y - y_0)\delta(t); \quad \text{with B.C.:} \quad \frac{\partial \Pi}{\partial y} = 0 \quad \text{at } y = \pm Y; \tag{11.75}$$

and thus it represents the probability that a tracer particle initially located at position y_0 diffuses and is found at position y at time t. Accordingly, the B-field, $B(y)$, represents the mean axial displacement (referred to its Vt mean) at long times of a Brownian particle located at y within the pipe cross section, assuming that its initial locations y_0 are all equally probable. At the end, we find the following expression for the effective diffusivity:

$$D^* = D + \int_0^\infty C(t) \, dt, \tag{11.76}$$

where

$$C(t) = \frac{1}{2Y} \int_{-Y}^{Y} dy \int_{-Y}^{Y} dy_0 \big[v(y) \Pi(y, t|y_0) v(y_0) \big] \tag{11.77}$$

is the average covariance of the fluid velocity at the points occupied by a Brownian tracer at times $t = 0$ and t, assuming that its initial positions are all equally probable.[2] In fact, since the joint probability that the Brownian particle is located at y_0 at time $t = 0$ and at y at time t is $\Pi(y, t; y_0) = \Pi(y, t|y_0) \Pi(y_0)$, with $\Pi(y_0) = 1/2Y$, Eq. (11.77) can be written as:

$$C(t) = \langle v(t) \, v(0) \rangle. \tag{11.78}$$

This expression shows clearly that the enhanced diffusivity results from the nonhomogeneity of the velocity field, thus justifying the scaling proposed at the beginning of this section. In fact, Taylor dispersion is perhaps the simplest case where the stretching induced by convection can be described in terms of an effective diffusivity.

II. Lagrangian Approach

The result (11.76) can be interpreted also using a Lagrangian approach where, for sake of convenience, we adopt a reference frame moving at constant velocity, V. In fact, consider the Langevin equation, stating that the instantaneous axial velocity of a Brownian tracer is the sum of two uncorrelated processes, namely a zero-mean convection drift, v, and a Wiener diffusion $\dot{w}(t)$,

$$\frac{dx}{dt} = v + \dot{w}, \tag{11.79}$$

with $x(t) = z(t) - Vt$ and $v(y) = V(y) - V$, while,

$$\langle \dot{w}(t) \rangle = 0; \qquad \langle \dot{w}(t') \dot{w}(t'') \rangle = 2D\delta(t' - t''). \tag{11.80}$$

Assuming a uniform initial particle distribution,[3] we obtain:

$$\left\langle \frac{dx}{dt}(t) \right\rangle = 0; \qquad \left\langle \frac{dx}{dt}(t') \frac{dx}{dt}(t'') \right\rangle = 2D\delta(t' - t'') + C(t' - t''), \tag{11.81}$$

---

[2]Since the Brownian particles sample all positions in the cross section with the same probability, irrespectively of their initial location, the same value of the effective diffusivity is obtained for any initial probability distribution; see [8].

[3]In general, Eq. (11.79) should be coupled to a free diffusion equation along the y-direction, which for long times, $t \ll Y^2/D$ determine a uniform probability distribution in the transverse direction.

with $C(t' - t'')$ denoting the velocity autocorrelation (11.78). Integrating this equation in time, with $x(0) = 0$, and averaging, we obtain:

$$\langle x(t)\rangle = 0; \qquad \langle x^2(t)\rangle = 2Dt + \int_0^t dt' \int_0^t dt'' C(t' - t''). \qquad (11.82)$$

At the end, considering that,

$$\lim_{t \to \infty} \frac{d}{dt} \int_0^t dt' \int_0^t dt'' C(t' - t'') = 2 \int_0^\infty d\tau C(\tau),$$

we see that Eq. (11.76) is equivalent to:

$$D^* = \frac{1}{2} \lim_{t \to \infty} \frac{d}{dt} \langle x^2(t)\rangle. \qquad (11.83)$$

This shows that D^* is the coefficient of self-diffusivity, as it equals (one half of) the growth rate of the mean square displacement of a tracer particle immersed in a uniform concentration field. Therefore, since D^* describes both self-diffusion and gradient diffusion (as it also describes the ratio between mass flux and concentration gradient), we see that the fluctuation-dissipation theorem is satisfied, as one would expect, since in this case the constitutive equation for the mass flux is linear (i.e. D^* is independent of the composition).

III. Eigenvalue Expansion

Consider the eigenvalue problem associated with Eq. (11.75), i.e.,

$$D\frac{d^2\phi_n}{dy^2} + \lambda_n\phi_n = 0, \quad \text{with B.C.:} \quad \frac{d\phi_n}{dy} = 0 \quad \text{at } y = \pm Y, \qquad (11.84)$$

where $\lambda_n \geq 0$ are the eigenvalues, while $\phi_n(y)$ are the orthonormalized eigenfunctions, i.e.[4]

$$\langle \phi_m(y)\phi_n(y)\rangle = \frac{1}{2Y} \int_{-Y}^{Y} \phi_m(y)\phi_n(y)\, dy = \delta_{m,n}. \qquad (11.85)$$

Clearly, in our case, we have:

$$\phi_n(y) = \sqrt{2}\cos\left(n\pi\frac{y}{Y}\right); \qquad \lambda_n = D\left(\frac{n\pi}{Y}\right)^2. \qquad (11.86)$$

---

[4]The eigenfunctions are orthogonal to each other because the operator ∇^2 is self-adjoint; see, for example, [5].

Considering that $\{\phi_n\}$ is a complete set of eigenfunctions, we have:

$$\delta(y' - y'') = \frac{1}{2Y} \sum_{n=0}^{\infty} \phi_n(y')\phi_n(y''),$$

(11.87)

so that it is easy to verify that the solution of Eq. (11.75) can be written as:

$$\Pi(y', t|y'') = \frac{1}{2Y} \sum_{n=0}^{\infty} \phi_n(y')\phi_n(y'')e^{-\lambda_n t}.$$

(11.88)

This result is a particular case of Eq. (4.48) with $\phi_n^{\dagger} = \phi_n$, considering that Eq. (11.75) coincides with (4.35), with $\mathcal{L} = -Dd^2/dy^2$.

Now, expand $v(y)$ in Fourier series, considering that its mean value (i.e. its zeroth-order expansion term) is identically zero:

$$v(y) = \sum_{n=1}^{\infty} v_n\phi_n(y).$$

(11.89)

Substituting (11.88) and (11.89) into (11.77) we obtain:

$$C(t) = \sum_{n=1}^{\infty} v_n^2 e^{-\lambda_n t}.$$

(11.90)

Finally, from (11.76) we obtain:

$$D^* = D + \sum_{n=1}^{\infty} \frac{v_n^2}{\lambda_n}.$$

(11.91)

The solution (11.91) is valid for any unidirectional flow field,[5] in particular for Poiseuille and Couette flow fields.

(A) **Poiseuille Flow in a Channel** In this case, when $V(y) = \frac{3}{2}V(1 - y^2/Y^2)$, with $-Y \le y \le Y$, we can expand the propagator and the velocity field in cosine Fourier series (11.86), obtaining,

$$v_n = \frac{3\sqrt{2}}{\pi^2}V\frac{(-1)^{n+1}}{n^2}.$$

Therefore, considering that $\sum_1^{\infty} n^{-6} = \pi^6/945$, substituting this result into (11.91) we obtain:

$$D^* = D + \frac{2}{105}\frac{V^2Y^2}{D},$$

(11.92)

which coincides with (11.69).

---

[5]See [7].

(B) Poiseuille Flow in a Pipe   The eigenvalue problem associated with Eq. (11.75) is,

$$D \frac{1}{r}\frac{d}{dr}\left(r\frac{d\phi_n}{dr}\right) + \lambda_n \phi_n = 0, \quad \text{with B.C.:} \quad \frac{d\phi_n}{dr} = 0 \quad \text{at } r = \pm R, \qquad (11.93)$$

whose solution is

$$\phi_n(r) = \frac{J_0(\gamma_n r/R)}{J_0(\gamma_n)}; \qquad \lambda_n = \gamma_n^2 \frac{D}{R^2}, \qquad (11.94)$$

where γ_n are the solution of the transcendental equation $J_1(\gamma_n) = 0$, with $J_0(x)$ and $J_1(x)$ denoting the Bessel functions of zeroth and first order, respectively.[6] Accordingly, we can expand the velocity field through (11.89), with $v_n = 8V/\gamma_n^2$, so that, considering that $\sum_{n=1}^{\infty} \gamma_n^{-6} = 1/3072$ [20], Eq. (11.91) yields:

$$D^* = D + 64\frac{V^2 R^2}{D}\sum_{n=1}^{\infty}\frac{1}{\gamma^6} = D + \frac{1}{48}\frac{V^2 R^2}{D}, \qquad (11.95)$$

which coincides with (11.73).

11.3 Method of Homogenization

The method of homogenization is the study of partial differential equations with rapidly oscillating coefficients [1, 19]. Therefore, it is the natural application of the multiple scale method to derive coarse grained effective equations. Clearly, that requires the problem being characterized by two lengthscales ℓ and L, indicating typical linear dimensions of the microscale and of the macroscale, respectively, with $\epsilon = \ell/L \ll 1$ denoting the small parameter of the perturbation analysis that follows. In fact, as it is customary in these cases, our primary interest is not the detailed knowledge of the microscale process, but rather its description on a coarse scale, where we expect that it is described through an effective-medium equation and constitutive relation in terms of effective parameters (such as the effective heat and mass diffusivities), which depend on the global characteristics of the microscale velocity field.

The homogenization procedure can be summarized as a three-stage recipe. In the first stage, each physical quantity is assumed to be representable by a locally random function, that is to depend separately on the macroscopic position vector

---

[6]The Bessel eigenfunctions are orthonormal since

$$\int_0^1 \frac{J_0(\gamma_m x)J_0(\gamma_n x)}{J_0(\gamma_n)J_0(\gamma_n)} 2x\,dx = \delta_{m,n}.$$

$\mathbf{r} = L\mathbf{x}$, with $|\mathbf{x}| = O(1)$, and on the stretched coordinate, $\ell\mathbf{y}$, with $|\mathbf{y}| = O(1)$, in such a way that any function $f(\mathbf{y})$ is stationary random. In the second stage, all quantities, as well as their space derivatives, are expanded as regular perturbations of the small parameter ϵ. Finally, in the third stage, coefficients of like powers of ϵ are equated, producing a set of boundary value problems to determine the effective coefficients appearing in the final, effective equations.

In the following, we will consider the transport of a Brownian tracer through a random, infinitely extended flow field $\mathbf{v}$. Applications can be found in the mass (and heat as well) transport in packed beds or in turbulent mixers, where the randomness of the velocity field is due, in the first case, to the random distribution of the bed particles and, in the second case, to the turbulent nature of the flow. First, we will assume that the system is statistically homogeneous (i.e. $\mathbf{v}$ depends on $\mathbf{r}$ only), determining a constant effective diffusivity that corresponds to the well known turbulent diffusion; then, we will show that, when the statistical properties of the system change over the macroscale (i.e. $\mathbf{v}$ depends both on $\mathbf{r}$ and on $\mathbf{R}$), then a drift velocity appears in the macroscale effective constitutive equation, thus obtaining the Fokker-Planck equation that was obtained in Sect. 5.3 using Stratonovich's approach to stochastic integration.

11.3.1 Transport in Homogeneous Random Velocity Fields

Consider the convection of Brownian tracers in a random incompressible velocity field $\mathbf{V}(\mathbf{r}, t)$, which is statistically homogeneous and infinitely extended. Neglecting inertia and all interactions among the particles, the tracer molar concentration $c(\mathbf{r}, t)$ at location $\mathbf{r}$ and time t satisfies the following convection-diffusion equation:

$$\frac{\partial c}{\partial t} + \nabla \cdot (\mathbf{V}c) - D\nabla^2 c = 0, \tag{11.96}$$

where D is the tracer molecular diffusivity, to be solved with a given initial condition,

$$c(\mathbf{r}, 0) = c_0(\mathbf{r}). \tag{11.97}$$

Boundary conditions are implicitly included in the statistical properties of the velocity field. In alternative, we can add periodic or no-flux boundary conditions, depending on whether the randomness of the velocity field is due to turbulence or the random distribution of the bed particles, respectively.

Clearly, this problem, in principle, could be solved exactly, provided that the velocity field were known. In reality, we are not interested in the detailed knowledge of the microscale process, but rather in its description on a coarse scale and, in addition, the velocity field is known only statistically. In fact, $\mathbf{V}(\mathbf{r}, t)$ is solenoidal (i.e. the fluid is incompressible), with a constant mean value, $\mathbf{V}$,

$$\nabla \cdot \mathbf{V} = 0; \qquad \langle \mathbf{V}(\mathbf{r}, t) \rangle_0 = \mathbf{V}, \tag{11.98}$$

and with a known Lagrangian velocity autocorrelation function, referred to the mean velocity,

$$\langle \mathbf{v}(\mathbf{r}, t) \, \mathbf{v}(\mathbf{r} + \Delta \mathbf{r}, t + \Delta t) \rangle_0 = \mathbf{C}(\mathbf{r}, \Delta t), \tag{11.99}$$

with $\mathbf{v}(\mathbf{r}, t) = \mathbf{V}(\mathbf{r}, t) - \mathbf{V}$, where the brackets indicate ensemble averaging, while $\mathbf{r} + \Delta \mathbf{r}$ is the position, at time $t + \Delta t$, of the fluid particle which, time t, is located at $\mathbf{r}$. Equations (11.98) and (11.99) indicate that the velocity field is stationary in time and homogeneous in space. Clearly, although the Lagrangian velocity auto-correlation can be extracted from numerical simulations of fluid flows in turbulent mixers or in packed beds, a direct experimental measurement of this function is not feasible, in general. However, it should be remarked that, since the Eulerian and La-grangian probability distribution functions of the velocity fluctuations are related to one another [18], it is possible to determine the Lagrangian velocity autocorrelation function from its Eulerian counterpart, which is more easily measurable.

From an elementary dimensional analysis, we see that the effective dispersion coefficient is $D^* \propto V^2/\tau$, where V is the mean longitudinal velocity, while τ is the characteristic time that is necessary for the solute to sample all velocities. Therefore, for large Peclet number, i.e. when $N_{Pe} = V\ell/D \gg 1$, since the different positions within the microstructure are sampled by convection, so that $\tau = \ell/V$, we easily obtain:

$$D^* \propto D\, N_{Pe}. \tag{11.100}$$

Therefore, compared to the Taylor dispersion case, now we have a quite different scaling.[7]

As mentioned above, the main idea of the homogenization procedure is that the effective equation is expected to arise naturally from Eq. (11.96) through a regular perturbation analysis in terms of the small parameter $\epsilon = \ell/L$, expressing the ratio between micro- and macro-scale variables. Accordingly, as a first step, we define the following non-dimensional quantities:

$$\tilde{\mathbf{y}} = \frac{\mathbf{r}}{\ell}; \qquad \tilde{\mathbf{x}} = \frac{\mathbf{r}}{L}; \qquad \tilde{\tau} = \frac{t}{L^2/D}; \qquad \tilde{t} = \frac{t}{L/V}; \qquad \tilde{\mathbf{V}} = \frac{\mathbf{V}}{V}. \tag{11.101}$$

Now, we assume that each quantity can be represented *separately* in terms of the macroscale variables $\tilde{\mathbf{x}}$ and $\tilde{\tau}$ and their microscale counterparts, $\tilde{\mathbf{y}}$ and $\tilde{t}$, so that:

$$c = c(\epsilon, \tilde{\mathbf{x}}, \tilde{\mathbf{y}}, \tilde{\tau}, \tilde{t}); \qquad \tilde{\mathbf{V}} = \tilde{\mathbf{V}}(\tilde{\mathbf{y}}, \tilde{t}). \tag{11.102}$$

Note that, since $\mathbf{v}$ represents a stationary random field, it depends on $\tilde{\mathbf{y}}$ and $\tilde{t}$ only, as it does not vary over the macroscale.

---

[7]As noted in [3], for periodic porous materials, when the incoming fluid velocity has the same direction as one of the principle axes of the microstructure, tracers sample the positions within the unit cell by diffusion, and therefore the same scaling as in Taylor dispersion is obtained, with $D^* \propto D N_{Pe}^2$.

In the second stage of the homogenization, the gradient operator ∇ can be expanded in terms of ϵ as:

$$L\nabla = \nabla_x + \frac{1}{\epsilon}\nabla_y, \tag{11.103}$$

where $\nabla_x = \partial/\partial\widetilde{x}$ and $\nabla_y = \partial/\partial\widetilde{y}$. In the same way, the time derivative operator can be expanded as:

$$\frac{L^2}{D}\frac{\partial}{\partial t} = \frac{\partial}{\partial\widetilde{\tau}} + \frac{1}{\epsilon}N_{Pe}\frac{\partial}{\partial\widetilde{t}}. \tag{11.104}$$

It can be shown [14] that expanding the time derivative in terms of ϵ is equivalent to referring the problem to a moving reference frame, as we did in the Taylor dispersion case [see Eq. (11.54)].

Then, the governing equation (11.96) becomes:

$$\frac{\partial c}{\partial\widetilde{\tau}} + \frac{1}{\epsilon}N_{Pe}\frac{\partial c}{\partial\widetilde{t}} + \frac{1}{\epsilon}N_{Pe}\widetilde{\mathbf{V}}\cdot\left(\nabla_x + \frac{1}{\epsilon}\nabla_y\right)c = \left(\nabla_x + \frac{1}{\epsilon}\nabla_y\right)^2 c, \tag{11.105}$$

where $N_{Pe} = \ell V_c/D_0$ is the microscale Peclet number. In the following, we will assume that $N_{Pe} \doteq O(1)$, which means that convection and diffusion balance each other at the microscale and therefore convection dominates diffusion at the macroscale, as $Pe_L = LV_c/D_0 = O(1/\epsilon)$. Now, let us expand the concentration field as a uniformly valid power series,

$$c(\epsilon, \widetilde{\mathbf{x}}, \widetilde{\mathbf{y}}, \widetilde{\tau}, \widetilde{t}) = \sum_{n=0}^{\infty} \epsilon^n c_n(\widetilde{\mathbf{x}}, \widetilde{\mathbf{y}}, \widetilde{\tau}, \widetilde{t}), \tag{11.106}$$

where each term c_n in this expansion is assumed to be locally ergodic, that is expressible as the product of an ergodic, $\widetilde{y}$-dependent function by an $\widetilde{x}$-dependent part. In addition, the c_n functions, with $n > 0$, are defined within an arbitrary additive constant, which here we choose so that the macroscopic behavior of the system is described in terms of c_0, i.e. $\langle c_0 \rangle = \langle c \rangle \equiv \overline{c}$, which means:

$$\langle c_n \rangle = \overline{c}\delta_{n,0}. \tag{11.107}$$

At leading order, we find:

$$N_{Pe}\widetilde{\mathbf{V}}\cdot\nabla_y c_0 = \nabla_y^2 c_0, \tag{11.108}$$

yielding: $c_0 = c_0(\widetilde{\mathbf{x}}, \widetilde{\tau}, \widetilde{t})$, so that $c_0 = \overline{c}$.

At the next order, considering that $\nabla_x \cdot \nabla_y c_0 = 0$, we obtain:

$$N_{Pe}\frac{\partial c_0}{\partial\widetilde{t}} + \nabla_y \cdot (N_{Pe}\widetilde{\mathbf{V}}c_1 - \nabla_y c_1) = -N_{Pe}\widetilde{\mathbf{V}}\cdot\nabla_x c_0. \tag{11.109}$$

Now, imposing that the ensemble average of the LHS equals that of the RHS, i.e. applying the solvability conditions, we see that the above equation requires:

$$\frac{\partial \overline{c}}{\partial \widetilde{t}} = -\mathbf{1}_v \cdot \nabla_x \overline{c}, \tag{11.110}$$

where $\mathbf{1}_v = \langle \widetilde{\mathbf{V}} \rangle$ is a unit vector along the mean velocity. Now, substituting,

$$c_1(\widetilde{\mathbf{x}}, \widetilde{\mathbf{y}}, \widetilde{\tau}, \widetilde{t}) = \widetilde{\mathbf{B}}(\widetilde{y}) \cdot \nabla_x \overline{c}(\widetilde{\mathbf{x}}, \widetilde{\tau}, \widetilde{t}), \tag{11.111}$$

into (11.109), we see that the Brenner's $\widetilde{\mathbf{B}}$-field[8] satisfies the following problem:

$$\widetilde{\mathbf{V}} \cdot \nabla_y \widetilde{\mathbf{B}} - N_{Pe}^{-1} \nabla_y^2 \widetilde{\mathbf{B}} = -\widetilde{\mathbf{v}}, \tag{11.112}$$

where $\widetilde{\mathbf{v}} = \widetilde{\mathbf{V}} - \langle \widetilde{\mathbf{V}} \rangle$. Here, the $\widetilde{\mathbf{B}}$-field is defined within an arbitrary constant, which can be determined applying (11.107) with $n = 1$, i.e.

$$\langle \widetilde{\mathbf{B}} \rangle = \mathbf{0}. \tag{11.113}$$

At $O(1)$ we obtain:

$$\frac{\partial c_0}{\partial \widetilde{\tau}} + N_{Pe} \frac{\partial c_1}{\partial \widetilde{t}} + \nabla_y \cdot (N_{Pe} \widetilde{\mathbf{V}} c_2 - \nabla_y c_2) = -N_{Pe} \widetilde{\mathbf{V}} \cdot \nabla_x c_1 + 2\nabla_x \cdot \nabla_y c_1 + \nabla_x^2 c_0. \tag{11.114}$$

Applying the solvability condition and considering that c_1, c_2 and $\widetilde{\mathbf{B}}$, with their y-gradients, are zero-average locally ergodic functions, we obtain:

$$\frac{\partial \overline{c}}{\partial \widetilde{t}} = \widetilde{\mathbf{D}} : \nabla_x \nabla_x \overline{c}, \tag{11.115}$$

where $\widetilde{\mathbf{D}}$ is a non-dimensional effective diffusivity,

$$\widetilde{\mathbf{D}} = \mathbf{I} - N_{Pe} \langle \widetilde{\mathbf{v}} \widetilde{\mathbf{B}} \rangle. \tag{11.116}$$

In dimensional variables, applying (11.104), we have:

$$\frac{\partial \overline{c}}{\partial t} + \nabla \cdot \mathbf{J} = 0, \tag{11.117}$$

where the material flux $\mathbf{J}$ satisfies the following constitutive equation,

$$\mathbf{J} = \mathbf{V} \overline{c} - \mathbf{D}^* \cdot \nabla \overline{c}, \tag{11.118}$$

with $\mathbf{D}^* = D\widetilde{\mathbf{D}}$ denoting the effective diffusivity.

---

[8]Note that, unlike in the Taylor dispersion case, here the B-field is defined without Peclet number.

11.3.2 Dispersion in Nonhomogeneous Random Fields

Consider the case where the velocity field is not homogeneous, but it varies along its macroscopic coordinate, so that $\widetilde{\mathbf{V}} = \widetilde{\mathbf{V}}(\widetilde{\mathbf{x}}, \widetilde{\mathbf{y}}, \widetilde{\tau}, \widetilde{t})$. In addition, let us assume that the fluid is quasi incompressible, which means that the velocity field is solenoidal at the microscale, i.e.,

$$\nabla_y \cdot \widetilde{\mathbf{V}} = 0. \tag{11.119}$$

Proceeding as in the previous section, we see that at leading order, we find again Eq. (11.108), yielding: $c_0 = c_0(\widetilde{\mathbf{x}}, \widetilde{\tau}, \widetilde{t})$, so that $c_0 = \bar{c}$.

At the next order, Eq. (11.109) becomes:

$$N_{Pe}\frac{\partial c_0}{\partial \widetilde{t}} + \nabla_y \cdot (N_{Pe}\widetilde{\mathbf{V}}c_1 - \nabla_y c_1) = -N_{Pe}\nabla_x \cdot (\widetilde{\mathbf{V}}c_0). \tag{11.120}$$

Now, applying the solvability conditions, we see that the above equation requires:

$$\frac{\partial \bar{c}}{\partial \widetilde{t}} = -\nabla_x \cdot \left(\langle\widetilde{\mathbf{V}}\rangle\bar{c}\right), \tag{11.121}$$

where the mean non-dimensional velocity $\langle\widetilde{\mathbf{V}}\rangle$ depends on the macrovariables, while in the case of random homogeneous fields it is a constant.

Now, substituting,

$$c_1(\widetilde{\mathbf{x}}, \widetilde{\mathbf{y}}, \widetilde{\tau}, \widetilde{t}) = \nabla_x \cdot \left[\widetilde{\mathbf{B}}(\widetilde{\mathbf{x}}, \widetilde{\mathbf{y}})\bar{c}(\widetilde{\mathbf{x}}, \widetilde{\tau}, \widetilde{t})\right], \tag{11.122}$$

into (11.120), we see that the $\widetilde{\mathbf{B}}$-field satisfies the following problem:

$$\widetilde{\mathbf{V}} \cdot \nabla_y \widetilde{\mathbf{B}} - N_{Pe}^{-1}\nabla_y^2 \widetilde{\mathbf{B}} = -\widetilde{\mathbf{v}}, \tag{11.123}$$

where $\widetilde{\mathbf{v}} = \widetilde{\mathbf{V}} - \langle\widetilde{\mathbf{V}}\rangle$. Again, the $\widetilde{\mathbf{B}}$-field is defined within an arbitrary constant, so that,

$$\langle\widetilde{\mathbf{B}}\rangle = \mathbf{0}. \tag{11.124}$$

At $O(1)$ we obtain:

$$\frac{\partial c_0}{\partial \widetilde{\tau}} + N_{Pe}\frac{\partial c_1}{\partial \widetilde{t}} + \nabla_y \cdot (N_{Pe}\widetilde{\mathbf{V}}c_2 - \nabla_y c_2) = -N_{Pe}\nabla_x \cdot (\widetilde{\mathbf{V}}c_1) + 2\nabla_x \cdot \nabla_y c_1 + \nabla_x^2 c_0. \tag{11.125}$$

Applying the solvability condition and considering that c_1, c_2 and $\widetilde{\mathbf{B}}$, with their y-gradients, are zero-average locally ergodic functions, we obtain:

$$\frac{\partial \bar{c}}{\partial \widetilde{t}} = \nabla_x^2 \bar{c} - N_{Pe}\nabla_x \cdot \left(\widetilde{\mathbf{V}}\nabla_x \cdot (\widetilde{\mathbf{B}}\bar{c})\right). \tag{11.126}$$

that is:[9]

$$\frac{\partial \bar{c}}{\partial \tilde{t}} + \nabla_x \cdot \tilde{\mathbf{J}} = 0. \tag{11.127}$$

Here, the macroscale material flux $\tilde{\mathbf{J}}$ is given by the following constitutive relation:

$$\tilde{\mathbf{J}} = \tilde{\mathbf{V}}_d \bar{c} - \tilde{\mathbf{D}} \cdot \nabla_x \bar{c}, \tag{11.128}$$

where $\tilde{\mathbf{D}}$ is the effective diffusivity,

$$\tilde{\mathbf{D}} = \mathbf{I} - N_{Pe} \langle \tilde{\mathbf{v}} \tilde{\mathbf{B}} \rangle, \tag{11.129}$$

while $\tilde{\mathbf{V}}_d$ is a drift velocity,

$$\tilde{\mathbf{V}}_d = N_{Pe} \langle \tilde{\mathbf{v}} \nabla_x \cdot \tilde{\mathbf{B}} \rangle. \tag{11.130}$$

In dimensional variables, applying (11.104), we obtain:

$$\frac{\partial \bar{c}}{\partial t} + \nabla \cdot \mathbf{J} = 0, \tag{11.131}$$

with,

$$\mathbf{J} = \mathbf{V}_E \bar{c} - \mathbf{D}^* \cdot \nabla \bar{c}, \tag{11.132}$$

where $\mathbf{V}_E = \langle \mathbf{V} \rangle + \mathbf{V}_d$ is the mean particle velocity, $\mathbf{V}_d = V \tilde{\mathbf{V}}_d$ is the drift velocity and $\mathbf{D}^* = D \tilde{\mathbf{D}}$ is the effective diffusivity.

Since the effective diffusivity has the same expression as in the case of homogeneous random fields, we obtain again Eq. (11.138), i.e.,

$$\mathbf{D}^* = D\mathbf{I} + \int_0^\infty \langle \mathbf{v}(0)\,\mathbf{v}(t) \rangle \, dt, \tag{11.133}$$

so that the drift velocity becomes:

$$\mathbf{V}_d = -\int_0^\infty \langle \mathbf{v}(0)\,\nabla \cdot \mathbf{v}(t) \rangle \, dt. \tag{11.134}$$

We may conclude that the flux of passive tracers can be expressed through a Fickian constitutive relation, characterized by an effective diffusivity and an Eulerian mean tracer velocity, the latter being equal to the sum of the mean fluid velocity, the ballistic tracer velocity (see below) and the particle drift velocity. In particular, the drift velocity takes into account the coupling between the non-solenoidal character of the flow field and its non-homogeneity at the macroscale, so that it vanishes whenever the flow field is homogeneous at the macroscale. Finally, we should note that the non solenoidal character of the flow field is not necessarily due to the fluid

---

[9]Decomposing $\tilde{\mathbf{V}} = \langle \tilde{\mathbf{V}} \rangle + \tilde{\mathbf{v}}$, we see that $\langle \tilde{\mathbf{V}} \rangle$ gives no contributions to Eq. (11.126).

compressibility. In fact, particle-particle interactions and particle inertial forces can cause the particle velocity to differ from the local fluid velocity, so that the particle velocity field can be non-solenoidal even when the fluid is incompressible [15].

Comment When the assumption of quasi incompressibility is removed, the result is very similar [16], the only difference being that all averaging must include a weighting factor, which is equal to the leading order concentration field. Consequently, the mean tracer velocity $\langle \mathbf{V} \rangle$ does not coincide with the mean fluid velocity any more, as it includes a so-called ballistic tracer velocity, accounting for the fact that tracer particles tend to concentrate more in the regions where the fluid velocity divergence is negative [24].

11.3.3 Lagrangian Approach

I. Velocity Covariance

The solution of Eq. (11.112) and (11.123) (in its dimensional form) can also be expressed as:

$$\widetilde{\mathbf{B}}(r) = \int_0^\infty dt \int_\mathbf{r} \Pi(\mathbf{r}, t | \mathbf{r}_0) \mathbf{v}(\mathbf{r}_0) \, d\mathbf{r}_0, \qquad (11.135)$$

where $\mathbf{v}(\mathbf{r}, t) = \mathbf{V}(\mathbf{r}, t) - \mathbf{V}$. Here, $\Pi(\mathbf{r}, t | \mathbf{r}_0)$ is the propagator (or Green function) of the convection-diffusion equation,

$$\frac{\partial \Pi}{\partial t} + \mathbf{V} \cdot \nabla \Pi - D \nabla^2 \Pi = \delta(\mathbf{r} - \mathbf{r}_0) \delta(t); \qquad (11.136)$$

it represents the probability that a tracer particle initially located at position $\mathbf{r}_0$ is found at position $\mathbf{r}$ at time t. At the end, proceeding as in the Taylor dispersion case, we find the following expression for the effective diffusivity [10, 14, 17]:

$$\mathbf{D}^* = D\mathbf{I} + \int_0^\infty \mathbf{C}(t) \, dt, \qquad (11.137)$$

where

$$\mathbf{C}(t) = \langle \mathbf{v}(t) \, \mathbf{v}(0) \rangle \qquad (11.138)$$

is the average covariance of the fluid velocity at the points occupied by a Brownian tracer at times $t = 0$ and t, assuming that its initial positions are all equally probable.

Note that, since the process is locally ergodic, proceeding as in Sect. 2.3 we can show that $\langle \mathbf{v}(t) \, \mathbf{v}(0) \rangle = \langle \mathbf{v}(0) \, \mathbf{v}(t) \rangle$, so that the effective diffusivity is a symmetric dyadic. In particular, when the inclusions within the porous material can be modeled as a dilute suspension of spheres, the effective diffusivity describing the flow of a solute is $D = \frac{3}{4} V a$, where V is the mean solvent velocity and a is the radius of the inclusions [10].

II. Homogeneous Random Velocity Fields

The result (11.137) can be interpreted also using a Lagrangian approach. First, assuming that the random velocity field is homogeneous, consider the Langevin equation, stating that the instantaneous velocity of a Brownian tracer is the sum of two uncorrelated processes, namely a zero-mean convection drift, $\mathbf{v}(\mathbf{r}, t) = \mathbf{V}(\mathbf{r}, t) - \mathbf{V}$, and a Wiener diffusion process, $\dot{\mathbf{w}}(t)$,

$$\frac{d\mathbf{z}}{dt} = \mathbf{v} + \dot{\mathbf{w}}, \tag{11.139}$$

where $\mathbf{z}(t) = \mathbf{Z}(t) - \mathbf{V}t$ is the position of the tracer particle, referred to a frame moving with constant velocity $\mathbf{V}$, while,

$$\langle \dot{\mathbf{w}}(t) \rangle = \mathbf{0}; \qquad \langle \dot{\mathbf{w}}(t') \dot{\mathbf{w}}(t'') \rangle = 2D\mathbf{I}\delta(t' - t''). \tag{11.140}$$

Assuming a uniform initial particle distribution, and proceeding as in the Taylor dispersion case [cf. Eqs. (11.82)–(11.83)], we obtain:

$$\left\langle \frac{d\mathbf{z}}{dt}(t) \right\rangle = \mathbf{0}; \qquad \left\langle \frac{d\mathbf{z}}{dt}(t') \frac{d\mathbf{z}}{dt}(t'') \right\rangle = 2D\mathbf{I}\delta(t' - t'') + \mathbf{C}(t' - t''), \tag{11.141}$$

with $\mathbf{C}(t' - t'') = \langle \mathbf{v}(\mathbf{r}') \mathbf{v}(\mathbf{r}'') \rangle$ denoting the velocity autocorrelation (11.138), where $\mathbf{r}'$ and $\mathbf{r}''$ are the locations of a Brownian particle at time t' and t'', respectively. At the end, we see that Eq. (11.137) is equivalent to:

$$\mathbf{D}^* = \frac{1}{2} \lim_{t \to \infty} \frac{d}{dt} \langle \mathbf{z}(t) \, \mathbf{z}(t) \rangle, \tag{11.142}$$

confirming that the effective diffusivity tensor is symmetric. This shows that $\mathbf{D}^*$ is the coefficient of self-diffusivity, as it equals (one half of) the growth rate of the mean square displacement of a tracer particle immersed in a uniform concentration field. Therefore, as for the Taylor dispersion case, since $\mathbf{D}^*$ describes both self-diffusion and gradient diffusion (as it also describes the ratio between mass flux and concentration gradient), we may conclude that the fluctuation-dissipation theorem is satisfied, as one would expect, since in this case the constitutive equation for the mass flux is linear (i.e. $\mathbf{D}^*$ is independent of the composition).

III. Non-homogeneous Random Velocity Fields

Here we intend to show that the same convection-diffusion equation (11.131)–(11.132) could be obtained, assuming that the trajectory $\mathbf{Z}(t)$ of any tracer particle is a random variable satisfying the following generalized non-linear Langevin equation,

$$\frac{d\mathbf{Z}}{dt}(\mathbf{r}, t) = \overline{\mathbf{V}}(\mathbf{r}, t) + \mathbf{v}(\mathbf{r}, t), \tag{11.143}$$

Here $\overline{\mathbf{V}}(\mathbf{r}, t)$ is a smoothly varying mean tracer velocity, while $\mathbf{v}(\mathbf{r}, t)$ is its random component, including both the random velocity field and the Wiener process, as examined above (in other words, now $\mathbf{v}$ can be identified with $d\mathbf{z}/dt$, as defined in (11.139)). Unlike the homogeneous case, however, now $\mathbf{v}$ depends explicitly on its position $\mathbf{r}$ and so it is described through the following white stochastic process:

$$\langle \mathbf{v}(\mathbf{r}, t) \rangle_0 = \mathbf{0}, \tag{11.144}$$

$$\langle \mathbf{v}(\mathbf{r}_1, t) \mathbf{v}(\mathbf{r}_2, t + \Delta t) \rangle_0 = 2\mathbf{D}^T (\mathbf{r}_1, \mathbf{r}_2, t)\delta(\Delta t), \tag{11.145}$$

with the angular brackets indicating ensemble averaging and $\mathbf{D}(\mathbf{r}_1, \mathbf{r}_2, t)$ denoting, by definition, the time integral of the Lagrangian velocity cross correlation dyadic. In fact, integrating Eq. (11.143) for a short time interval Δt, we obtain:

$$\Delta \mathbf{r} = \overline{\mathbf{V}}(\mathbf{r}, t)\Delta t + \int_t^{t+\Delta t} \mathbf{v}\big[\mathbf{r}(t_1), t_1\big] dt_1 + o(\Delta t), \tag{11.146}$$

where $\Delta \mathbf{r} = \mathbf{r}(t + \Delta t) - \mathbf{r}(t)$ and we have considered that $\overline{\mathbf{V}}$ changes in time much slower than $\mathbf{v}$. Now expand[10]

$$\mathbf{v}\big[\mathbf{r}(t_1), t_1\big] = \mathbf{v}\big[\mathbf{r}(t), t_1\big] + \Delta \mathbf{r}_1 \cdot \frac{\partial}{\partial \mathbf{r}}\mathbf{v}\big[\mathbf{r}(t), t_1\big] + o(\Delta t), \tag{11.147}$$

where

$$\Delta \mathbf{r}_1 = \mathbf{r}(t_1) - \mathbf{r}(t) = \int_t^{t_1} \mathbf{v}\big[\mathbf{r}(t), t_2\big] dt_2. \tag{11.148}$$

Consequently, we can define the mean Lagrangian particle velocity, $\mathbf{V}_L$, and the particle self diffusivity $\mathbf{D}^s$ as,

$$\mathbf{V}_L(\mathbf{r}, t) = \lim_{\Delta t \to 0} \frac{\langle \Delta \mathbf{r} \rangle}{\Delta t}; \qquad \mathbf{D}^s(\mathbf{r}, t) = \frac{1}{2}\lim_{\Delta t \to 0} \frac{\langle (\Delta \mathbf{r})^2 \rangle}{\Delta t} \tag{11.149}$$

obtaining:

$$\mathbf{V}_L(\mathbf{r}, t) = \overline{\mathbf{V}}(\mathbf{r}, t) + \left[\frac{\partial}{\partial \mathbf{r}_1} \cdot \mathbf{D}(\mathbf{r}_1, \mathbf{r}, t) \right]_{\mathbf{r}_1 = \mathbf{r}} \tag{11.150}$$

and

$$\mathbf{D}^s(\mathbf{r}, t) = \mathbf{D}(\mathbf{r}, \mathbf{r}, t). \tag{11.151}$$

These quantities constitute the convective and diffusive part of the particle flux appearing in the Fokker-Planck equation (4.22), with

$$\mathbf{J} = \mathbf{V}_L \overline{c} - \nabla \cdot (\mathbf{D}^s \overline{c}), \tag{11.152}$$

---

[10]This expansion is equivalent to Stratonovich stochastic integration as seen in Sect. 5.3. See [12].

which is equivalent to Eq. (11.132) when $\mathbf{D}^s = \mathbf{D}^*$, as the Lagrangian mean particle velocity is related to its Eulerian counterpart, $\mathbf{V}_E$ through the relation:

$$\mathbf{V}_L = \mathbf{V}_E + \nabla \cdot \mathbf{D}^s. \tag{11.153}$$

Therefore, we may conclude that the convection-diffusion equation (11.131)–(11.132) is equivalent to the non-linear stochastic process (11.143)–(11.145). In particular, that means that (a) the smoothly varying mean tracer velocity $\overline{\mathbf{V}}$ appearing in the random process is the microscale mean tracer velocity (which in turn is the sum of the mean fluid velocity and the ballistic tracer velocity); (b) the effective diffusivity $\mathbf{D}^s$ appearing in the Fokker-Planck equation is a self-diffusion dyadic, as it equals the time derivative of the Lagrangian velocity autocorrelation function. An identical conclusion was reached for the constitutive relations of the volumetric flux of a suspension of rigid particles immersed in a viscous fluid [15].

11.4 Problems

Problem 11.1 Determine the mean velocity and the Taylor dispersivity of small spherical solute particles of radius a immersed in a Poiseuille flow in a microchannel of width $2Y$.

Problem 11.2 Study the Taylor dispersion in a Couette flow, using (a) Eulerian approach; (b) Lagrangian approach.

Problem 11.3 Determine the permeability of a porous material using the method of homogenization.

References

1. Bensoussan, A., Lions, J.L., Papanicolaou, G.: Asymptotic Analysis for Periodic Structures. North-Holland, Amsterdam (1978)
2. Brenner, H.: PhysicoChem. Hydrodyn. **1**, 91 (1980)
3. Brenner, H.: Philos. Trans. R. Soc. Lond. A **297**, 81 (1980)
4. Brenner, H., Edwards, D.A.: Macrotransport Processes. Butterworth-Heinemann, Stoneham (1993)
5. Brown, J.W., Churchill, R.V.: Fourier Series and Boundary Value Problems. McGraw-Hill, New York (1993)
6. Childress, S.: J. Chem. Phys. **56**, 2527 (1972)
7. Dill, L.H., Brenner, H.: J. Colloid Interface Sci. **93**, 343 (1983)
8. Haber, S., Mauri, R.: J. Fluid Mech. **190**, 201 (1988)
9. Kim, S., Russel, W.B.: J. Fluid Mech. **154**, 269 (1985)
10. Koch, D.L., Brady, J.F.: J. Fluid Mech. **154**, 399 (1985)
11. Koch, D.L., Brady, J.F.: J. Fluid Mech. **180**, 387 (1987)
12. Kubo, R., Toda, M., Hashitsume, N.: Statistical Physics II. Nonequilibrium Statistical Mechanics. Springer, Berlin (1991), Chap. 2.4

13. Lundgren, T.S.: J. Fluid Mech. **51**, 273 (1972)
14. Mauri, R.: J. Eng. Math. **29**, 77 (1995)
15. Mauri, R.: Phys. Fluids **15**, 1888 (2003)
16. Mauri, R.: Phys. Rev. E **68**, 066306 (2003)
17. Monin, A.S., Yaglom, A.M.: Statistical Fluid Mechanics: Mechanics of Turbulence, vol. 1. MIT Press, Cambridge (1971), pp. 540–547, 591–606
18. Pope, S.B.: Turbulent Flows, pp. 480–482. Cambridge University Press, Cambridge (2000)
19. Sanchez-Palencia, E.: Non-Homogeneous Media and Vibration Theory. Springer, Berlin (1980)
20. Sneddon, I.N.: Proc. Glasg. Math. Assoc. **4**(144), 151 (1960)
21. Taylor, G.I.: Proc. R. Soc. A **219**, 186 (1953)
22. Taylor, G.I.: Proc. R. Soc. A **223**, 446 (1953)
23. Taylor, G.I.: Proc. R. Soc. A **225**, 473 (1954)
24. Vergassola, M., Avellaneda, M.: Physica D **106**, 148 (1997)

Appendix A
Review of Probability Distribution

In this appendix we review some fundamental concepts of probability theory. After deriving in Sect. A.1 the binomial probability distribution, in Sect. A.2 we show that for systems described by a large number of random variables this distribution tends to a Gaussian function. Finally, in Sect. A.3 we define moments and cumulants of a generic probability distribution, showing how the central limit theorem can be easily derived.

A.1 Binomial Distribution

Consider an ideal gas at equilibrium, composed of N particles, contained in a box of volume V. Isolating within V a subsystem of volume v, as the gas is macroscopically homogeneous, the probability to find a particle within v equals $p = v/V$, while the probability to find it in a volume $V - v$ is $q = 1 - p$. Accordingly, the probability to find any *one* given configuration (i.e. assuming that particles could all be distinguished from each other), with n molecules in v and the remaining $(N - n)$ within $(V - v)$, would be equal to $p^n q^{N-n}$, i.e. the product of the respective probabilities.[1] However, as the particles are identical to each other, we have to multiply this probability by the Tartaglia pre-factor, that us the number of possibilities of

---

[1] Here, we assume that finding a particle within v is not correlated with finding another, which is true in ideal gases, that are composed of point-like molecules. In real fluids, however, since molecules do occupy an effective volume, we should consider an excluded volume effect.

R. Mauri, *Non-Equilibrium Thermodynamics in Multiphase Flows*,
Soft and Biological Matter, DOI 10.1007/978-94-007-5461-4,
© Springer Science+Business Media Dordrecht 2013

choosing n particles within a total of N identical particles, obtaining:[2]

$$\Pi_N(n) = \binom{N}{n} p^n q^{N-n} = \frac{N!}{n!(N-n)!} \left(\frac{v}{V}\right)^n \left(\frac{V-v}{V}\right)^{N-n}. \qquad (A.1)$$

This is the binomial distribution, expressing the probability to find n particles in the volume v and the remaining $(N-n)$ within $(V-v)$.

Normalization of the Binomial Distribution   Recall the binomial theorem:

$$(p+q)^N = \sum_{n=0}^{N} \frac{N!}{n!(N-n)!} p^n q^{N-n}. \qquad (A.2)$$

This shows that, as $p+q=1$, the binomial distribution is normalized,

$$\sum_{n=0}^{N} \Pi_N(n) = 1. \qquad (A.3)$$

Calculation of the Mean Value   From

$$\bar{n} = \sum_{n=0}^{N} n \Pi_N(n) = \sum_{n=0}^{N} \frac{N!}{n!(N-n)!} p^n q^{N-n} n, \qquad (A.4)$$

considering that

$$n p^n = p \frac{\partial}{\partial p} (p^n), \qquad (A.5)$$

we obtain:

$$\bar{n} = p \frac{\partial}{\partial p} \left[\sum_{n=0}^{N} \frac{N!}{n!(N-n)!} p^n q^{N-n} \right]. \qquad (A.6)$$

Then considering (A.2), we obtain:

$$\bar{n} = p \frac{\partial}{\partial p} (p+q)^N = pN(p+q)^{N-1}, \qquad (A.7)$$

---

[2]The number of ways with which N distinguishable objects can fill N slots is equal to $N!$. In fact, if we imagine to fill the slots one by one, the first particle will have N free slots to choose from, the second particle will have $(N-1)$, the third $(N-2)$, etc., until the last particle will find only one free slot. Now, suppose that of the N objects, n are, say, white and indistinguishable (this is equivalent, in our case, to having them located within v), while the remaining $(N-n)$ are black, i.e. they are located within $(V-v)$. That means that we must divide $N!$ by the number of permutations among the n white particles and by the number of permutations among the $(N-n)$ black particles, to account for the fact that if any white (or black) particles exchange their places we find a configuration that is indistinguishable from the previous one. So, at the end, we find the factor $N!/n!(N-n)!$, which correspond to the Tartaglia prefactors, appearing in the Tartaglia triangle.

and finally, as $p + q = 1$, we conclude:

$$\bar{n} = Np. \tag{A.8}$$

Calculation of the Variance Considering that,

$$\overline{(\Delta n)^2} = \overline{(n - \bar{n})^2} = \overline{n^2} - \bar{n}^2, \tag{A.9}$$

we see that we need to compute $\overline{n^2}$, defined as

$$\overline{n^2} = \sum_{n=0}^{N} n^2 \Pi_N(n) = \sum_{n=0}^{N} \frac{N!}{n!(N-n)!} p^n q^{N-n} n^2. \tag{A.10}$$

Considering that

$$n^2 p^n = n \left(p \frac{\partial}{\partial p} \right)(p^n) = \left(p \frac{\partial}{\partial p} \right)^2 p^n, \tag{A.11}$$

we obtain:

$$\overline{n^2} = \left(p \frac{\partial}{\partial p} \right)^2 \left[\sum_{n=0}^{N} \frac{N!}{n!(N-n)!} p^n q^{N-n} \right] = \left(p \frac{\partial}{\partial p} \right)^2 (p+q)^N, \tag{A.12}$$

where we have considered (A.2) and (A.7). Then, we obtain:

$$\overline{n^2} = p \frac{\partial}{\partial p} \left[pN(p+q)^{N-1} \right] = pN \left[(p+q)^{N-1} + p(N-1)(p+q)^{N-2} \right], \tag{A.13}$$

and finally, as $p + q = 1$, we conclude:

$$\overline{n^2} = Np(1 + Np - p) = (Np)^2 + Npq, \tag{A.14}$$

so that:

$$\overline{(\Delta n)^2} = Npq. \tag{A.15}$$

As expected, defining the mean root square dispersion, we find as expected:

$$\frac{\delta n}{\bar{n}} = \frac{1}{\sqrt{N}} \sqrt{\frac{q}{p}}, \quad \text{where } \delta n \equiv \sqrt{\overline{(\Delta n)^2}}. \tag{A.16}$$

In particular, when $p = q = 1/2$, we have: $\delta n / \bar{n} = 1/\sqrt{n}$.

A.2 Poisson and Gaussian Distribution

Let us consider the case when $v \ll V$, i.e. $\bar{n} \ll N$. Then $N! \cong (N - n)! N^n$ and the binomial distribution becomes:

$$\Pi_N(n) = \frac{\bar{n}^n}{n!} \left(1 - \frac{\bar{n}}{N} \right)^N, \tag{A.17}$$

where $\bar{n} = Np$. Now, letting $N \to \infty$ and considering that

$$\lim_{x \to \infty} \left(1 - \frac{a}{x} \right)^x = e^{-a}, \tag{A.18}$$

we obtain the so-called *Poisson distribution*,[3]

$$\Pi(n) = \frac{\bar{n}^n e^{-\bar{n}}}{n!}. \tag{A.19}$$

It is easy to see that, predictably, the Poisson distribution gives the same results as the binomial distribution in terms of normalization condition, mean value and dispersion, i.e.

$$\sum_{n=0}^{\infty} \Pi(n) = 1; \qquad \sum_{n=0}^{\infty} n \, \Pi(n) = \bar{n}; \qquad \sum_{n=0}^{\infty} (n - \bar{n})^2 \Pi(n) = \bar{n}. \tag{A.20}$$

When fluctuations are small and $n \gg 1$, the Poisson distribution can be simplified applying Stirling's formula, $n! = \sqrt{2\pi n} \, n^n e^{-n}$, valid when $n \gg 1$, obtaining:

$$\Pi(n) = \frac{1}{\sqrt{2\pi n}} \left(\frac{\bar{n}}{n} \right)^n e^{\Delta n}, \tag{A.21}$$

where $\Delta n = n - \bar{n}$. Now, considering that

$$\left(\frac{\bar{n}}{n} \right)^n = \exp\left[-n \ln\left(1 + \frac{\Delta n}{\bar{n}} \right) \right] = \exp\left\{ -n \left[\frac{\Delta n}{\bar{n}} + \frac{1}{2}\left(\frac{\Delta n}{\bar{n}} \right)^2 + \cdots \right] \right\}, \tag{A.22}$$

we see that at leading order it reduces to the following *Gaussian distribution*:

$$\Pi(n) = \frac{1}{\sqrt{2\pi n}} \exp\left[-\frac{(n - \bar{n})^2}{2\bar{n}} \right]. \tag{A.23}$$

Often, instead of a discrete probability distribution $\Pi(n)$, it is convenient to define a continuous distribution, $\Pi(x)$, defined so that $\Pi(x)\,dx$ is the probability that

---

[3]Consider that in this case we keep $Np = \bar{n}$ fixed as we let $N \to \infty$, so that $q = 1$.

x assumes values between x and $x + dx$. In our case, this can be done considering the limit of $x = n/N$ as $N \to \infty$,

$$\Pi(x)\,dx = \lim_{N \to \infty} \Pi(n)\frac{dn}{N}, \qquad (A.24)$$

where $dn = 1$ and $dx = \lim_{N \to \infty}(1/N)$. Predictably, $\Pi(x)$ is again a Gaussian distribution,

$$\Pi(x) = \frac{1}{\sqrt{2\pi}\sigma} \exp\left[-\frac{(x - \bar{x})^2}{2\sigma^2}\right], \qquad (A.25)$$

where $\sigma = \delta x$.

A.3 Moments and Cumulants

The rth *central moment*, or moment about the mean, of a random variable X with probability density $\Pi(x)$ is defined as the mean value of $(x - \mu)^r$, i.e.,

$$\mu^{(r)} = \langle (x - \mu)^r \rangle = \int (x - \mu)^r \Pi(x)\,dx, \quad r = 0, 1, \ldots, \qquad (A.26)$$

where $\mu = \int x \Pi(x)\,dx$ is the mean value, provided that the integrals converge.[4] Note that $\mu^{(0)} = 1$ because of the normalization condition, while $\mu^{(1)} = 0$ by definition. An easy way to combine all of the moments into a single expression is to consider the following Fourier transform of the probability density,

$$\widehat{\Pi}(k) = \int e^{ik(x-\mu)} \Pi(x)\,dx = \langle e^{ik(x-\mu)} \rangle, \qquad (A.27)$$

which is generally referred to as the characteristic (or moment generating) function. Now, since $\widehat{\Pi}$ coincides with the mean value of $e^{ik(x-\mu)}$, expanding it in a Taylor series about the origin, we easily find:

$$\widehat{\Pi}(k) = 1 + \langle (x - \mu) \rangle(ik) + \frac{1}{2}\langle (x - \mu)^2 \rangle(ik)^2 + \cdots = \sum_{r=0}^{\infty} \frac{1}{r!}\mu^{(r)}(ik)^r. \quad (A.28)$$

Therefore, we see that the r-th moment is the r-th derivative of the characteristic function with respect to (ik) at $k = 0$.

Sometimes it is more convenient to consider the logarithm of the probability density. In that case, proceeding as before, we can define the cumulant generating function,

$$\Phi(k) = \ln \widehat{\Pi}(k) = \sum_{r=1}^{\infty} \frac{1}{r!}\kappa^{(r)}(ik)^r, \qquad (A.29)$$

---

[4] In general, the rth moment of a random variable is defined as the mean values of x^r.

where $\kappa^{(r)}$ are called *cumulants*, with obviously $\kappa^{(0)} = \kappa^{(1)} = 0$. The first three cumulants are equal to the respective central moments, that is $\kappa^{(1)} = \mu^{(1)}$ is the mean value,[5] $\kappa^{(2)} = \mu^{(2)}$ is the variance, and $\kappa^{(3)} = \mu^{(3)}$ is the skewness, measuring the lopsidedness of the distribution. Instead, higher order moments and cumulants are different from each other. For example, $\kappa^{(4)} = \mu^{(4)} - 3(\mu^{(2)})^2$, where the forth central moment is a measure of whether the distribution is tall and skinny or short and squat, compared to the normal distribution of the same variance.[6]

When the random variable x is the sum of many mutually independent random variables x_i, we know that the global probability is the product of the probabilities of the individual variables, and therefore the cumulant generating function (A.29) is the sum of the individual functions, i.e. $\Phi(k) = \sum_i \Phi_i(k)$. Therefore, from (A.29) we see that,

$$\kappa^{(r)} = \sum_i \kappa_i^{(r)}. \tag{A.30}$$

This is the additive property of cumulants, stating that the cumulants of a sum of random variables equals the sum of the individual cumulants.

For a binomial distribution, when the independent variable is rescaled as $x = n/N$, we found: $\kappa^{(2)} = \sigma^2 = p(1-p)/N$ and $\kappa^{(r)} = O(N^{-r+1})$, so that, when $N \gg 1$, the expansion (A.29) becomes at leading order:

$$\Phi(k) = -\frac{1}{2}\sigma^2 k^2, \quad \text{i.e.,} \quad \hat{\Pi}(k) = e^{-\frac{1}{2}\sigma^2 k^2}. \tag{A.31}$$

Antitransforming, as in Eq. (4.78)–(4.79), we find the Gaussian distribution (A.25). Clearly, this shows that the Gaussian distribution has the property that all cumulants of higher order than 2 vanish identically.

This is basically the *central limit theorem*, stating that the mean of a sufficiently large number of independent random variables (each with finite mean and variance) will be approximately *normally distributed*.

---

[5]This equality is true even when we consider moments referred to the origin, instead of moments referred to the mean.

[6]This characteristic is measured through a non-dimensional parameter, called *kurtosis*, or "peakedness" in Greek, which is defined as the ratio between the fourth cumulant and the square of the second cumulant.

Appendix B
Review of Statistical Thermodynamics

In this appendix a brief review of statistical thermodynamics is presented. First, starting from some intuitive postulates at the microscale, in Sect. B.1 we derive Boltzmann's fundamental definition of entropy, showing from that how to define some basic macroscopic quantities, such as temperature and pressure. Then, in Sect. B.2, we see the connection between thermodynamics and statistical mechanics on a deeper level, showing the equivalence between free energy and partition function. Finally, in Sect. B.3, we derive an expression for the free energy of a fluid that is used in Chap. 9 within the context of the diffuse interface model.

B.1 Introduction

The basic postulate of statistical mechanics is that at equilibrium an isolated system is equally likely to be found in any of its accessible states. Accordingly, the probability $\Pi^{eq}(\mathcal{C})$ that at equilibrium an isolated system is found in a given configuration $\mathcal{C}$ equals the ratio between the number of accessible states associated with $\mathcal{C}$, $\Omega(\mathcal{C})$, and the total number of accessible states, Ω_{tot}, i.e.,

$$\Pi_{\mathcal{C}}^{eq} = \frac{\Omega(\mathcal{C})}{\Omega_{tot}}. \tag{B.1}$$

As an example, consider a system composed of four spheres, indicated as a, b, c and d, which are to be placed in two containers, indicated as A and B. Then, the accessible states associated with having two spheres in A and two in B correspond to having in A the spheres ab, ac, ad, bc, bd or cd, so that $\Omega = 6$, while $\Omega_{tot} = 2^4 = 16$, yielding $\Pi_4^{eq}(2) = 3/8$. Generalizing this result to N spheres, we find:

$$\Omega(\mathcal{C}) = \binom{N}{N/2}, \qquad \Omega_{tot} = 2^N; \qquad \Pi_N^{eq}\left(\frac{N}{2}\right) = \binom{N}{N/2}\frac{1}{2^N}, \tag{B.2}$$

R. Mauri, *Non-Equilibrium Thermodynamics in Multiphase Flows*,
Soft and Biological Matter, DOI 10.1007/978-94-007-5461-4,
© Springer Science+Business Media Dordrecht 2013

where

$$\binom{N}{n} = \frac{N!}{n!(N-n)!} \tag{B.3}$$

is the Tartaglia prefactor (see Appendix A).

Now, consider an ideal gas at equilibrium, composed of N molecules and contained in a box of volume V. Isolating within V a subsystem of volume v, since at equilibrium the gas is homogeneous, the probability to find a particle within the volume v equals $p = v/V$, while the probability to find it in a volume $V - v$ is $q = 1 - p$. If the particles cannot be distinguished from each other, the probability to find n molecules in the volume v and the remaining $(N - n)$ within $(V - v)$ is equal to the binomial distribution (see Appendix A)

$$\Pi_N^{eq}(n) = \binom{N}{n} p^n q^{N-n} = \frac{N!}{n!(N-n)!} \left(\frac{v}{V}\right)^n \left(\frac{V-v}{V}\right)^{N-n}. \tag{B.4}$$

In Appendix A we show that the binomial distribution is normalized, with the following values of the mean and dispersion values of n,

$$\sum_{n=0}^{N} \Pi_N^{eq}(n) = 1; \tag{B.5}$$

$$\bar{n} = \sum_{n=0}^{N} n \Pi_N^{eq}(n) = Np; \tag{B.6}$$

$$(\delta n)^2 = \overline{(\Delta n)^2} = \sum_{n=0}^{N} (\Delta n)^2 \Pi_N^{eq}(n) = Npq, \tag{B.7}$$

where $\Delta n = n - \bar{n}$.

When we plot Π_N^{eq} as a function of n/N for a given value of v/V, we see that, as N increases, the binomial distribution p becomes sharper an sharper around its mean, equilibrium value, $\bar{n}/N = p$, as $\delta n/N = \sqrt{pq/N}$, thus confirming what we saw in Sect. 1.1. In addition, letting $N \to \infty$, with $\rho = x = \lim_{N\to\infty} n/N$ denoting a continuous density function, we can define a continuous distribution $\Pi^{eq}(x)\,dx = \lim_{N\to\infty} \frac{1}{N}\Pi_N^{eq}(n)$, seeing that the binomial curve tends to a Gaussian distribution, centered around $\bar{x} = \bar{n}/N$ and having a dispersion, or variance, equal to $\sigma^2 = \overline{(\Delta x)^2} = (\delta x)^2 = pq/N$, [see Eqs. (A.24)–(A.25)]

$$\Pi^{eq}(x) = \frac{1}{\sqrt{2\pi}\sigma} \exp\left[-\frac{(x-\bar{x})^2}{2\sigma^2}\right]. \tag{B.8}$$

In general, we may conclude that for any thermodynamic quantity x, there is a probability $\Pi^{eq}(\tilde{x})$ that A will be equal to a certain value, $\tilde{x}$, defined as the ratio between the number of times that we measure $x = \tilde{x}$ and the total number of measurements, as this latter goes to infinity. This probability is maximum when $\tilde{x}$ coincides with the equilibrium value, $x = \bar{x}$, while it decays rapidly as we move away from it. In fact, $\Pi^{eq}(\tilde{x})$ is a sharp function whose width tends to zero as $N \to \infty$, i.e. for classical thermodynamics, where we assume that all the variables describing the state of a system are constant.

Now, in isolated systems, the number of accessible states Ω (and the equilibrium probability distribution as well) depends on its state variables, but it cannot depend on a particular reference frame. Accordingly, Ω may depend only on the invariants of the system, namely its energy, E, momentum, $\mathbf{Q}$, and angular momentum, $\mathbf{M}$. To understand which type of relations we expect to exist among these quantities, let us suppose that our system is composed of two separate subsystems, 1 and 2 [3, Chaps. 4 and 9]. Then, the accessible states $\Omega_{1+2}(\mathbf{x}_1 + \mathbf{x}_2)$ associated with subsystem 1 having a given configuration $\mathbf{x}_1$ and subsystem 2 another given configuration $\mathbf{x}_2$, equals the product of the accessible states $\Omega_1(\mathbf{x}_1)$ and $\Omega_2(\mathbf{x}_2)$ of the single subsystems, i.e. $\Omega_{1+2}(\mathbf{x}_1 + \mathbf{x}_2) = \Omega_1(\mathbf{x}_1) \times \Omega_2(\mathbf{x}_2)$. On the other hand, E, $\mathbf{Q}$ and $\mathbf{M}$ are extensive variables, so that $E = E_1 + E_2$, $\mathbf{Q} = \mathbf{Q}_1 + \mathbf{Q}_2$ and $\mathbf{M} = \mathbf{M}_1 + \mathbf{M}_2$. So, we may conclude:

$$\ln \Omega = \alpha + \beta E + \mathbf{C}_1 \cdot \mathbf{Q} + \mathbf{C}_2 \cdot \mathbf{M}, \tag{B.9}$$

where α, β, $\mathbf{C}_1$ and $\mathbf{C}_2$ are characteristics of the system. Therefore, when our system is isotropic, $\mathbf{C}_1 = \mathbf{C}_2 = \mathbf{0}$, so that the probability that the system is in a state k is:

$$\Omega_k = \Omega(E_k) = C e^{\beta E_k}, \tag{B.10}$$

where E_k is the energy of the state k and C a constant, showing that $\Omega(E)$ is a very rapidly increasing function of E, compatible with all the other constraints that the system must satisfy.

Let us consider an isolated macroscopic system with constant volume, divided into two parts through a fixed, thermally conducting partition, therefore assuming that the two subsystems are free to exchange heat, but not work. Accordingly, although the total energy, $E_{tot} = E_1 + E_2$, will remain constant, the energy of subsystem 1 and 2, E_1 and E_2, will fluctuate around their equilibrium value, due to the heat $\delta Q = d E_1$, that is exchanged between the two subsystems. As we saw, the number of accessible states Ω_1 available to subsystem 1 is an exponentially increasing function of its energy E_1, the same happening to subsystem 2. Accordingly, just as we saw in the case of an ideal gas contained in two subsystems [cf. Eq. (B.8) at $x = \bar{x}$ when $N \gg 1$], the total number of accessible states $\Omega = \Omega_1 \Omega_2$ (and therefore its logarithm as well) will exhibit an extremely sharp maximum when $E_1 = \bar{E}_1$, which is the equilibrium value of the energy of subsystem 1, (obviously, the equilibrium value of subsystem 2 is given by $\bar{E}_2 = E_{tot} - \bar{E}_1$) [4, Chap. 3.3]. Therefore,

$$\frac{d \ln \Omega}{d E_1} = \frac{d \ln \Omega_1}{d E_1} + \frac{d \ln \Omega_2}{d E_1} = 0, \quad \text{i.e.} \quad \frac{d \ln \Omega_1}{d E_1} = \frac{d \ln \Omega_2}{d E_2}, \tag{B.11}$$

where we have considered that $d E_1 = -d E_2$. So, defining

$$\beta = \frac{d \ln \Omega}{d E}, \tag{B.12}$$

we obtain:

$$\beta_1 = \beta_2, \tag{B.13}$$

indicating that at thermal equilibrium the property β of the two subsystems are equal to each other. In fact, as β has the units of an inverse energy and is obviously connected with the thermodynamic temperature, it is convenient to define it so that it agrees with all previous definition of temperature as

$$\beta = \frac{1}{kT}, \tag{B.14}$$

where k is *Boltzmann constant*, $k = 1.38 \times 10^{-23} \, J/K$.

Therefore, considering that from thermodynamics,

$$\frac{1}{T} = \left(\frac{\partial S}{\partial E}\right)_V, \tag{B.15}$$

we may conclude that:

$$S = k \ln \Omega. \tag{B.16}$$

This is Boltzmann's fundamental definition of entropy, which allows to connect microscopic statistical mechanics with macroscopic thermodynamics. In fact, as we indicate below, this result will lead directly to the thermodynamic second law, as stated in Eq. (B.22). In addition, considering that, according to the postulate (B.1), the number of accessible states is proportional to the probability at local equilibrium, Eq. (B.16) can be rewritten as,

$$\Pi^{eq}(x) = C \exp\left[\frac{1}{k} S_{tot}(x)\right], \tag{B.17}$$

which is generally referred to as *Einstein's fluctuation formula* [1].

Generalizing this analysis, subdividing the system in as many subsystems as possible, we see that at equilibrium the energy of a system tends to distribute equally among all its degrees of freedom, with each degree of freedom having an average energy equal to $\frac{1}{2}kT$. This is the so-called *equipartition principle*, and is equivalent to the second law of thermodynamics [2]. In fact, with this assumption, the definition of temperature of classical thermodynamics can be obtained as follows. Consider one mole of a very dilute gas[1] contained in a volume V. From elementary considerations of kinetic theory, we see that the N_A gas particles exert on the wall a force per unit surface (i.e. a pressure P) equal to $\frac{1}{3}\frac{N_A}{V}m\langle v^2\rangle$, where m is the mass of each particle. Then, considering that from the equipartition principle $m\langle v^2\rangle = 3kT$ (as each particle has three degrees of freedom), we obtain the well known ideal

---

[1]Remind that in 1 mole of an ideal gas there are N_A molecules, where $N_A = 6.022 \times 10^{23}$ mole^{-1} is the Avogadro number.

gas law, $PV = RT$, where $R = N_A k$ is the gas constant. This coincides with the definition of temperature in classical thermodynamics, i.e.

$$T = \lim_{P \to 0} \frac{PV}{R}.$$ (B.18)

Naturally, we could carry over the same analysis assuming that the two subsystems are separated by a moving, thermally insulating, partition, so that they can exchange mechanical work, but not heat. In that case, imposing that the equilibrium volume V_1 will maximize the total number of accessible states Ω, we can derive the condition of mechanical equilibrium, as

$$\frac{d \ln \Omega}{d V_1} = \frac{d \ln \Omega_1}{d V_1} + \frac{d \ln \Omega_2}{d V_1} = 0, \quad \text{i.e.} \quad \frac{d \ln \Omega_1}{d V_1} = \frac{d \ln \Omega_2}{d V_2},$$ (B.19)

where we have considered that $dV_1 = -dV_2$. Therefore, we obtain the condition that at equilibrium the two pressures are equal, defining

$$P = \frac{1}{\beta} \frac{d \ln \Omega}{dV}.$$ (B.20)

Clearly, when our system is not at equilibrium, with the two subsystems having different temperatures, T_1 and T_2 and pressures P_1 and P_2, then a heat Q and a work W will be exchanged between the two subsystems (from higher to lower temperature and pressure) in order to equilibrate their temperature and their pressure, thus maximizing the total entropy $S_1 + S_2$.

In general, considering that,

$$d \ln \Omega = \frac{\partial \ln \Omega}{\partial E} dE + \frac{\partial \ln \Omega}{\partial V} dV,$$ (B.21)

and substituting (B.15), (B.16) and (B.20), we obtain the well known following expression:

$$dE = TdS - PdV,$$ (B.22)

where both the first and the second laws of thermodynamics are manifested.

B.2 Connection with Thermodynamics

Consider again an isolated system, with total energy E_0, composed of two subsystems in thermal interaction with one another, assuming that one of them, say A, is much smaller than the other, say R, which we will therefore call a heat reservoir, having temperature T. If A is in one definite state k, with energy E_k, the number of states accessible to the whole system is just the number of states accessible to R, say $\Omega_R(E_0 - E_k)$. However, considering that $E_k \ll E_0$, we can write:[2]

$$\ln \Omega_R(E_0 - E_k) = \ln \Omega_R(E_0) - \left(\frac{\partial \ln \Omega_R}{\partial E}\right)_0 E_k + \cdots \simeq \alpha - \beta E_k, \qquad (B.23)$$

where $\alpha = \ln \Omega_R(E_0)$ is a constant independent of k, while $\beta = 1/kT$. Accordingly, since the probability of A being in a state k is proportional to the number of accessible states under these conditions, we obtain:

$$\Pi_k^{eq} = \Pi^{eq}(E_k) = \frac{1}{Z} e^{-\beta E_k}. \qquad (B.24)$$

Here Z is a normalization constant, which can be evaluated imposing that the sum of all probabilities taken over all the accessible states, irrespective of their energy, must be equal to 1,

$$\sum_k \Pi^{eq}(E_k) = 1, \qquad (B.25)$$

obtaining:

$$Z = \sum_k e^{-\beta E_k}. \qquad (B.26)$$

Z is called "partition function" and is a fundamental quantity in statistical mechanics.[3]

Now we intend to show that all physical quantities can be expressed in terms of the partition function, Z, defined in (B.26). So, for example, the mean energy of a subsystem in thermal contact with a heat reservoir can be calculated as follows:[4]

$$\langle E \rangle = \overline{E} = \sum_k E_k \Pi_k^{eq} = \frac{\sum_k E_k e^{-\beta E_k}}{\sum_k e^{-\beta E_k}}. \qquad (B.27)$$

Accordingly, we find:

$$\overline{E} = -\frac{1}{Z}\frac{\partial Z}{\partial \beta} = -\frac{\partial \ln Z}{\partial \beta} = -T^2 \frac{\partial (F/T)}{\partial T}, \qquad (B.28)$$

where

$$F = -kT \ln Z = -kT \ln \sum_n \exp\left(-\frac{E_n}{kT}\right) \qquad (B.29)$$

---

[2] See [4, pp. 202–206].

[3] The letter Z is used because the German name is *Zustandsumme*.

[4] In general, the mean energy of a system is identified with its internal energy, provided that the contribution of the overall kinetic and potential energies can be neglected.

is the Helmholtz free energy. This equality can be easily seen considering that, as $S = -\partial F/\partial T$, Eq. (B.28) leads to the free energy definition,

$$F = \overline{E} - TS \tag{B.30}$$

and, conversely,

$$S = \frac{\overline{E}}{T} + k \ln Z. \tag{B.31}$$

This is another way to express the connection between mechanics (i.e. the energy of the states of the system) and thermodynamics. In particular, substituting (B.29) into (B.24), we obtain:

$$kT \ln \Pi_k^{eq} = F - E_k. \tag{B.32}$$

This last relation is important, as it allows us to express Boltzmann's fundamental definition of entropy, Eq. (B.16), in an equivalent form. In fact, considering that,

$$S = \frac{1}{T}(\overline{E} - F) = \frac{1}{T}\sum_k (E_k - F)\Pi_k^{eq}, \tag{B.33}$$

where we have applied (B.27), together with the normalization condition (B.25), and substituting (B.32), we obtain:

$$S = -k \sum_k \Pi_k^{eq} \ln \Pi_k^{eq}, \tag{B.34}$$

where, we remind, $\Pi_k^{eq} = \Pi^{eq}(E_k)$ is the probability (B.24) that the system has energy E_k. Therefore, considering that $\ln \Pi^{eq}(E)$ is a linear function of E, this last relation can also be written as,

$$S = -k\langle \ln \Pi^{eq}(E)\rangle = -k \ln \Pi^{eq}(\overline{E}). \tag{B.35}$$

Note that in this derivation the system is assumed to have T and V constant. Therefore, the partition function and the free energy are defined as a function of T and V, i.e., $F = F(T, V)$. Then, once we know F, all other thermodynamic functions can be determined from it, such as $P = (\partial F/\partial V)_T$, $S = -(\partial F/\partial T)_V$ and $G = F + PV$.

Proceeding in the same way, we find:

$$P = -\left\langle \frac{\partial E}{\partial V}\right\rangle = \sum_n \frac{\partial E_n}{\partial V}e^{-\beta E_n} \Big/ \sum_n e^{-\beta E_n}, \tag{B.36}$$

that is:

$$P = \frac{1}{\beta}\frac{\partial \ln Z}{\partial V}. \tag{B.37}$$

Therefore, considering that

$$d \ln Z = \frac{\partial \ln Z}{\partial V} dV + \frac{\partial \ln Z}{\partial \beta} d\beta, \tag{B.38}$$

substituting (B.28) and (B.37), we find:

$$d \ln Z = \beta P dV - E d\beta, \qquad d(\ln Z + \beta E) = \beta(\delta W + dE) = \beta \delta Q, \tag{B.39}$$

where E denotes the mean energy $\overline{E}$. Here $\delta W = -PdV$ is the work done by (and therefore exiting) the system, and we have used the definition (i.e. the first law of thermodynamics) $\delta Q = dE + \delta W$ of the heat Q entering the system. Finally, substituting (B.31) into (B.39), we obtain the second law of thermodynamics in the form,

$$\delta Q = T dS, \tag{B.40}$$

showing that although δQ is not an exact differential, an exact differential results when it is divided by T.

B.3 Free Energy of Fluids

Let us consider a fluid composed of N identical molecules[5] enclosed in a container of volume V and having a temperature T. The starting point is the expression (B.29) for the free energy,

$$F = -kT \ln \sum_n \exp\left(-\frac{E_n}{kT}\right), \tag{B.41}$$

where the sum is taken over all the possible states of the system. In general, as the total energy E is the sum of the energies of each of the N molecules, i.e. $E = \sum_k \epsilon_k$ and considering that the molecules are indistinguishable from each other, we have:

$$\sum_n \exp\left(-\frac{E_n}{kT}\right) = \frac{1}{N!}\left[\sum_k \exp\left(-\frac{\epsilon_k}{kT}\right)\right]^N, \tag{B.42}$$

where $N!$ indicates the number of permutations of the N molecules, while k represents the states of the molecules. Therefore, we obtain the following expression for the free energy:

$$F = kT \ln N! - NkT \ln \sum_k \exp\left(-\frac{\epsilon_k}{kT}\right). \tag{B.43}$$

---

[5]See [3, Chap. 74].

Since $N \gg 1$, applying Sterling's approximation formula, $\ln N! = N \ln N - N$, we write:

$$F = -NkT \ln \left[\frac{e}{N} \sum_k \exp\left(-\frac{\epsilon_k}{kT}\right) \right]. \tag{B.44}$$

In general, the energy of a molecule is the sum of a kinetic and a potential part,

$$\epsilon = \frac{p^2}{2m} + \psi, \tag{B.45}$$

where $\mathbf{p}$ is the momentum and ψ the potential energy.

First, let us consider the case of an ideal gas, where $\psi = 0$, i.e. the molecules do not interact with each other. Converting the summation into an integral we apply the following rule,

$$\sum_k (\ldots) = \frac{1}{h^3} \int_V \int_{-\infty}^{+\infty} (\ldots) \, d^3\mathbf{p} \, d^3\mathbf{r}, \tag{B.46}$$

where the volume integral is taken over the volume occupied by the fluid (in this case, it is the molar volume), while the momentum integral is taken over all the possible momenta, i.e. from $-\infty$ to $+\infty$. Here h is a normalization constant, representing the volume, in phase space, occupied by a single state[6] having the units of an action, which, as we will see, has no importance for our purposes. In order to determine it, though, we must apply quantum mechanics, finding that it coincides with Planck's constant. Then we find:

$$F_{id} = -NkT \ln \frac{e}{Nh^3} \int \int \exp\left(-\frac{p^2}{2mkT}\right) d^3\mathbf{p} \, d^3\mathbf{r}, \tag{B.47}$$

that is,

$$F_{id} = -NkT \ln \frac{eV}{N} \left(\frac{2\pi mkT}{h^2}\right)^{3/2} = -NkT \ln V + \text{funct}(T). \tag{B.48}$$

Clearly, from here we obtain the equation state of an ideal gas,

$$P = -\left(\frac{\partial f_{id}}{\partial V}\right)_T = \frac{NkT}{V}. \tag{B.49}$$

For real gases, adding the contribution of the potential energy, $\psi = \psi(\mathbf{r}_1, \ldots, \mathbf{r}_N)$ and denoting by N_A the Avogadro number, we obtain:

$$F = F_{id} - kT \ln \left\{ C N_A^N \int_V \cdots \int_V e^{-\psi/kT} \rho(\mathbf{r}_1) \, d^3\mathbf{r}_1 \ldots \rho(\mathbf{r}_N) \, d^3\mathbf{r}_N \right\}, \tag{B.50}$$

---

[6]In case of degeneracy, we simply multiply the result by a constant.

where ρ is the molar density, so that $N_A \rho(\mathbf{r}) d^3 \mathbf{r}$ is the number of molecules in the volume $d^3 \mathbf{r}$, while $\mathcal{C}$ is an appropriate constant that avoids to count twice the same molecule. Now, adding and subtracting 1, and assuming pairwise interactions, i.e. the potential energy ψ depends only on the distance r between two molecules, we obtain:

$$F = F_{id} - kT \ln \left\{ \frac{N_A^2}{2} \int_V \int_V \left[e^{-\psi(r)/kT} - 1 \right] \rho(\mathbf{r}_1) \rho(\mathbf{r}_2) \, d\mathbf{r}_1 \, d\mathbf{r}_2 + 1 \right\}, \quad (B.51)$$

where $r = |\mathbf{r}_1, -\mathbf{r}_2|$, while $N_A^2/2 \simeq N_A(N_A - 1)/2$ is the number of ways we can choose a couple within a sample of N_A molecules. At this point, assuming that the integral is much smaller than 1, since $\ln(1 + \epsilon) \simeq \epsilon$, we obtain:

$$F = F_{id} + \frac{kT N_A^2}{2} \int_V \int_V \left[1 - e^{-\psi(r)/kT} \right] \rho(\mathbf{r}_1) \rho(\mathbf{r}_2) \, d\mathbf{r}_1 \, d\mathbf{r}_2. \quad (B.52)$$

Finally, denoting $\mathbf{x} = \mathbf{r}_1$ and $\mathbf{r} = \mathbf{r}_2 - \mathbf{r}_1$, and defining the local molar free energy f as

$$F = \int_V f(\mathbf{x}) \, d^3 \mathbf{x}, \quad (B.53)$$

we obtain:

$$f(\mathbf{x}) = f_{id}(\mathbf{x}) + \frac{1}{2} RT N_A \int_V \left[1 - e^{-\psi(r)/kT} \right] \rho(\mathbf{x} + \mathbf{r}) \, d^3 \mathbf{r}, \quad (B.54)$$

where

$$f_{id}(\mathbf{x}) = RT \ln \rho(\mathbf{x}) + \text{funct}(T). \quad (B.55)$$

The expression (B.54) is the starting point to derive the diffuse interface model of multiphase flows.

References

1. Einstein, A.: Ann. d. Physik **33**, 1275 (1910)
2. Hatsopoulos, G.N., Keenan, J.H.: Principles of General Thermodynamics. Wiley, New York (1965)
3. Landau, L.D., Lifshitz, E.M.: Statistical Physics, Part I. Pergamon, Elmsford (1980)
4. Reif, F.: Fundamentals of Statistical and Thermal Physics. McGraw Hill, New York (1965)

Appendix C
Principle of Causality

The causality principle states that an effect may never precede in time its cause. In this appendix, we see how this trivial statement leads us to derive a general relation (i.e. the Kramers-Kronig relations) that must be satisfied by any susceptibility.

C.1 Correlation Functions

Denote by $\widehat{\sigma}_{ik}^{xx}(\omega)$ the Fourier transform of the correlation function $\sigma_{ik}^{xx}(t)$, defined in (2.15), i.e.,

$$\widehat{\sigma}_{ik}^{xx}(\omega) = \int_{-\infty}^{\infty} \sigma_{ik}^{xx}(t)e^{i\omega t}\,dt, \tag{C.1}$$

where

$$\sigma_{ik}^{xx}(t) = \langle x_i(t)x_k(0)\rangle = \int_{-\infty}^{\infty} \widehat{\sigma}_{ik}^{xx}(\omega)e^{-i\omega t}\,\frac{d\omega}{2\pi}. \tag{C.2}$$

From the definition (C.1) and considering that $\sigma_{ik}^{xx}(t)$ is real, we see that

$$\widehat{\sigma}_{ik}^{xx}(\omega) = \widehat{\sigma}_{ki}^{xx*}(\omega) = \widehat{\sigma}_{ik}^{xx}(-\omega). \tag{C.3}$$

where the asterisk denotes complex conjugate. In addition, from the condition (2.16) of microscopic reversibility, i.e. $\sigma_{ik}^{xx}(t) = \sigma_{ik}^{xx}(-t)$, we obtain the following relation:

$$\widehat{\sigma}_{ik}^{xx}(\omega) = \widehat{\sigma}_{ki}^{xx}(\omega). \tag{C.4}$$

Now, from Eqs. (C.3)–(C.4) we see that $\widehat{\sigma}_{ik}^{xx}(\omega)$ is a real and even (i.e. its imaginary and odd parts are zero), symmetric tensor, i.e.,

$$\widehat{\sigma}_{ik}^{xx}(\omega) = \widehat{\sigma}_{ik}^{xx*}(\omega) = \widehat{\sigma}_{ik}^{xx}(-\omega) = \widehat{\sigma}_{ki}^{xx}(\omega). \tag{C.5}$$

R. Mauri, *Non-Equilibrium Thermodynamics in Multiphase Flows*,
Soft and Biological Matter, DOI 10.1007/978-94-007-5461-4,
© Springer Science+Business Media Dordrecht 2013

Accordingly,

$$\sigma_{ik}^{xx}(\tau) = 2 \int_0^\infty \widehat{\sigma}_{ik}^{xx}(\omega) \cos(\omega\tau) \frac{d\omega}{2\pi}. \tag{C.6}$$

In particular,

$$\sigma_{ik}^{xx}(0) = \langle x_i x_k \rangle = k g_{ik}^{-1} = 2 \int_0^\infty \widehat{\sigma}_{ik}^{xx}(\omega) \frac{d\omega}{2\pi}. \tag{C.7}$$

C.2   Kramers-Kronig Relations

As we saw in the previous chapter, near equilibrium generalized forces, $\mathbf{X}$ and generalized displacements $\mathbf{x}$ are linearly related. For time-dependent processes, this relation can be generalized as the following convolution integral:

$$x_i(t) = \sum_{k=1}^{n} \int_0^\infty \kappa_{ik}(t - t') X_k(t') \, dt', \tag{C.8}$$

where $\kappa_{ik}(t)$ is a time dependent generalized susceptibility. Since the effect may not precede in time its cause, we have:

$$\kappa_{ik}(t) = 0 \quad \text{for } t < 0. \tag{C.9}$$

Now, when we consider the Fourier transform of $\mathbf{x}$ and $\mathbf{X}$, the convolution integral (C.8) becomes the simple relation,

$$\widehat{x}_i(\omega) = \sum_{k=1}^{n} \widehat{\kappa}_{ik}(\omega) \widehat{X}_k(\omega). \tag{C.10}$$

At steady state, we know that [cf. Eq. (1.17)]

$$\widehat{\kappa}_{ik}(0) = -g_{ik}^{-1}. \tag{C.11}$$

Let us see what effect the causality condition (C.9) has on the susceptibility matrix. For this purpose, we extend the definition of Fourier transform (C.1) to complex values $\omega = \omega^{(r)} + i\omega^{(i)}$ of the argument. Since $\kappa_{ik}(t)$ vanishes for negative times, the Fourier integral (C.1) can be rewritten between 0 and ∞ as:

$$\kappa_{ik}(\omega) = \int_0^\infty \kappa_{ik}(t) e^{i\omega t} \, dt = \int_0^\infty \kappa_{ik}(t) e^{i\omega^{(r)}t - \omega^{(i)}t} \, dt. \tag{C.12}$$

For positive values of $\omega^{(i)}$, this integral exists and is finite, and it tends to zero as $\omega^{(i)} \to \infty$. Therefore, the statement equivalent to the causality condition (C.9) is:

The function $\widehat{\kappa}(\omega)$, with $\omega = \omega^{(r)} + i\omega^{(i)}$, has no poles (i.e. no singular points)
in the upper half of the complex plane and tends to zero in the limit as $\omega^{(i)} \to \infty$.

Now we apply Cauchy's theorem, stating that, given a closed contour in the complex plane, any function $f(\omega)$ having no poles inside that contour satisfies the relation:

$$\oint f(\omega)\,d\omega = 0, \tag{C.13}$$

where the integral is taken along the contour (e.g. in counter clockwise direction). We shall apply this theorem to the function,

$$f(\omega) = \frac{\widehat{\kappa}_{ik}(\omega)}{\omega - u}, \tag{C.14}$$

where u is real. The contour we choose include the whole upper half of the complex plane (in the lower half there might be poles). It extends along the whole real axis, avoiding the point $w = u$ (which is a pole of the function $f(\omega)$) with a semi-circle of radius r in the upper half of the complex plane and is closed by a semi-infinite semi-circle, also in the upper half of the complex plane. Inside this contour, the function $f(\omega)$ has no poles and therefore we can apply Cauchy's theorem,

$$\int_{-\infty}^{u-r} \frac{\widehat{\kappa}_{ik}(\omega)}{\omega - u}\,d\omega + \int_{u+r}^{\infty} \frac{\widehat{\kappa}_{ik}(\omega)}{\omega - u}\,d\omega + \int_{C_r} \frac{\widehat{\kappa}_{ik}(\omega)}{\omega - u}\,d\omega = 0, \tag{C.15}$$

where C_r is the small semicircle with radius r around the pole $\omega = u$, passed in clock-wise direction, and we have considered that the line integral along the infinite semicircle vanishes, since $\widehat{\kappa}_{ik}(\omega)$ tends exponentially to zero as $\omega^{(i)} \to \infty$. In the limit as $r \to 0$, the first two integrals together reduce to the so-called principal part of the integral from $-\infty$ to ∞. The last integral can be easily evaluated considering that $\omega = u + re^{i\theta}$ (here θ denotes the angle between the radius of the circle and the positive side of the real axis) as follows,

$$\int_{C_r} \frac{\widehat{\kappa}_{ik}(\omega)}{\omega - u}\,d\omega = \lim_{r \to 0} \int_{\pi}^{0} \widehat{\kappa}_{ik}(u + re^{i\theta})\,i\,d\theta = -i\pi\widehat{\kappa}_{ik}(u). \tag{C.16}$$

Therefore we obtain:

$$\widehat{\kappa}_{ik}(u) = \frac{1}{i\pi}\mathbb{P}\int_{-\infty}^{\infty} \frac{\widehat{\kappa}_{ik}(\omega)}{\omega - u}\,d\omega, \tag{C.17}$$

where the symbol $\mathbb{P}$ denotes the principal value integral. Equation (C.17) is referred to as the Kramers-Kronig relation, which can be considered as a direct consequence of the causality condition (C.9).

Applying again Cauchy's theorem to the function,

$$f(\omega) = \frac{e^{i\omega|t|}\widehat{\kappa}_{ik}(\omega)}{\omega - u}, \tag{C.18}$$

we obtain the similar relation:

$$e^{iu|t|}\widehat{\kappa}_{ik}(u) = \frac{1}{i\pi}\mathbb{P}\int_{-\infty}^{\infty}\frac{e^{i\omega|t|}\widehat{\kappa}_{ik}(\omega)}{\omega - u}\,d\omega. \tag{C.19}$$

For $t = 0$, (C.19) reduces to (C.17). In particular, when $u = 0$, this relation becomes:

$$\widehat{\kappa}_{ik}(0) = \frac{1}{i\pi}\mathbb{P}\int_{-\infty}^{\infty}\frac{e^{i\omega|t|}\widehat{\kappa}_{ik}(\omega)}{\omega}\,d\omega. \tag{C.20}$$

Now, let us see what happens when a constant driving force F_i is applied at times $t < 0$, while it is lifted at $t = 0$, i.e. when,

$$X_i(t) = F_i H(-t), \tag{C.21}$$

where $H(t)$ is Heaviside's step function, i.e.

$$H(t) = \begin{Bmatrix} 1 & \text{when } t > 0 \\ 0 & \text{when } t < 0 \end{Bmatrix}. \tag{C.22}$$

From the definition of the Fourier integral, we obtain:

$$\widehat{X}_i(\omega) = F_i\left[\pi\delta(\omega) + \mathbb{P}\frac{1}{i\omega}\right]. \tag{C.23}$$

Now substitute these results into

$$x_i(t) = \sum_{k=1}^{n}\int_{-\infty}^{\infty}e^{-i\omega t}\widehat{\kappa}_{ik}(\omega)\widehat{X}_k(\omega)\frac{d\omega}{2\pi}, \tag{C.24}$$

and obtain:

$$x_i(t) = \frac{1}{2}\sum_{k=1}^{n}\left[\widehat{\kappa}_{ik}(0) + \frac{1}{i\pi}\mathbb{P}\int_{-\infty}^{\infty}e^{-i\omega t}\frac{\widehat{\kappa}_{ik}(\omega)}{\omega}\,d\omega\right]F_k. \tag{C.25}$$

Now, for negative times, applying (C.20), this relation becomes:

$$x_i(t) = \sum_{k=1}^{n}\widehat{\kappa}_{ik}(0)F_k \quad (t < 0). \tag{C.26}$$

This is a trivial result, indicating that for $t < 0$ the constant driving force induces a constant response. On the other hand, for positive times, substituting (C.20) into (C.24) we find:

$$x_i(t) = \frac{1}{i\pi}\sum_{k=1}^{n}\mathbb{P}\int_{-\infty}^{\infty}\cos(\omega t)\frac{\widehat{\kappa}_{ik}(\omega)}{\omega}\,d\omega F_k \quad (t > 0). \tag{C.27}$$

Therefore, in this case, we can define a generalized time-dependent susceptibility, $\chi_{ik}(t)$, such that

$$x_i(t) = \sum_{k=1}^{n} \chi_{ik}(t) F_k,$$ (C.28)

with

$$\chi_{ik}(t) = \begin{cases} \widehat{\kappa}_{ik}(0) & \text{if } t < 0 \\ \frac{1}{i\pi} \mathbb{P} \int_{-\infty}^{\infty} \cos(\omega t) \frac{\widehat{\kappa}_{ik}(\omega)}{\omega} d\omega & \text{if } t > 0 \end{cases}.$$ (C.29)

Formally, that means that for practical purposes we can write:

$$\widehat{\chi}_{ik} = \frac{2}{i\omega} \widehat{\kappa}_{ik}(\omega).$$ (C.30)

Appendix D
Review of Analytical Mechanics

Analytical mechanics was developed in the 19th century as an elegant generalization of Newton's mechanics. In Sects. D.1, D.3 and D.4 of this appendix, we see that in both Lagrangian and Hamiltonian approaches energy, instead of force, is the fundamental quantity. This allows to see clearly the connection between invariance with respect to a certain variable and the conservation of the associated momentum (Sect. D.2). Finally, in Sect. D.5, we present a generalization to non conservative systems.

D.1 Lagrangian Approach

Consider an isolated, conservative system composed of N point particles. The motion of any particle i can be described through *Newton*'s equation,

$$\mathbf{F}_i = m\ddot{\mathbf{r}}_i, \tag{D.1}$$

where $\mathbf{r}_i$ is the position vector of point i, referred to a fixed reference frame, $\ddot{\mathbf{r}}_i = d^2\mathbf{r}_i/dt^2$ is the acceleration, while $\mathbf{F}_i = -\partial V/\partial \mathbf{r}_i$ is the force, which here is expressed as the gradient of a position-dependent potential energy $V(\mathbf{r})$. Newton's equations consist of second-order ordinary differential equations, which can be integrated by knowing two conditions. Therefore if, at any instant of time, for each particle we know the position and the velocity, then we can determine their trajectories in the future and in the past.

Now, multiply Eq. (D.1) by any, so called, virtual displacement $\delta \mathbf{r}_i$ of the system, describing a movement that is consistent with all the constraints.[1] Therefore, we

---

[1] A virtual displacement is an infinitesimal change of the system coordinates occurring while time is held constant; it is called virtual rather than real since no actual displacement can take place without the passage of time.

R. Mauri, *Non-Equilibrium Thermodynamics in Multiphase Flows*,
Soft and Biological Matter, DOI 10.1007/978-94-007-5461-4,
© Springer Science+Business Media Dordrecht 2013

obtain the so-called *d'Alembert*'s principle of virtual works:[2]

$$\sum_{i=1}^{N} \mathbf{F}_i \cdot \delta \mathbf{r}_i = -\delta\mathcal{V} = \sum_{i=1}^{N} m_i \ddot{\mathbf{r}}_i \cdot \delta \mathbf{r}_i, \tag{D.2}$$

where the negative change of potential energy, $-\delta\mathcal{V}$, equals the virtual work of all active forces.

When the system is subjected to time independent constraints (e.g. the distance between two particles is constant), we will have only $n < 3N$ degrees of freedom, corresponding to n independent generalized coordinates, q_j, with $j = 1, 2, \ldots, n$, and we can write

$$\mathbf{r}_i = \mathbf{r}_i(q_1, q_2, \ldots, q_n); \qquad \delta \mathbf{r}_i = \sum_{j=1}^{n} \frac{\partial \mathbf{r}_i}{\partial q_j} \delta q_j. \tag{D.3}$$

Therefore, substituting (D.3) into (D.2) we have:

$$-\delta\mathcal{V} = -\sum_{j=1}^{n} \frac{\partial \mathcal{V}}{\partial q_j} \delta q_j = \sum_{i=1}^{N} \sum_{j=1}^{n} m_i \ddot{\mathbf{r}}_i \cdot \frac{\partial \mathbf{r}_i}{\partial q_j} \delta q_j, \tag{D.4}$$

and considering that the virtual generalized displacements δq_j are arbitrary, we obtain:

$$\sum_{i=1}^{N} m_i \ddot{\mathbf{r}}_i \cdot \frac{\partial \mathbf{r}_i}{\partial q_j} + \frac{\partial \mathcal{V}}{\partial q_j} = 0. \tag{D.5}$$

The first term on the LHS of the above equation can be rewritten as,

$$\sum_{i=1}^{N} m_i \ddot{\mathbf{r}}_i \cdot \frac{\partial \mathbf{r}_i}{\partial q_j} = \left[\frac{d}{dt} \left(\frac{\partial \mathcal{T}}{\partial \dot{q}_j} \right) - \frac{\partial \mathcal{T}}{\partial q_j} \right], \tag{D.6}$$

where $\mathcal{T}$ is the kinetic energy of the system,

$$\mathcal{T} = \frac{1}{2} \sum_{i=1}^{N} m_i |\dot{\mathbf{r}}_i|^2. \tag{D.7}$$

The proof of this statement can be broken down in three steps. First, substituting (D.3), we have:

$$\frac{\partial \mathcal{T}}{\partial \dot{q}_j} = \sum_{i=1}^{N} m_i \dot{\mathbf{r}}_i \cdot \frac{\partial \dot{\mathbf{r}}_i}{\partial \dot{q}_j}. \tag{D.8}$$

---

[2] When applied to non-conservative systems with time-dependent constraints, d'Alembert principle is not as trivial as it is in this case.

Then, considering that the operators of virtual variation, δ, and that of time differentiation, d/dt, are commutable, as they are independent from each other, we can write:

$$\delta \dot{\mathbf{r}}_i = \frac{d}{dt}\delta \mathbf{r}_i = \sum_{j=1}^{n} \frac{d}{dt}\left(\frac{\partial \mathbf{r}_i}{\partial q_j}\right)\delta q_j + \sum_{j=1}^{n} \frac{\partial \mathbf{r}_i}{\partial q_j}\delta \dot{q}_j, \qquad (D.9)$$

and therefore:

$$\frac{\partial \dot{\mathbf{r}}_i}{\partial q_j} = \frac{d}{dt}\left(\frac{\partial \mathbf{r}_i}{\partial q_j}\right); \qquad \frac{\partial \dot{\mathbf{r}}_i}{\partial \dot{q}_j} = \frac{\partial \mathbf{r}_i}{\partial q_j}. \qquad (D.10)$$

Finally, substituting this last relation into (D.8) and time differentiating, we obtain:

$$\frac{d}{dt}\left(\frac{\partial \mathcal{T}}{\partial \dot{q}_j}\right) = \sum_{i=1}^{N} m_i \ddot{\mathbf{r}}_i \cdot \frac{\partial \mathbf{r}_i}{\partial q_j} + \sum_{i=1}^{N} m_i \dot{\mathbf{r}}_i \cdot \frac{\partial \dot{\mathbf{r}}_i}{\partial q_j}, \qquad (D.11)$$

where we have considered that the operators d/dt and $\partial/\partial q_j$ commute with each other, as seen in (D.10). At the end, substituting the equality

$$\frac{\partial \mathcal{T}}{\partial q_j} = \sum_{i=1}^{N} m_i \dot{\mathbf{r}}_i \cdot \frac{\partial \dot{\mathbf{r}}_i}{\partial q_j}. \qquad (D.12)$$

into Eq. (D.11), we obtain Eq. (D.6).

The previous findings, Eqs. (D.4) and (D.6), can be reinterpreted defining a new function, $\mathcal{L}$, called *Lagrangian*, as

$$\mathcal{L}(\mathbf{q}, \dot{\mathbf{q}}) = \mathcal{T}(\mathbf{q}, \dot{\mathbf{q}}) - \mathcal{V}(\mathbf{q}), \qquad (D.13)$$

obtaining:

$$\frac{d}{dt}\left(\frac{\partial \mathcal{L}}{\partial \dot{q}_j}\right) - \frac{\partial \mathcal{L}}{\partial q_j} = 0. \qquad (D.14)$$

These are the equations of *Lagrange*. They constitute a system of n second-order differential equations and are expressed in terms of the coordinates q_j ($j = 1, 2, \ldots, n$). Although Lagrange's equation is equivalent to Newton's equation of motion,[3] in many problems it is more useful, as it is easier to write down an expression for the energy of a system instead of recognizing all the various forces (see Comment D.3 below).

Comment D.1 The Lagrangian is defined within the time derivative of an arbitrary function of time and position, $\dot{f}(\mathbf{q}, t)$. This can be easily seen, considering that

$$\frac{d}{dt}\left(\frac{\partial \dot{f}}{\partial \dot{q}_j}\right) - \frac{\partial \dot{f}}{\partial q_j} = \frac{d}{dt}\left(\frac{\partial f}{\partial q_j}\right) - \frac{\partial \dot{f}}{\partial q_j} = 0, \qquad (D.15)$$

where Eq. (D.10) has been applied.

---

[3]Trivially, with no constrains, when $\mathbf{q} = \mathbf{r}$, applying (D.14) we obtain again (D.1).

Comment D.2 When we express the "old" coordinates in terms of the generalized coordinates, applying Eq. (D.3) we see that the kinetic energy becomes:

$$T(\mathbf{q}, \dot{\mathbf{q}}) = \frac{1}{2} \sum_{j,k=1}^{n} a_{jk}(\mathbf{q}) \dot{q}_j \dot{q}_k; \quad a_{jk}(\mathbf{q}) = \sum_{i=1}^{N} m_i \frac{\partial \mathbf{r}_i}{\partial q_j} \cdot \frac{\partial \mathbf{r}_i}{\partial q_k}. \tag{D.16}$$

This shows that, while the potential energy is only a function of the generalized coordinates, the kinetic energy (a) is a quadratic function of the generalized velocities; (b) depends on the generalized coordinates as well.

Comment D.3 As an example, consider the 2D motion of a single point particle in radial coordinates (r, ϕ). We see that $|\dot{\mathbf{r}}|^2 = \dot{r}^2 + r^2 \dot{\phi}^2$. Therefore, denoting $q_1 = r$ and $q_2 = \phi$, we obtain the expression (D.16) for the kinetic energy, with $a_{11} = m$, $a_{12} = a_{21} = 0$ and $a_{22} = mr^2$. Now, if the particle has mass m and is immersed in a Coulomb force field, with $V(r) = -K/r$, where K is a constant, the Lagrangian is:

$$\mathcal{L} = \frac{m}{2}\left(\dot{r}^2 + r^2 \dot{\phi}^2\right) + \frac{K}{r}. \tag{D.17}$$

Therefore, Lagrange's equations (D.14) give:

$$\frac{d}{dt}\left(\frac{\partial \mathcal{L}}{\partial \dot{r}}\right) - \frac{\partial \mathcal{L}}{\partial r} = 0 \quad \Rightarrow \quad \frac{d}{dt}(m\dot{r}) = mr\dot{\phi}^2 - \frac{K}{r^2}, \tag{D.18}$$

and

$$\frac{d}{dt}\left(\frac{\partial \mathcal{L}}{\partial \dot{\phi}}\right) - \frac{\partial \mathcal{L}}{\partial \phi} = 0 \quad \Rightarrow \quad \frac{d}{dt}(mr^2\dot{\phi}) = 0. \tag{D.19}$$

This latter equation shows that $mr^2\dot{\phi}$, denoting angular momentum, is a conserved quantity.

Clearly, if we try to derive (D.18) and (D.19) from Newton's equation of motion, we would have a much more difficult time.

D.2 Conservation Laws

In general, integration of the Lagrange's equations of motion require $2n$ constants. Expressing them in terms of $\mathbf{q}$ and $\dot{\mathbf{q}}$ at any time, we see that during the evolution of our system there are $2n$ quantities that are conserved. Among them, though, energy and momentum conservation are particularly important as they derive from symmetry considerations.

Conservation of Energy   The conservation of energy derives from the homogeneity of time.

Imposing that the Lagrangian does not change if time t is shifted by a constant amount $a = \delta t$, i.e.

$$\delta \mathcal{L} = \frac{\partial \mathcal{L}}{\partial t} a = 0 \quad \Rightarrow \quad \frac{\partial \mathcal{L}}{\partial t} = 0, \tag{D.20}$$

wee see that $\mathcal{L}$ cannot depend explicitly on time. Accordingly, considering that

$$\frac{d\mathcal{L}}{dt} = \frac{\partial \mathcal{L}}{\partial t} + \sum_j \left(\frac{\partial \mathcal{L}}{\partial q_j} \dot{q}_j + \frac{\partial \mathcal{L}}{\partial \dot{q}_j} \ddot{q}_j \right), \tag{D.21}$$

and substituting (D.20), together with Lagrange's equations (D.14), we obtain:

$$\frac{d\mathcal{L}}{dt} = \sum_j \left(\dot{q}_j \frac{d}{dt} \frac{\partial \mathcal{L}}{\partial \dot{q}_j} + \ddot{q}_j \frac{\partial \mathcal{L}}{\partial \dot{q}_j} \right) = \sum_j \frac{d}{dt} \left(\dot{q}_j \frac{\partial \mathcal{L}}{\partial \dot{q}_j} \right). \tag{D.22}$$

Therefore,

$$\frac{d\mathcal{H}}{dt} = 0; \tag{D.23}$$

where[4]

$$\mathcal{H} = \sum_j \left(\dot{q}_j \frac{\partial \mathcal{L}}{\partial \dot{q}_j} - \mathcal{L} \right), \tag{D.24}$$

revealing that $\mathcal{H}$ is conserved, i.e. it remains invariant during the evolution of the system.

The function $\mathcal{H}$ is called *Hamiltonian* and coincides with the total energy of the system. In fact, considering that the kinetic energy is a quadratic function of the generalized velocities [cf. Eq. (D.16)], we obtain:[5]

$$\sum_j \dot{q}_j \frac{\partial \mathcal{L}}{\partial \dot{q}_j} = \sum_j \dot{q}_j \frac{\partial T}{\partial \dot{q}_j} = 2T, \tag{D.25}$$

and therefore we see that

$$\mathcal{H} = T + V, \tag{D.26}$$

i.e. the Hamiltonian is the sum of the kinetic and potential energy.

Conservation of Momentum and Angular Momentum   The conservation of momentum and angular momentum derive from the homogeneity and isotropy of space.

---

[4]With the symbol $\mathcal{H}$ here we indicate also the Hamiltonian of the system (see next section) which coincides with this function, but it is expressed in terms of generalized coordinates and momenta (instead of velocities).

[5]This can be seen as a result of Euler's theorem.

Imposing that, as a consequence of space homogeneity, the Lagrangian does not change if the origin of our reference frame is shifted by a constant amount $\mathbf{a} = \delta\mathbf{r}$, we obtain:

$$\delta\mathcal{L} = \sum_{i=1}^{N} \frac{\partial\mathcal{L}}{\partial\mathbf{r}_i} \cdot \mathbf{a} = 0 \quad \Rightarrow \quad \sum_{i=1}^{N} \frac{\partial\mathcal{L}}{\partial\mathbf{r}_i} = 0. \tag{D.27}$$

Considering that $\partial\mathcal{L}/\partial\mathbf{r}_i = -\partial V/\partial\mathbf{r}_i = \mathbf{F}_i$ represents the force acting on the i-th particle, the expression (D.27) indicates that the sum of all the forces acting on the particles of an isolated system must vanish, i.e.,

$$\sum_{i=1}^{N} \mathbf{F}_i = \mathbf{0}. \tag{D.28}$$

Now, substituting Lagrange's equations into (D.27), we obtain:

$$\sum_{i=1}^{N} \frac{d}{dt} \frac{\partial\mathcal{L}}{\partial\dot{\mathbf{r}}_i} = \frac{d}{dt} \sum_{i=1}^{N} \frac{\partial\mathcal{L}}{\partial\dot{\mathbf{r}}_i} = 0. \tag{D.29}$$

In our case, though, $\partial\mathcal{L}/\partial\dot{\mathbf{r}}_i = m_i\dot{\mathbf{r}}_i$ is the momentum $\mathbf{p}_i$ of the i-th particle. Therefore, (D.29) can be rewritten as:

$$\frac{d\mathbf{P}}{dt} = 0; \quad \mathbf{P} = \sum_{i=1}^{N} \mathbf{p}_i; \quad \mathbf{p}_i = \frac{\partial\mathcal{L}}{\partial\dot{\mathbf{r}}_i}, \tag{D.30}$$

indicating that the total momentum of the system is conserved. This relation also reveals that momentum is an additive quantity, as the total momentum is the sum of the momenta of each particle.

The same procedure can be applied imposing that, as a consequence of space isotropy, the Lagrangian does not change if our reference frame is rotated by a constant angle $\delta\phi$, so that each point i is shifted by an amount $\delta\mathbf{r}_i = \mathbf{r}_i \times \delta\phi$. At the end, in place of (D.28), we obtain:

$$\sum_{i=1}^{N} \mathbf{\Gamma}_i = \mathbf{0}; \quad \mathbf{\Gamma}_i = \mathbf{r}_i \times \mathbf{F}_i, \tag{D.31}$$

indicating that the sum of all the torques acting on the particles of an isolated system must vanish. Also, considering that $\partial\mathcal{L}/\partial\dot{\phi}_i = \mathbf{r}_i \times m_i\dot{\mathbf{r}}_i$ is the angular momentum $\mathbf{m}_i$ of the i-th particle, in place of Eq. (D.30) we obtain:

$$\frac{d\mathbf{M}}{dt} = 0; \quad \mathbf{M} = \sum_{i=1}^{N} \mathbf{m}_i; \quad \mathbf{m}_i = \frac{\partial\mathcal{L}}{\partial\dot{\phi}_i}, \tag{D.32}$$

indicating that the total angular momentum of the system is conserved.

Conservation of a Generalized Momentum When the system is described in terms of generalized coordinates q_j, we can define the generalized momentum p_j,

$$p_j = \frac{\partial \mathcal{L}}{\partial \dot{q}_j},$$ (D.33)

and the generalized force F_j,

$$F_j = \frac{\partial \mathcal{L}}{\partial q_j}.$$ (D.34)

Accordingly, Lagrange's equations (D.14) can be rewritten as:

$$\dot{p}_j = F_j.$$ (D.35)

Therefore, when the Lagrangian does not depend explicitly on a coordinate q_k, i.e. $F_k = \partial \mathcal{L}/\partial q_k = 0$, then we see that $dp_k/dt = 0$, that is the conjugated momentum p_k is conserved. Conservation of energy, momentum and angular momentum can be considered as particular cases of this statement. This property is generally referred to as Noether's theorem.

In general, from (D.16) we see that, although the generalized momentum p_j is a linear function of the generalized velocities $\dot{q}_j$, it does not always reduce to a simple product of a mass by a velocity. In fact, as generalized coordinates and momenta can be transformed through canonical transformations [2], denoting q_j as coordinate and p_j as momentum is only a question of nomenclature.

D.3 Hamiltonian Approach

When the Lagrangian $\mathcal{L}$ is not an explicit function of time, we obtain:

$$d\mathcal{L} = \sum_j \left(\frac{\partial \mathcal{L}}{\partial q_j} dq_j + \frac{\partial \mathcal{L}}{\partial \dot{q}_j} d\dot{q}_j \right),$$ (D.36)

that is, applying (D.33)–(D.35),

$$d\mathcal{L} = \sum_j (\dot{p}_j dq_j + p_j d\dot{q}_j).$$ (D.37)

Now, consider the definition (D.24) of the total energy of a system,

$$\mathcal{H} = \sum_j p_j \dot{q}_j - \mathcal{L},$$ (D.38)

where we have applied the definition (D.33) of momentum. The function $\mathcal{H}$ is called the Hamiltonian of a system and is expressed in terms of coordinates and momenta, i.e. $\mathcal{H} = \mathcal{H}(q_j, p_j)$. Differentiating (D.38) and substituting (D.37) we obtain:

$$d\mathcal{H} = \sum_j (-\dot{p}_j dq_j + \dot{q}_j dp_j). \tag{D.39}$$

From this expression we derive *Hamilton's equations*,

$$\dot{q}_j = \frac{\partial \mathcal{H}}{\partial p_i}; \qquad \dot{p}_j = -\frac{\partial \mathcal{H}}{\partial q_i}. \tag{D.40}$$

Hamilton's equations are expressed in terms $2n$ variables, q_j and p_j ($j = 1, 2, \ldots, n$) and constitute a system of $2n$ first-order differential equations. From a practical point of view, in most cases they are not as convenient to implement as Lagrange's equations which, as we saw, constitute a set of n second-order differential equations. However, due to their symmetry and elegance, they provide a very important insight into the intrinsic laws of mechanics. For that reason, they are often called *canonical* equations.

In the most general case where $\mathcal{H} = \mathcal{H}(q_j, p_j, t)$ and $\mathcal{L} = \mathcal{L}(q_j, \dot{q}_j, t)$ depend explicitly on time, we have,

$$\frac{d\mathcal{H}}{dt} = \frac{\partial \mathcal{H}}{\partial t} + \sum_j \left(\frac{\partial \mathcal{H}}{\partial q_j} \dot{q}_j + \frac{\partial \mathcal{H}}{\partial p_j} \dot{p}_j \right), \tag{D.41}$$

and applying the canonical equations (D.40) we obtain:

$$\frac{d\mathcal{H}}{dt} = \frac{\partial \mathcal{H}}{\partial t}. \tag{D.42}$$

In addition, in the expression (D.37) we would have the additional term, $(\partial \mathcal{L}/\partial t)dt$, which would be subtracted from (D.39), so that at the end we obtain:

$$\left(\frac{\partial \mathcal{H}}{\partial t} \right)_{q,p} = -\left(\frac{\partial \mathcal{L}}{\partial t} \right)_{q,\dot{q}}. \tag{D.43}$$

Here, we have stressed that the time derivative on the LHS must be performed with q_j and p_j constant, while that on the RHS have q_j and $\dot{q}_j$ constant.

The procedure above can be generalized to any function $F(q_j, p_j, t)$, i.e.,

$$\frac{dF}{dt} = \frac{\partial F}{\partial t} + \sum_j \left(\frac{\partial F}{\partial q_j} \dot{q}_j + \frac{\partial F}{\partial p_j} \dot{p}_j \right), \tag{D.44}$$

and applying the canonical equations (D.40) we obtain:

$$\frac{dF}{dt} = \frac{\partial F}{\partial t} + \{\mathcal{H}, F\}, \tag{D.45}$$

where

$$\{\mathcal{H}, F\} = \sum_j \left(\frac{\partial \mathcal{H}}{\partial p_j} \frac{\partial F}{\partial q_j} - \frac{\partial \mathcal{H}}{\partial q_j} \frac{\partial F}{\partial p_j} \right). \tag{D.46}$$

This is called *Poisson's parenthesis* between $\mathcal{H}$ and F. Therefore, we wee that a dynamical variable $F = F(q_j, p_j)$ that does not depend explicitly on time is an integral of motion, i.e. $dF/dt = 0$, only when $\{\mathcal{H}, F\} = 0$, i.e. its Poisson's parenthesis with the Hamiltonian is zero. In addition, from the definition (D.46) we obtain:

$$\{q_i, q_k\} = 0; \qquad \{p_i, p_k\} = 0; \qquad \{q_i, p_k\} = \delta_{ik}, \tag{D.47}$$

whose quantum mechanical interpretation leads to the Heisenberg uncertainty principle.

A particularly important application of this theory is the theorem of *Liouville*, where we consider systems composed of a very large number of degrees of freedom, N. In this case, as solving the system of $2N$ differential canonical equations is impossible,[6] we study the evolution of the probability density $\Pi_N(q_j, p_j, t)$ in the phase space (q_j, p_j). Since Π_N is conserved, it must remain constant during the evolution of the system, i.e.,[7]

$$\frac{d\Pi_N}{dt} = 0, \tag{D.48}$$

so that, applying (D.45), we obtain Liouville's equation,

$$\frac{\partial \Pi_N}{\partial t} + \{\mathcal{H}, \Pi_N\} = 0. \tag{D.49}$$

In particular, at steady state, when $\partial \Pi_N/\partial t = 0$, we reach an equilibrium condition, with $\Pi_N = \Pi_N^{eq}$, so that:

$$\{\mathcal{H}, \Pi_N^{eq}\} = 0, \tag{D.50}$$

that is Π_N^{eq} must be an integral of motion. In particular, Eq. (D.50) is satisfied when Π_N^{eq} is an arbitrary function of the total energy $E = \mathcal{H}$, although this is by no means the only solution.

D.4   Principle of Minimum Action

In the cases that are of interest here, i.e. systems subjected to conservative forces and with ideal (i.e. non dissipative) constraints, Lagrange's equation can be derived from a very general and beautiful principle. The idea is to study the evolution of a

---

[6]The main reason of this impossibility is that we do not know the initial conditions for all the N degrees of freedom.

[7]This statement is proven formally in Appendix E; see Eqs. (E.1)–(E.5).

system, composed of particles having positions $\mathbf{r}_i(t)$, from an initial configuration $\mathbf{r}_i(t_1)$ at time t_1 to a final configuration $\mathbf{r}_i(t_2)$ at time t_2 by perturbing the trajectories $\mathbf{r}_i(t)$ by virtual (i.e. compatible with all constraints) displacements $\delta\mathbf{r}_i(t)$, where, by definition,

$$\delta\mathbf{r}_i(t_1) = \delta\mathbf{r}_i(t_2) = 0. \tag{D.51}$$

First, take the time integral of Eq. (D.2):

$$\int_{t_1}^{t_2} \left[\delta V + \sum_{i=1}^{N} m_i \ddot{\mathbf{r}}_i \cdot \delta\mathbf{r}_i \right] dt = 0. \tag{D.52}$$

Integrating by parts the second term on the LHS we obtain:

$$\int_{t_1}^{t_2} \sum_{i=1}^{N} m_i \frac{d\dot{\mathbf{r}}_i}{dt} \cdot \delta\mathbf{r}_i \, dt = \sum_{i=1}^{N} [m_i \dot{\mathbf{r}}_i \cdot \delta\mathbf{r}_i]_{t_1}^{t_2} - \int_{t_1}^{t_2} \sum_{i=1}^{N} m_i \dot{\mathbf{r}}_i \cdot \delta\dot{\mathbf{r}}_i. \tag{D.53}$$

Therefore applying (D.51), observing that the last term above is the variation of the kinetic energy (D.7), $\delta\mathcal{T}$, and considering that the operator of virtual variation, δ, and that of time integration, $\int dt$ (just like time differentiation, d/dt) are independent from each other and therefore are commutable, we obtain:

$$\delta\mathcal{S} = 0, \tag{D.54}$$

where $\mathcal{S}$, called *action*, is the time integral of the Lagrangian, $\mathcal{L} = \mathcal{T} - \mathcal{V}$,[8]

$$\mathcal{S} = \int_{t_1}^{t_2} \mathcal{L} \, dt. \tag{D.55}$$

Equation (D.54) is the principle of minimum action and reveals that a system evolves in such a way to minimize the action.[9]

As mentioned in the previous section, when the system is subjected to time independent constraints we will have only n independent generalized coordinates, q_j, with $j = 1, 2, \ldots, n$, and we can write $\mathcal{L}(\mathbf{q}, \dot{\mathbf{q}}) = \mathcal{T}(\mathbf{q}, \dot{\mathbf{q}}) - \mathcal{V}(\mathbf{q})$. Accordingly, from Eq. (D.54) we can obtain again Lagrange's equation. In fact, we have:

$$\delta\mathcal{L} = \sum_{j=1}^{n} \frac{\partial\mathcal{L}}{\partial q_j}\delta q_j + \sum_{j=1}^{n} \frac{\partial\mathcal{L}}{\partial\dot{q}_j}\delta\dot{q}_j, \tag{D.56}$$

---

[8]Being a scalar that depends on a function, i.e. the trajectory of the system, the action $\mathcal{S}$ is a functional.

[9]Clearly, Eq. (D.54) could also mean that the action is maximized. This is, however, impossible, as one could always find a path with a larger action by slowing up the movement of the system for most of the time (keeping the same trajectory) and then accelerating it at the end.

and therefore, since $\delta \dot{q}_j = d(\delta q_j)/dt$, Eq. (D.54) becomes:

$$\int_{t_1}^{t_2} \sum_{j=1}^{n} \left[\frac{\partial \mathcal{L}}{\partial \dot{q}_j} \frac{d}{dt}(\delta q_j) + \frac{\partial \mathcal{L}}{\partial q_j} \delta q_j \right] dt = 0. \tag{D.57}$$

Integrating by parts the first term and considering that $\delta q_j = 0$ at $t = t_1$ and $t = t_2$, we finally obtain:

$$\int_{t_1}^{t_2} \sum_{j=1}^{n} \left[\frac{d}{dt}\left(\frac{\partial \mathcal{L}}{\partial \dot{q}_j} \right) - \frac{\partial \mathcal{L}}{\partial q_j} \right] \delta q_j \, dt = 0. \tag{D.58}$$

Now, since the virtual displacements δq_j are arbitrary and since this equality must hold for any t_1 and t_2, we conclude that the integrand must be zero, yielding Lagrange's equation (D.14), i.e.,

$$\frac{d}{dt}\left(\frac{\partial \mathcal{L}}{\partial \dot{q}_j} \right) - \frac{\partial \mathcal{L}}{\partial q_j} = 0. \tag{D.59}$$

According to the principle of minimum action, only one of the different trajectories connecting initial and final configurations refers to the real motion of the system, and that corresponds to the trajectories that minimizes the S. Now, instead of keeping the lower and upper limit of (D.55) fixed, let us compare the different values of S corresponding to trajectories having the same initial configuration, but with variable final configurations, i.e. when $\delta \mathbf{r}_i(t_1) = \mathbf{0}$ and $\delta \mathbf{r}_i(t_2) = \delta \mathbf{r}_i$. In this case, the first term on the RHS of Eq. (D.53) does not vanish, yielding $\sum_i m_i \dot{\mathbf{r}}_i \cdot \delta \mathbf{r}_i$, so that at the end, instead of (D.54), we obtain, in generalized coordinates:

$$\delta S = \sum_j p_j \delta q_j. \tag{D.60}$$

Therefore, expressing S is a function of the final configuration, q_j, we have:

$$\frac{\partial S}{\partial q_j} = p_j. \tag{D.61}$$

In the same way, the action can be seen as a function the time, i.e. we consider the trajectories starting from the same initial configuration and ending at the same final configuration, but at a different time, $t_2 = t$. From its definition (D.55), we see that the total derivative of S is:

$$\frac{dS}{dt} = \mathcal{L}. \tag{D.62}$$

Therefore, considering that the action (D.55) is a function of both coordinates and time of the upper limit, $S = S(q_j, t)$, we find:

$$\mathcal{L} = \frac{dS}{dt} = \frac{\partial S}{\partial t} + \sum_j \frac{\partial S}{\partial q_j} \dot{q}_j = \frac{\partial S}{\partial t} + \sum_j p_j \dot{q}_j, \tag{D.63}$$

thus obtaining the equation of Hamilton-Jacobi,[10]

$$\frac{\partial S}{\partial t} + \mathcal{H}\left(q_j, \frac{\partial S}{\partial q_j}, t\right) = 0, \tag{D.64}$$

where $\mathcal{H} = \sum_j p_j \dot{q}_j - \mathcal{L}$ is the Hamiltonian and the momenta have been expressed as gradient of the action through (D.61). The equation of Hamilton-Jacobi, together with Lagrange's equations and the canonical equations, can be considered as one of the governing equations of motion. Being a first-order partial differential equation, however, solving the Hamilton-Jacobi equation is in general more complicated than solving a system of ordinary differential equations, unless we can identify one or more integral of motion [1].

D.5 Dissipative Terms

In general, forces can be conservative (i.e. potential) and non conservative. Near local equilibrium, non conservative forces are linearly related to velocities (i.e. they are viscous forces), leading to the phenomenological equations (2.21),

$$\dot{q}_i = \sum_{i=1}^{n} L_{ij} Q_j, \quad \text{i.e.,} \quad Q_i = \sum_{i=1}^{n} L_{ij}^{-1} \dot{q}_j, \tag{D.65}$$

where $Q_i = k^{-1} \partial S / \partial q_i$ is a generalized force, while the coefficients L_{ij} satisfy the Onsager reciprocity relations, $L_{ij} = L_{ji}$. Accordingly, the entropy production rate is determined as [see Eqs. (2.34) and (2.35)],

$$k_B^{-1} \dot{S} = k_B^{-1} \sum_{i=1}^{n} \frac{\partial S}{\partial q_i} \dot{q}_i = \sum_{i=1}^{n} Q_i \dot{q}_i, \tag{D.66}$$

obtaining:

$$k_B^{-1} \dot{S} = \sum_{i=1}^{n} \sum_{j=1}^{n} L_{ij} Q_i Q_j = \sum_{i=1}^{n} \sum_{j=1}^{n} L_{ij}^{-1} \dot{q}_i \dot{q}_j. \tag{D.67}$$

Now, defining the drag coefficients, $\zeta_{ij} = kT L_{ij}^{-1}$, this last relation can be more conveniently written defining the following dissipation function (an energy per unit time), first defined by Rayleigh,

$$\mathcal{F} = \frac{1}{2} T \dot{S} = \frac{1}{2} \sum_{i=1}^{n} \sum_{j=1}^{n} \zeta_{ij} \dot{q}_i \dot{q}_j. \tag{D.68}$$

---

[10]It was discovered by Hamilton, but Jacobi pointed out its importance in determining solutions of the canonical equations.

Accordingly, the total forces (both conservative and non conservative) can be written as:

$$F_i = \frac{\partial \mathcal{L}}{\partial q_i} - \frac{\partial \mathcal{F}}{\partial \dot{q}_i}, \tag{D.69}$$

so that the Lagrange equations (D.59) can be rewritten as:

$$\frac{d}{dt}\left(\frac{\partial \mathcal{L}}{\partial \dot{q}_j}\right) - \frac{\partial \mathcal{L}}{\partial q_j} = -\frac{\partial \mathcal{F}}{\partial \dot{q}_i}. \tag{D.70}$$

Obviously, when $\mathcal{T} = \frac{1}{2}\dot{\mathbf{x}} \cdot \mathbf{m}\dot{\mathbf{x}}$, $\mathcal{V} = \frac{1}{2}\mathbf{x} \cdot \mathbf{A} \cdot \mathbf{x}$ and $\mathcal{F} = \frac{1}{2}\dot{\mathbf{x}} \cdot \boldsymbol{\zeta} \cdot \dot{\mathbf{x}}$, the equation of motion becomes:

$$\mathbf{m} \cdot \dot{\mathbf{x}} + \boldsymbol{\zeta} \cdot \dot{\mathbf{x}} + \mathbf{A} \cdot \mathbf{x} = \mathbf{0}. \tag{D.71}$$

References

1. Arnold, V.I.: Mathematical Methods of Classical Mechanics, pp. 255–270. Springer, Berlin (1978)
2. Landau, L.D., Lifshits, E.M.: Mechanics. Pergamon, Elmsford (1976)

Appendix E
Microscopic Balance Equations

In this appendix, the thermodynamic expression of the entropy production rate is derived from the basic equation of statistical mechanics; in particular, in Sect. E.1 Boltzmann's H theorem is derived, explaining the apparent paradox between microscopic reversibility and macroscopic irreversibility. Then, in Sect. E.2, we see that the balance equations that we have derived in Sect. 7 can also be obtained by coarse graining the fundamental equations of classical mechanics, describing the motion of all the particles that constitute our system.

E.1 H theorem

Let us consider a classical N-body system and define the ensemble probability function $\Pi_N(\mathbf{x}_1, \mathbf{x}_2, \ldots, \mathbf{x}_N, t)$, where $\mathbf{x}_i = (\mathbf{r}_i, \mathbf{p}_i)$ are the six-dimensional vectors consisting of the space coordinates $\mathbf{r}_i$ and momentum $\mathbf{p}_i$. $\Pi_N(\mathbf{X}, t)d\mathbf{X}$ is the fraction of the ensemble of replicates of the system to be found at time t in the volume element $d\mathbf{X}$ about the point $\mathbf{X} = (\mathbf{x}_1, \mathbf{x}_2, \ldots, \mathbf{x}_N)$ in the $6N$-dimensional space. Therefore, Π_N is normalized as:

$$\int_{\mathbf{X}} \Pi_N(\mathbf{X}) \, d\mathbf{X} = 1. \tag{E.1}$$

The trajectories of the N particles can be determined through Hamilton's equation of motion (D.40):

$$\dot{\mathbf{r}}_i = \nabla_{\mathbf{p}_i} \mathcal{H}_N; \quad \dot{\mathbf{p}}_i = -\nabla_{\mathbf{r}_i} \mathcal{H}_N, \tag{E.2}$$

where $\nabla_{\mathbf{r}_i} = \partial/\partial \mathbf{r}_i$ and $\nabla_{\mathbf{p}_i} = \partial/\partial \mathbf{p}_i$, the dot indicates time derivative and $\mathcal{H}_N$ is the Hamiltonian of the system.

Since particles are neither created nor destroyed, Π_N is a density in the $\mathbf{X}$-space, so that it obeys the continuity equation,

$$\dot{\Pi}_N + \nabla_{\mathbf{X}} \cdot (\dot{\mathbf{X}} \Pi_N) = 0. \tag{E.3}$$

R. Mauri, *Non-Equilibrium Thermodynamics in Multiphase Flows*,
Soft and Biological Matter, DOI 10.1007/978-94-007-5461-4,
© Springer Science+Business Media Dordrecht 2013

Now, considering that, applying (E.2),

$$\nabla_{\mathbf{X}} \cdot \dot{\mathbf{X}} = \sum_{i=1}^{N} \nabla_{\mathbf{r}_i} \cdot \dot{\mathbf{r}}_i + \nabla_{\mathbf{p}_i} \cdot \dot{\mathbf{p}}_i = \sum_{i=1}^{N} \left(\frac{\partial^2 \mathcal{H}_N}{\partial \mathbf{r}_i \mathbf{p}_i} - \frac{\partial^2 \mathcal{H}_N}{\partial \mathbf{p}_i \mathbf{r}_i} \right) = 0, \qquad \text{(E.4)}$$

we obtain the Liouville equation (D.49),

$$\dot{\Pi}_N + \dot{\mathbf{X}} \cdot \nabla_{\mathbf{X}} \Pi_N = 0, \qquad \text{(E.5)}$$

i.e.,

$$\frac{\partial \Pi_N}{\partial t} + \sum_{i=1}^{N} \dot{\mathbf{r}}_i \cdot \nabla_{\mathbf{r}_i} \Pi_N + \dot{\mathbf{p}}_i \cdot \nabla_{\mathbf{p}_i} \Pi_N = 0. \qquad \text{(E.6)}$$

The entropy of the system is given by Eq. (B.34),

$$S_N = -k \int_{\mathbf{X}} \Pi_N \ln \Pi_N \, d\mathbf{X} = -k \langle\!\langle \ln \Pi_N \rangle\!\rangle, \qquad \text{(E.7)}$$

where we have defined the overall average $\langle\!\langle A \rangle\!\rangle = \int A \Pi_N \, d\mathbf{X}$. Taking the time derivative and considering the normalization (E.1) we obtain:

$$-k^{-1} \frac{d S_N}{dt} = \left\langle\!\!\left\langle \frac{1}{\Pi_N} \frac{d \Pi_N}{dt} \right\rangle\!\!\right\rangle = \int_{\mathbf{X}} \frac{d \Pi_N}{dt} \, d\mathbf{X} = \frac{d}{dt} \int_{\mathbf{X}} \Pi_N \, d\mathbf{X} = 0, \qquad \text{(E.8)}$$

showing that, when the motion of all particles is monitored exactly, with unlimited precision, any process is iso-entropic, or reversible. In fact, entropy increase is a consequence of a loss of information, which may be caused by coarse-graining or particle indistinguishability.

Now, as the time derivative of the position is the velocity and the time derivative of the momentum is the force, assuming that the particles interact *via* a two-body potential we obtain:

$$\dot{\mathbf{r}}_i = \mathbf{p}_i / m = \mathbf{v}_i; \quad \dot{\mathbf{p}}_i = \mathbf{F}_i + \sum_{j=1}^{N} \mathbf{f}_{ij}, \qquad \text{(E.9)}$$

where $\mathbf{F}_i(\mathbf{r}_i)$ is the external force exerted on particle i, which depends only on the position of particle i, while $\mathbf{f}_{ij} = \nabla_{\mathbf{r}_i} \psi_{ij}$ is the interparticle force, with $\psi_{ij} = \psi(|\mathbf{r}_i - \mathbf{r}_j|)$ denoting a potential, that depends only on the interparticle distance. Here we have assumed that particles are indistinguishable, having, in particular, the same mass m. For multicomponent systems, we can repeat this treatment for each chemical species.

Now let us keep one particle (anyone of them, as they are identical) fixed, and average the continuity equation (E.3) over positions and momenta of all the other $N - 1$ particles, defining,

$$\Pi_1(\mathbf{x}_1, t) = \int_{\mathbf{X}} \Pi_N(\mathbf{x}_1, \mathbf{x}_2, \ldots, \mathbf{x}_N, t) \, d\mathbf{x}_2, \ldots, \mathbf{x}_N. \qquad \text{(E.10)}$$

Applying the equations of motion (E.2) and considering that, since Π_N is normalized it must vanish very rapidly at infinity, we obtain:

$$\frac{\partial \Pi}{\partial t} + \mathbf{v} \cdot \nabla_{\mathbf{r}} \Pi + \mathbf{F} \cdot \nabla_{\mathbf{p}} \Pi = -N \int f_{12} \nabla_{\mathbf{p}} \Pi_2 \, d\mathbf{r}_2 \, d\mathbf{p}_2 = \frac{D_C \Pi}{Dt}, \qquad \text{(E.11)}$$

where the index 1 has been dropped for convenience. This is basically the Boltzmann equation, where its RHS indicates a loss of coherence and it needs some assumptions on Π_2 to be solved. In fact, as $D_C \Pi$ represents the difference between the probability distribution before and after a particle-particle interaction, Boltzmann proposed his celebrated assumption of molecular chaos (Stosszahl Ansatz) to estimate it. The role of this term is further clarified as we can easily show that

$$\frac{dS}{dt} > 0; \quad S = -k \langle\langle \ln \Pi \rangle\rangle, \qquad \text{(E.12)}$$

showing that the loss of coherence is the cause of the entropy monotonic increase predicted by the second law of thermodynamics. This is generally referred to as the H theorem, since S, apart from the minus sign, is also referred to as the H-function.[1]

Comment Equation (E.11) is the first of a series of coupled equations describing the evolution of $\Pi_k(\mathbf{x}_1, \mathbf{x}_2, \ldots, \mathbf{x}_k, t)$, with $k = 1, 2, \ldots, N$, which is generally referred to as the BBGKY hierarchy, after the work by Bogolyubov, Born, Green, Kirkwood and Yvon. In general, these equations have the form:

$$\frac{D\Pi_k}{Dt} = F(\Pi_{k+1}); \quad \Pi_k = \int \Pi_N \, d\mathbf{x}_{k+1} \ldots d\mathbf{x}_N,$$

that is, in order to calculate Π_k, we need to know Π_{k+1}. Therefore, we need a closure to truncate the series, and that is generally done by approximating, or ignoring, the second- or higher-order correlation function.

E.2 Balance Equations

Let us define the momentum average of any quantity $f(\mathbf{r}, \mathbf{p})$ as:

$$\langle f \rangle (\mathbf{r}, t) = \frac{\int f(\mathbf{r}, \mathbf{p}) \Pi(\mathbf{r}, \mathbf{p}, t) \, d\mathbf{p}}{\int \Pi(\mathbf{r}, \mathbf{p}, t) \, d\mathbf{p}} = \frac{N}{n(\mathbf{r}, t)} \int f(\mathbf{r}, \mathbf{p}) \Pi(\mathbf{r}, \mathbf{p}, t) \, d\mathbf{p}, \qquad \text{(E.13)}$$

where n is the number density, i.e. the ratio between mass density and the mass of a single particle,

$$n(\mathbf{r}, t) = \frac{1}{m} \rho(\mathbf{r}, t) = N \int \Pi(\mathbf{r}, \mathbf{p}, t) \, d\mathbf{p}. \qquad \text{(E.14)}$$

---

[1] Therefore, $dH/dt < 0$. See [3, Chap. 40].

From the Boltzmann equation (E.11) we obtain:

$$\int f \frac{D_t \Pi}{Dt} \, d\mathbf{p} = \int f \frac{D_C \Pi}{Dt} \, d\mathbf{p}, \tag{E.15}$$

where $D_t/Dt = \partial/\partial t + \mathbf{v} \cdot \nabla_\mathbf{r} + \mathbf{F} \cdot \nabla_\mathbf{p}$ is a material derivative. Considering the physical meaning of the RHS of this equation, we see that it is identically zero when f is a conserved quantity. In that case we find:

$$\int f(\mathbf{r}, \mathbf{p}) \left(\frac{\partial}{\partial t} + \mathbf{v} \cdot \nabla_\mathbf{r} + \mathbf{F} \cdot \nabla_\mathbf{p} \right) \Pi(\mathbf{r}, \mathbf{p}, t) \, d\mathbf{p} = 0. \tag{E.16}$$

Now, consider the equalities:

$$f\mathbf{v} \cdot \nabla_\mathbf{r} \Pi = \nabla_\mathbf{r} \cdot (\mathbf{v} f \Pi) - \Pi \mathbf{v} \cdot \nabla_\mathbf{r} f; \qquad f\mathbf{F} \cdot \nabla_\mathbf{p} \Pi = \nabla_\mathbf{p} \cdot (\mathbf{F} f \Pi) - \Pi \mathbf{F} \cdot \nabla_\mathbf{p} f \tag{E.17}$$

where we have taken into account the fact that $\mathbf{v}$ is an independent variable, while $\mathbf{F}$ does not depend on $\mathbf{p}$. Substituting (E.17) into (E.16) and applying the divergence theorem, considering that Π decays exponentially fast as $|\mathbf{p} \to \infty|$, we obtain:

$$\frac{\partial}{\partial t} \int f \Pi \, d\mathbf{p} + \nabla_\mathbf{r} \cdot \int \mathbf{v} f \Pi \, d\mathbf{p} - \int \Pi \mathbf{v} \cdot \nabla_\mathbf{r} f \, d\mathbf{p} - \int \Pi \mathbf{F} \cdot \nabla_\mathbf{p} f \, d\mathbf{p} = 0, \tag{E.18}$$

where we have considered that f is not an explicit function of t. Finally we can rewrite the general balance equation for any conserved quantity f in the following form,[2]

$$\frac{\partial}{\partial t} \left(n \langle f \rangle \right) + \nabla_\mathbf{r} \cdot \left(n \langle \mathbf{v} f \rangle \right) = n \langle \mathbf{v} \cdot \nabla_\mathbf{r} f \rangle + n \mathbf{F} \cdot \langle \nabla_\mathbf{p} f \rangle, \tag{E.19}$$

where $\langle nf \rangle = n \langle f \rangle$ because n is independent of $\mathbf{p}$.

Assume that the system is composed of monatomic molecules. Then, the independent conserved quantities are mass, momentum and energy (we could also include the electric charge, but this extension is trivial). Accordingly, we set successively, $f = m$ (mass), $f = mv_i$, with $i = 1, 2, 3$ (momentum) and $f = \frac{1}{2} m |\mathbf{v} - \mathbf{u}|^2$ (thermal energy), where $\mathbf{u}(\mathbf{r}, t) = \langle \mathbf{v} \rangle$ is the mean velocity.

Mass Balance   For $f = m$, since $mn = \rho$ is the mass density [see Eq. (E.14)], we find immediately:

$$\frac{\partial \rho}{\partial t} + \nabla_\mathbf{r} \cdot (\rho \mathbf{u}) = 0, \tag{E.20}$$

which coincides with (7.7).

---

[2]A more complete treatment can be found, for example, in [1], and [2].

Momentum Balance   For $f = p_i = mv_i$, we obtain:

$$\frac{\partial}{\partial t}\langle \rho v_i \rangle + \nabla_{\mathbf{r}} \cdot (\rho v_i \mathbf{v}) = \rho F_i', \tag{E.21}$$

where $\mathbf{F}' = \mathbf{F}/m$ is the force per unit mass. Now, applying the obvious identity $\langle v_i v_j \rangle = u_i u_j + \langle \tilde{v}_i \tilde{v}_j \rangle$, where $\tilde{v}_i = v_i - u_i$, with $\langle \tilde{v}_i \rangle = 0$, and substituting (E.20), we obtain:

$$\rho\left(\frac{\partial \mathbf{u}}{\partial t} + \mathbf{u} \cdot \nabla \mathbf{u}\right) = \rho \mathbf{F}' - \nabla \cdot \mathbf{P}, \tag{E.22}$$

where,

$$P_{ij} = \rho\langle \tilde{v}_i \tilde{v}_j \rangle, \tag{E.23}$$

is the symmetric pressure tensor, that is the diffusive momentum flux tensor. This equation coincides with (7.25).

Energy Balance   For $f = \frac{1}{2}m\tilde{v}^2$, we obtain:

$$\frac{\partial}{\partial t}\left\langle \frac{1}{2}\rho\tilde{v}^2 \right\rangle + \nabla_{\mathbf{r}} \cdot \left(\frac{1}{2}\rho\mathbf{v}\tilde{v}^2\right) = \frac{1}{2}\rho\langle \mathbf{v} \cdot \nabla_{\mathbf{r}}\tilde{v}^2 \rangle. \tag{E.24}$$

Now, define the temperature, T,

$$kT = \frac{1}{3}m\langle \tilde{v}^2 \rangle, \tag{E.25}$$

and the diffusive heat flux vector, $\mathbf{J}^{(q)}$,

$$\mathbf{J}^{(q)} = \frac{1}{2}\rho\langle \tilde{v}^2 \tilde{\mathbf{v}} \rangle. \tag{E.26}$$

At the end, we obtain:

$$\frac{3}{2}k\left(\frac{\partial(\rho T)}{\partial t} + \nabla_{\mathbf{r}} \cdot (\rho T\mathbf{u})\right) = -\frac{\partial J_i^{(q)}}{\partial r_i} - mP_{ij}\frac{\partial u_j}{\partial r_i}, \tag{E.27}$$

where $\mathbf{P}$ is the pressure tensor. Finally, substituting the continuity equation (E.20) and dividing by m we obtain:

$$\rho c\left(\frac{\partial T}{\partial t} + \mathbf{u} \cdot \nabla_{\mathbf{r}}T\right) = -\nabla_{\mathbf{r}} \cdot \mathbf{J}^{(q)} - \mathbf{P} : \nabla\mathbf{u}, \tag{E.28}$$

where $c = 3k/2m$ is the specific heat per unit mass of monatomic molecules. The last term expresses the viscous heat dissipation; as $\mathbf{P}$ is symmetric, only the symmetric part of the velocity gradient contributes to this term.

Therefore, starting from first principles, conservation equations can be derived exactly. In particular, note that the diffusive momentum and heat fluxes, $\mathbf{P}$ and $\mathbf{J}^{(q)}$, depend on the velocity fluctuations and arise from coarse graining of the fluctuating part of the total fluxes. Identical considerations lead to defining Taylor dispersivity (see Sect. 11.2.2) and turbulent fluxes (see Sect. 11.3.3).

Comment If particles do not interact with each other, the fluctuating part of the particle velocities satisfies the Maxwellian distribution, so that the pressure tensor, **P**, results to be isotropic (i.e. there is pressure but not shear stresses) and the diffusive heat flux $\mathbf{J}^{(q)}$ is identically zero.[3] So, as expected, dissipation can be accounted for only when particle collision is considered.

References

1. de Groot, S.R., Mazur, P.: Non-Equilibrium Thermodynamics. Dover, New York (1984), Chap. IX
2. Kreuzer, H.J.: Non-Equilibrium Thermodynamics and Its Statistical Foundations. Clarendon, Oxford (1981), Chap. 7
3. Landau, L.D., Lifshitz, E.M.: Statistical Physics, Part I. Pergamon, Elmsford (1980)

---

[3]In fact, we find: $\langle \tilde{v}_i \tilde{v}_j \rangle = kT/m$; therefore, since $\rho = mN_A/V_w$, where N_A is the Avogadro number and V_w is the molar volume, we see that $P_{ij} = P\delta_{ij}$, where $P = RT/V_w$ is the pressure of an ideal gas and $R = N_A k$ is the gas constant.

Appendix F
Some Results of Transport Phenomena

In this appendix we review some fundamental results of transport phenomena that we utilize in Chaps. 10 and 11. First, in Sect. F.1, we express the solution of the Laplace equation in terms of vector harmonics, namely the fundamental harmonic functions and all their gradients. Then, in Sect. F.2, these results are generalized to determine the general solution of the Stokes equation, in particular the stokeslet, rotlet and stresslet. These results are applied in Sect. F.3 to derive the reciprocal theorems of transport phenomena and Faxen's laws. Finally, in Sect. F.4, the sedimentation velocity of dilute suspensions is determined as an example of George Batchelor's renormalization procedure.

F.1 Laplace Equation and Spherical Harmonics

Consider the temperature field $T(\mathbf{r})$ in a homogeneous medium, at steady state and in the absence of convection, where $\mathbf{r}$ is the position vector. As we know, $T(\mathbf{r})$ is a *harmonic function*, i.e. it satisfies the Laplace equation,

$$\nabla^2 T = 0, \qquad (F.1)$$

with appropriate boundary conditions. Harmonic functions can be expressed as linear combinations of the fundamental solutions of the Laplace equation, i.e. 1 and $1/|\mathbf{r}|$, together with all their gradients. These are tensorial harmonic functions named *vector harmonics*, that can be written in invariant notation, i.e. in a form that is independent of the coordinate system and involves only the position vector, $\mathbf{r}$, and its length, $r = |\mathbf{r}|$. Vector harmonics can be growing or decaying, i.e. they approach 0 as $r \to 0$ and as $r \to \infty$, respectively.

R. Mauri, *Non-Equilibrium Thermodynamics in Multiphase Flows*,
Soft and Biological Matter, DOI 10.1007/978-94-007-5461-4,
© Springer Science+Business Media Dordrecht 2013

F.1.1 Decaying Harmonics

Among the decaying harmonic functions, the first to be considered is $1/r$, which is often referred to as a *monopole* and represents the Green function, or propagator, of the Laplace equation, i.e. it is the solution of the equation: $\nabla^2 T = \delta(\mathbf{r})$. The monopole is the temperature field generated by an energy impulse, $\dot{Q}$, placed at the origin, that is a point source which radiates equally well in all directions. In fact, since at steady state the energy that crosses per unit time any closed surface that includes the origin is constant and equal to $\dot{Q}$, for a sphere of radius a we have,

$$\dot{Q} = \oint_{r=a} \mathbf{n} \cdot \mathbf{J}_U \, d^2\mathbf{r}_S; \quad \mathbf{J}_U = -k\nabla T, \tag{F.2}$$

where $\mathbf{r}_S$ is a position vector located on the sphere surface, $\mathbf{n} = \mathbf{r}_s/a$ is the outer unit vector, k is the heat conductivity, while $\mathbf{J}_U$ is the heat flux, so we easily find that a temperature distribution $T = 1/r$ induces a heat flux $\dot{Q} = 4\pi k$. Accordingly, the monopole temperature distribution is often indicated as:

$$T^{(m)}(\mathbf{r}; \alpha) = \frac{\alpha}{r}; \quad \alpha = \frac{\dot{Q}}{4\pi k}, \tag{F.3}$$

showing that the strength of the monopole, α, is directly related to the total heat flux.

Due to the linearity of the Laplace equation, all the gradients of the singular fundamental solution, r^{-1}, will be (singular) solutions as well. So, for example, ∇r^{-1}, which is often referred to as a *dipole* distribution, is the solution of the equation $\nabla^2 \nabla T = \nabla \delta(\mathbf{r})$, and therefore represents the temperature field generated by two monopole sources of infinitely large, equal, and opposite strengths, separated by an infinitesimal distance d, located at the origin;[1] therefore, as one monopole generates the same energy that is absorbed by the other, the dipole has no net energy release. In the same way, we can define a quadrupole, as two opposing dipoles, an octupole, as two opposing quadrupole, and so on, so that, in general, we see that any decaying temperature field is determined as a linear combination of decaying vector harmonics, which are defined (after multiplication by convenient constants) by the following n-th order tensorial functions,

$$\mathbf{H}^{[-(n+1)]}(\mathbf{r}) = \frac{(-1)^n}{1 \times 3 \times 5 \times \cdots \times (2n-1)} \underbrace{\nabla\nabla\cdots\nabla}_{n \ times}\left(\frac{1}{r}\right), \tag{F.4}$$

for $n = 1, 2, \ldots$; in particular,

$$H^{(-1)} = -\frac{1}{r}; \quad H_i^{(-2)} = \frac{r_i}{r^3}; \quad H_{ij}^{(-3)} = \frac{r_i r_j}{r^5} - \frac{\delta_{ij}}{3r^3};$$

---

[1]In fact, the dipole strength is the product of the strength of the two opposite impulses by their distance.

$$H_{ijk}^{(-4)} = \frac{r_i r_j r_k}{r^7} - \frac{x_i \delta_{jk} + x_j \delta_{ik} + x_k \delta_{ij}}{5r^5}.$$

Concretely, we have:

$$T(\mathbf{r}) = \sum_{n=0}^{\infty} \mathbf{C}_n \, (\cdot)^n \, \mathbf{H}^{[-(n+1)]}(\mathbf{r}), \tag{F.5}$$

where $\mathbf{C}_n$ are n-th order constant tensors, representing the strength of the monopole, dipole, quadrupole, etc., to be determined by satisfying the boundary conditions. Formally, this temperature distribution solves the following Laplace equation:

$$\nabla^2 T = \sum_{n=0}^{\infty} \mathbf{C}_n \, (\cdot)^n \, \frac{(-1)^n}{1 \times 3 \times 5 \times \cdots \times (2n-1)} \underbrace{\nabla \nabla \nabla \cdots \nabla}_{n \ times} \delta(r). \tag{F.6}$$

To understand the meaning of all these singular terms, consider the temperature field disturbance due to the presence of a particle. Clearly, each infinitesimal surface element $dS = d^2 \mathbf{r}_S$ located at position $\mathbf{r}_S$ on the particle surface will act as a point impulse of strength $d\alpha_S = f_S dS$, where f_S has to be determined, inducing a temperature disturbance $d\alpha_S |\mathbf{r} - \mathbf{r}_S|^{-1}$ at position $\mathbf{r}$. At the end, superimposing the effects of all surface elements, we obtain the following *boundary integral* equation:

$$T(\mathbf{r}) = \oint_S f_S(\mathbf{r}_S) |\mathbf{r} - \mathbf{r}_S|^{-1} d^2 \mathbf{r}_S, \tag{F.7}$$

where the impulse strength $f_S(\mathbf{r}_S)$ is determined by satisfying the boundary conditions, e.g. imposing a given temperature or flux distribution on the surface.

When we are far from the particle, i.e. when $|\mathbf{r}| \gg |\mathbf{r}_S|$, we can expand the Green function, obtaining:

$$|\mathbf{r} - \mathbf{r}_S|^{-1} = r^{-1} - \mathbf{r}_S \cdot \nabla r^{-1} + \frac{1}{2} \mathbf{r}_S \mathbf{r}_S : \nabla \nabla r^{-1} + \cdots + \frac{(-1)^n}{n!} \mathbf{r}_S^n (\cdot)^n \nabla^n r^{-1} + \cdots,$$

where $\mathbf{r}^n = \mathbf{r} \mathbf{r} \cdots \mathbf{r}$ (n times), and similarly for ∇^n and $(\cdot)^n$. Substituting this expression into (F.7) we obtain the multipole expansion (F.5).

F.1.2 General Solution

Growing harmonics, in contrast, grow with distance from the origin. They are composed of the fundamental uniform solution, $H^{(0)} = 1$, and of all its inverse gradients, that are harmonic tensorial functions whose gradients are equal to 1. For example,

$$\nabla_i^{-1} 1 = \frac{1}{3} r_i; \qquad \nabla_i^{-1} \nabla_j^{-1} 1 = \frac{1}{10} \left(r_i r_j - \frac{r^2}{3} \delta_{ij} \right) \cdots.$$

Multiplying them by appropriate constants, the growing harmonics can be obtained from the decaying harmonics as:

$$\mathbf{H}^{(n)}(\mathbf{r}) = r^{2n+1}\mathbf{H}^{[-(n+1)]}(\mathbf{r}), \tag{F.8}$$

and, in particular,

$$H^{(0)} = 1; \qquad H_i^{(1)} = r_i; \qquad H_{ij}^{(2)} = r_i r_j - \frac{r^2}{3}\delta_{ij};$$

$$H_{ijk}^{(3)} = r_i r_j r_k - \frac{r^2}{5}(x_i\delta_{jk} + x_j\delta_{ik} + x_k\delta_{ij}).$$

In general, the solution of the Laplace equation can be written as a linear combination of all spherical harmonics,

$$T(\mathbf{r}) = \sum_{n=-\infty}^{\infty}\left[\mathbf{C}_n\,(\cdot)^n\mathbf{H}^{(n)}(\mathbf{r})\right] = \sum_{n=0}^{\infty}\left[(\mathbf{C}_n' + r^{2n+1}\mathbf{C}_n'')(\cdot)^n\mathbf{H}^{[-(n+1)]}(\mathbf{r})\right], \tag{F.9}$$

where $\mathbf{C}_n'$ and $\mathbf{C}_n''$ are n-th order constant tensors, to be determined imposing that the boundary conditions be satisfied. In particular,

$$T(\mathbf{r}) = \left(C_0' + rC_0''\right)\frac{1}{r} + \left(\mathbf{C}_1' + r^3\mathbf{C}_1''\right)\cdot\frac{\mathbf{r}}{r^3} + \left(\mathbf{C}_2' + r^5\mathbf{C}_2''\right):\left(\frac{\mathbf{rr}}{r^5} - \frac{\mathbf{I}}{3r^3}\right) + \cdots$$

We will find these sets of harmonic tensorial functions very useful in constructing solutions that take advantage of the symmetry of the problem under consideration. In fact, there are only a limited number of ways in which the boundary conditions can be combined with the set of decaying or growing harmonics to generate tensors that constitute the desired solutions in invariant form.

For example, suppose that we want to determine the temperature distribution around a sphere induced by an imposed ΔT, i.e. $T = T_0$ at $r \to \infty$ and $T = 0$ at the surface $r = a$ of the sphere. Then, the temperature field must be the product of the scalar driving force, $\Delta T = T_0$, and a linear combination of the scalar harmonic functions, $H^{(0)}$ and $H^{(-1)}$. Therefore, $T(\mathbf{r}) = T_0(\lambda'/r + \lambda'')$, where λ' and λ'' are scalar constants to be determined through the boundary conditions, finding at the end the obvious solution:[2]

$$T(\mathbf{r}) = T_0(1 - a/r) = T_0 - T^{(m)}(\mathbf{r};\alpha), \tag{F.10}$$

where $\alpha = aT_0$. This shows that the temperature distribution around a sphere induced by an imposed ΔT is determined uniquely by a monopole at the center of the

---

[2]In this case, we could simply say that the temperature field must be spherically symmetric, i.e. only a function of r.

sphere. Accordingly, applying Eq. (F.3), we can easily determine the total heat flux leaving the sphere,

$$\dot{Q} = 4\pi ka\Delta T, \tag{F.11}$$

where $\Delta T = -T_0$ is the temperature difference between the sphere and infinity.

A more complex case is when we impose a constant temperature gradient, G_i, at infinity, i.e. $T(\mathbf{r}) = G_i r_i$ at $r \to \infty$, while, as before, $T = 0$ at the surface of the sphere $r = a$.[3] Then, the temperature field is the product of the driving force, i.e. the vector G_i, by a linear combination of the vectorial harmonic functions, $H_i^{(1)}$ and $H_i^{(-2)}$, obtaining:

$$T(\mathbf{r}) = G_i r_i \left(\lambda' \frac{1}{r^3} + \lambda'' \right) = G_i r_i \left(1 - \frac{a^3}{r^3} \right), \tag{F.12}$$

where the constants λ' and λ'' have been determined by imposing that the boundary conditions are satisfied. So, as expected, we have obtained the temperature distribution due to a dipole located at the center of the sphere and, therefore, no net heat flux is released, or absorbed.

In the most general case, if we impose a temperature distribution at infinity (and yet satisfying the Laplace equation),

$$T(\mathbf{r}) = \sum_{n=0}^{\infty} \mathbf{C}_n (\cdot)^n \mathbf{H}^{(n)}(\mathbf{r}) \quad \text{as } r \to \infty, \tag{F.13}$$

while $T = 0$ at the surface $r = a$ of a sphere, from (F.8) and (F.9) we obtain:

$$T(\mathbf{r}) = \sum_{n=0}^{\infty} \mathbf{C}_n (\cdot)^n \left(r^{2n+1} - a^{2n+1} \right) \mathbf{H}^{[-(n+1)]}(\mathbf{r}), \tag{F.14}$$

thus generalizing the two temperature distributions (F.10) and (F.12).

F.2 Stokes Equation

A similar procedure can be applied to find solutions of the Stokes equations,

$$\nabla p = \eta \nabla^2 \mathbf{v}; \qquad \nabla \cdot \mathbf{v} = 0, \tag{F.15}$$

where p and $\mathbf{v}$ are the pressure and the velocity fields, satisfying appropriate boundary conditions, while η is the fluid viscosity.

---

[3]Note that $\mathbf{r} = \mathbf{H}^{(1)}$ and so the imposed temperature at infinity satisfies, as it must, the Laplace equation.

Taking the divergence of the Stokes equation and considering that the velocity field is solenoidal, we see that the pressure field is harmonic, i.e. it satisfies the Laplace equation,

$$\nabla^2 p = 0. \tag{F.16}$$

In addition, it is easy to verify that the velocity field can be written as

$$\mathbf{v} = \frac{1}{2\eta}\mathbf{r}p + \mathbf{u}, \tag{F.17}$$

where $\mathbf{u}$ is a harmonic vectorial function,

$$\nabla^2 \mathbf{u} = 0. \tag{F.18}$$

Obviously, only three of the four harmonic functions deining p and $\mathbf{v}$ are independent, as the condition that the velocity field must be divergence free must be implemented, obtaining:

$$\nabla \cdot \mathbf{u} = -\frac{1}{2\eta}(3p + \mathbf{r} \cdot \nabla p) = 0. \tag{F.19}$$

So, the important idea here is that we are looking for solutions of the Laplace equation in one case for a scalar field (pressure) and in the other for a vector field (homogeneous solution for the velocity).

F.2.1 Stokeslet

The simplest application of these procedure arises when we calculate the propagator, or Green function, of the Stokes equation, that is the velocity and pressure fields induced in an unbounded and otherwise quiescent fluid by a point body force located at the origin, $\mathbf{f}(\mathbf{r}) = \mathbf{F}\delta(\mathbf{r})$, where $\mathbf{F}$ is a force. Then, both p and $\mathbf{u}$ are decaying harmonic function proportional to $\mathbf{F}$, i.e.,

$$p = F_k \lambda \frac{r_k}{r^3}, \tag{F.20}$$

and

$$u_i = \frac{F_k}{2\eta}\left[\lambda'\frac{\delta_{ik}}{r} + \lambda''\left(\frac{r_i r_k}{r^5} - \frac{\delta_{ik}}{3r^3}\right)\right]. \tag{F.21}$$

Now, since λ'' has the units of a square length, considering that there is no characteristic dimension in this problem, it must be $\lambda'' = 0$. Consequently, imposing that $\nabla_i v_i = 0$, we see that $\lambda' = \lambda$, so that we obtain:

$$v_i = \frac{F_k}{2\eta}\lambda\left(\frac{\delta_{ik}}{r} + \frac{r_i r_k}{r^3}\right). \tag{F.22}$$

The value of λ can be determined imposing that the total force applied to the fluid located *outside* a sphere of radius a must equal to $\mathbf{F}$, i.e.,

$$\oint_{r=a} n_i T_{ij} \, d^2\mathbf{r} = F_i, \tag{F.23}$$

where $n_i = -r_i/a$ is the normal unit vector, perpendicular to the surface of the sphere and directed inward, while T_{ij} is the stress tensor,

$$T_{ij} = -p\delta_{ij} + \eta(\nabla_i v_j + \nabla_j v_i) = -3\lambda F_k \frac{r_i r_j r_k}{r^5}. \tag{F.24}$$

Now, considering that

$$n_i T_{ij} = 3\lambda F_k \frac{r_j r_k}{a^4},$$

and using the identity,[4]

$$\oint_{r=a} r_i r_k \, d^2\mathbf{r} = \frac{4}{3}\pi a^4 \delta_{ik}, \tag{F.25}$$

we finally obtain: $\lambda = 1/4\pi$. Therefore, the Stokeslet has the following pressure and velocity fields:[5]

$$p^{(s)}(\mathbf{r}; \mathbf{F}) = F_k P_k^{(s)}(\mathbf{r}); \quad P_k^{(s)}(\mathbf{r}) = \frac{1}{4\pi} \frac{r_k}{r^3}, \tag{F.26}$$

$$v_i^{(s)}(\mathbf{r}; \mathbf{F}) = F_k V_{ik}^{(s)}(\mathbf{r}); \quad V_{ik}^{(s)}(\mathbf{r}) = \frac{1}{8\pi\eta r}\left(\delta_{ik} + \frac{r_i r_k}{r^2}\right), \tag{F.27}$$

with the associated stress tensor field as well,

$$T_{ik}^{(s)}(\mathbf{r}) = \frac{3}{4\pi} \frac{r_i r_j r_k}{r^5} F_k. \tag{F.28}$$

The multiplier $\mathbf{V}^{(s)}$ is called the *Oseen* tensor, and describes the disturbance to an unperturbed flow due to a point force.

This result could also be obtained taking the Fourier transform of the Stokes equations,

$$\nabla p - \eta\nabla^2\mathbf{v} = \mathbf{F}\delta(\mathbf{r}); \qquad \nabla \cdot \mathbf{v} = 0, \tag{F.29}$$

defining the Fourier transform $\widehat{f}(\mathbf{k})$ of any function $f(\mathbf{r})$,

$$\widehat{f}(\mathbf{k}) = \mathcal{F}\{f(\mathbf{r})\} = \int f(\mathbf{r})e^{i\mathbf{k}\cdot\mathbf{r}} \, d^3\mathbf{r}, \tag{F.30}$$

---

[4]It is evident that when $i \neq k$ this integral must be zero. Then, multiplying both member by δ_{ik} and considering that $\delta_{ik}\delta_{ik} = 3$, we easily verify the identity.

[5]Sometimes the strength of the Stokeslet is indicated as $\alpha = \mathbf{F}/8\pi\eta$.

and its corresponding antitransform,

$$f(\mathbf{r}) = \mathcal{F}^{-1}\{f(\mathbf{k})\} = \int \widehat{f}(\mathbf{k})e^{-i\mathbf{k}\cdot\mathbf{r}}\frac{d^3\mathbf{k}}{(2\pi)^3}. \tag{F.31}$$

Now, considering that $\mathcal{F}\{\delta(\mathbf{r})\} = 1$ and $\mathcal{F}\{\nabla f(\mathbf{r})\} = -i\mathbf{k}\widehat{f}(\mathbf{k})$, we obtain:

$$\widehat{p}(\mathbf{k}) = i\frac{k_j}{k^2}F_j, \tag{F.32}$$

and

$$\widehat{v}_i(\mathbf{k}) = \frac{1}{\eta k^2}\left(\delta_{ij} - \frac{k_ik_j}{k^2}\right)F_j. \tag{F.33}$$

Finally, antitransforming Eqs. (F.32) and (F.33), we find again Eqs. (F.26) and (F.27).

F.2.2 Uniform Flow Past a Sphere; Potential Doublet

In this case, the unperturbed velocity field is uniform, with $\mathbf{v}(\mathbf{r}) = \mathbf{U}$ as $r \to \infty$, while a sphere is kept fixed at the origin, i.e. $\mathbf{v}(\mathbf{r}) = \mathbf{0}$ at $r = a$. Then, proceeding as before, we obtain the same expressions (F.20)–(F.21),

$$p = a\eta U_k\lambda\frac{r_k}{r^3}, \tag{F.34}$$

and

$$v_i = U_i + \frac{aU_k}{2}\left[\lambda\left(\frac{r_ir_k}{r^3} + \frac{\delta_{ik}}{r}\right) + \lambda''\left(\frac{r_ir_k}{r^5} - \frac{\delta_{ik}}{3r^3}\right)\right], \tag{F.35}$$

where the divergence-free condition, $\nabla_i v_i = 0$, is satisfied identically (with $\lambda = \lambda'$). Note that now $\lambda'' \neq 0$, since we have a characteristic dimension a.

Imposing that $\mathbf{v}(\mathbf{r}) = \mathbf{0}$ at $r = a$ we find: $\lambda = -\lambda''/a^2 = -3/2$. So, at the end, we find:

$$p = -\frac{3}{2}a\eta U_k\frac{r_k}{r^3}, \tag{F.36}$$

and

$$v_i = U_i - \frac{3}{4}aU_k\left[\left(\frac{r_ir_k}{r^3} + \frac{\delta_{ik}}{r}\right) - a^2\left(\frac{r_ir_k}{r^5} - \frac{\delta_{ik}}{3r^3}\right)\right]. \tag{F.37}$$

At large r, the flow perceives only a point force F_k, so that the pressure and the velocity fields must reduce to the Stokeslet (F.26)–(F.27). Therefore, $F_k/4\pi = \frac{3}{2}a\eta U_k$, i.e.,

$$\mathbf{F} = 6\pi\eta a\mathbf{U}. \tag{F.38}$$

This is the Stokes law, establishing the drag force exerted on a sphere by a uniform fluid flow.

The last term in the expression (F.35) can be expressed in terms of the following singular solution of the Stokes equation,

$$v_i^{(d)}(\mathbf{r}; \mathbf{F}) = -\frac{1}{6}\nabla^2 v_i^{(s)}(\mathbf{r}; \mathbf{F}) = F_k\, V_{ik}^{(d)}(\mathbf{r}); \qquad p^{(d)}(\mathbf{r}; \mathbf{F}) = 0, \qquad (F.39)$$

where,

$$V_{ik}^{(d)}(\mathbf{r}) = \frac{1}{8\pi\eta\, r^3}\left(\frac{r_i r_k}{r^2} - \frac{1}{3}\delta_{ik}\right) = \frac{1}{8\pi\eta}H_{ik}^{(-3)}(\mathbf{r}). \qquad (F.40)$$

This term is the harmonic tensorial function (F.4) with $n = 2$, and therefore it is referred to as a *potential doublet*, since it is identical to a doublet in potential flow. Consequently, according to the d'Alambert paradox, the drag force exerted by the potential doublet (as well as by any potential flow) is equal zero.

Comparing the expression (F.27) and (F.40) with the solution (F.37) of the creeping flow past a sphere we see that the Stokes flow past a sphere is the sum of a Stokeslet and a potential doublet, i.e.,

$$\mathbf{v}(\mathbf{r}) = \mathbf{U} - \mathbf{F} \cdot \left[\mathbf{V}^{(s)}(\mathbf{r}) - a^2 \mathbf{V}^{(d)}(\mathbf{r})\right], \qquad (F.41)$$

where $\mathbf{F} = 6\pi\eta a\mathbf{U}$, that is,

$$\mathbf{v}(\mathbf{r}) = \mathbf{U} \cdot \left[\mathbf{I} - 6\pi\eta a\left(1 + \frac{a^2}{6}\nabla^2\right)\mathbf{V}^{(s)}(\mathbf{r})\right], \qquad (F.42)$$

while $p(\mathbf{r}) = -\mathbf{F} \cdot \mathbf{P}^{(s)}(\mathbf{r})$. When $\mathbf{r} = 0$, Eq. (F.42) becomes a particular case of Faxen's law (see Sect. F.3.2).

F.2.3 Shear Flow Past a Sphere

In this case, the unperturbed velocity field is a constant shear flow, $\mathbf{U}(\mathbf{r}) = \mathbf{G}^s \cdot \mathbf{r}$ as $r \to \infty$, with $\mathbf{I}:\mathbf{G}^s = 0$ and $G_{ij}^s = G_{ji}^s$ is a symmetric tensor, while a sphere of radius a is kept fixed at the origin, i.e. $\mathbf{v}(\mathbf{r}) = 0$ at $r = a$. Then, proceeding as before, we obtain,

$$p = a^3\eta G_{ij}^s\lambda\left(\frac{r_i r_j}{r^5} - \frac{1}{3}\frac{\delta_{ij}}{r^3}\right), \qquad (F.43)$$

and

$$u_i = a^5 G_{jk}^s\lambda'\left(\frac{r_i r_j r_k}{r^7} - \frac{\delta_{ik}r_j + \delta_{ij}r_k + \delta_{jk}r_i}{5r^5}\right). \qquad (F.44)$$

Considering that $G_{ii}^s = 0$, we see that the last term in (F.43) and the last term in (F.44) are identically zero. Then, substituting these results in (F.17) we obtain:

$$v_i = G_{ik}^s r_k \left(1 - \frac{2}{5}\lambda' \frac{a^5}{r^5}\right) + G_{jk}^s \frac{r_i r_j r_k}{2r^5} a^3 \left(\lambda + 2\lambda' \frac{a^2}{r^2}\right). \qquad (F.45)$$

Imposing that $\mathbf{v}(\mathbf{r}) = \mathbf{0}$ at $r = a$ we find: $\lambda = -5$ and $\lambda' = 5/2$, so that at the end we obtain:

$$p = -5a^3 \eta G_{ij}^s \frac{r_i r_j}{r^5}, \qquad (F.46)$$

and

$$v_i = G_{ik}^s r_k \left(1 - \frac{a^5}{r^5}\right) - \frac{5}{2} G_{jk}^s \frac{r_i r_j r_k}{r^5} a^3 \left(1 - \frac{a^2}{r^2}\right). \qquad (F.47)$$

F.2.4 Rotation Flow Past a Sphere

In this case, the unperturbed velocity field is a constant rigid body rotation, with $\mathbf{v}(\mathbf{r}) = \mathbf{G}^a \cdot \mathbf{r}$ as $r \to \infty$, with $G_{ij}^a = -G_{ji}^a$ is an antisymmetric tensor, while a sphere of radius a is kept fixed at the origin, i.e. $\mathbf{v}(\mathbf{r}) = \mathbf{0}$ at $r = a$. Then, since $p = 0$ out of symmetry, the velocity must be a harmonic function and we easily find:

$$v_i = G_{ik}^a r_k \left(1 - \frac{a^3}{r^3}\right); \qquad p = 0, \qquad (F.48)$$

where $\mathbf{G}^a \cdot \mathbf{r} = \mathbf{\Omega} \times \mathbf{r}$ and $\mathbf{\Omega} = \frac{1}{2}\boldsymbol{\epsilon} : \nabla \mathbf{v}$ is the angular velocity of the unperturbed flow. In this case, the torque experienced by the sphere can be easily determined, obtaining:

$$\mathbf{\Gamma} = 8\pi \eta a^3 \mathbf{\Omega}. \qquad (F.49)$$

F.2.5 Stokes Multipole Expansion

As for the Laplace equation, the disturbance of the flow field due to the presence of a submerged object can be generally expressed using a multipole expansion in terms of the gradients of the Stokeslet, as in (F.5), obtaining,

$$p(\mathbf{r}) = \sum_{n=0}^{\infty} \mathbf{F}^{(n+1)} (\cdot)^{n+1} \nabla^n \mathbf{P}^{(s)}, \qquad (F.50)$$

and,

$$\mathbf{v}(\mathbf{r}) = \sum_{n=0}^{\infty} \mathbf{F}^{(n+1)} (\cdot)^{n+1} \nabla^n \mathbf{V}^{(s)}, \tag{F.51}$$

where $\nabla^n = \nabla\nabla\nabla \cdots \nabla$ (n times), while $\mathbf{F}^{(n)}$ is an n-th order constant tensor expressing the strength of the n-th multipole, to be determined by satisfying the boundary conditions. Formally, in fact, this flow field distribution solves the following Stokes equation, as in (F.6),

$$\nabla p - \eta\nabla^2\mathbf{v} = \sum_{n=0}^{\infty} \mathbf{F}^{(n+1)} (\cdot)^n \underbrace{\nabla\nabla \cdots \nabla}_{n \text{ times}} \delta(r); \qquad \nabla \cdot \mathbf{v} = 0, \tag{F.52}$$

thus generalizing the Stokeslet definition (F.29). Thus, $\mathbf{F}^{(1)}$ is the force exerted by the object on the fluid, $\mathbf{F}^{(2)}$ is the corresponding moment of dipole, $\mathbf{F}^{(3)}$ the moment of quadrupole, etc. In turn, due to the linearity of the Stokes equation, these multipole strengths are proportional to the gradients of the unperturbed velocity $\mathbf{U}$ at the origin, that is the velocity field in the absence of the particle, i.e.,

$$\mathbf{F}^{(m)} = -\sum_{n} \mathbf{R}^{(mn)} (\cdot)^n \nabla^{n-1}\mathbf{U}|_{\mathbf{r}=0}, \tag{F.53}$$

where $\mathbf{R}^{(mn)}$ is a grand resistance matrix.

In particular, for an isolated sphere of radius a fixed at the origin and immersed in an arbitrary flow $\mathbf{U}(\mathbf{r})$, Faxen's law (F.88) yields:

$$\mathbf{F}^{(1)} = 6\pi a\eta\left[1 + \frac{a^2}{6}\right]\mathbf{U}|_{\mathbf{r}=0}, \tag{F.54}$$

so that

$$R_{ij}^{(11)} = 6\pi\eta a\delta_{ij}; \qquad R_{ijk}^{(12)} = 0; \qquad R_{ijk\ell}^{(13)} = \pi\eta a^3\delta_{i\ell}\delta_{jk}. \tag{F.55}$$

As for the dipole strength, $\mathbf{F}^{(2)}$, its contribution to (F.51) can be written as:

$$v_i = F_{jk}^{(2)}\nabla_k V_{ji}^{(s)} = F_{jk}^{(S)}\tfrac{1}{2}\left(\nabla_k V_{ji}^{(s)} + \nabla_j V_{ki}^{(s)}\right) + F_{jk}^{(R)}\tfrac{1}{2}\left(\nabla_k V_{ji}^{(s)} - \nabla_j V_{ki}^{(s)}\right), \tag{F.56}$$

where $\mathbf{F}^{(S)}$ and $\mathbf{F}^{(R)}$ are the symmetric and antisymmetric components of $\mathbf{F}^{(2)}$, respectively, while the superscript S and R stand for stresslet and rotlet (see next section). Now, consider that the gradient of the Stokeslet is:

$$\nabla_k V_{ji}^{(s)}(\mathbf{r}) = \frac{1}{8\pi\eta r^3}\left[\left(\delta_{jk} - 3\frac{r_j r_k}{r^2}\right)r_i + (\delta_{ij}r_k - \delta_{ik}r_j)\right], \tag{F.57}$$

and

$$\nabla_k P_i^{(s)}(\mathbf{r}) = \frac{1}{4\pi}\left(\frac{\delta_{ik}}{r^3} - 3\frac{r_i r_k}{r^5}\right). \tag{F.58}$$

Comparing (F.57) and (F.58) with (F.46)–(F.48) for $r \ll a$, and considering that $\mathbf{I}{:}\mathbf{G}^s = 0$, we obtain $3/(8\pi\eta)\mathbf{F}^{(S)} = -5/2a^3\mathbf{G}^s$ and $2/(8\pi\eta)\mathbf{F}^{(R)} = -a^3\mathbf{G}^a$, so that at the end we find:

$$\mathbf{F}^{(2)} = -\frac{20}{3}\pi\eta a^3 \left[\frac{1}{2}(\nabla\mathbf{U}+\nabla\mathbf{U}^+)\right]_{\mathbf{r}=0} - 4\pi\eta a^3 \left[\frac{1}{2}(\nabla\mathbf{U}-\nabla\mathbf{U}^+)\right]_{\mathbf{r}=0} \qquad (\text{F.59})$$

and therefore,

$$R_{ijk\ell}^{(22)} = \pi\eta a^3 \left(\frac{16}{3}\delta_{ik}\delta_{j\ell} + \frac{4}{3}\delta_{i\ell}\delta_{jk}\right). \qquad (\text{F.60})$$

Obviously, if the suspended sphere is neutrally buoyant, then $\mathbf{F}^{(1)} = \mathbf{F}^{(R)} = \mathbf{0}$, so that the only contribution that is left is the stresslet, yielding at the end:

$$R_{ijk\ell}^{(22)} = \frac{10}{3}\pi\eta a^3 (\delta_{ik}\delta_{j\ell} + \delta_{i\ell}\delta_{jk}). \qquad (\text{F.61})$$

F.2.6   Stresslet and Rotlet

As we saw in the previous section, the disturbance of the flow field due to the presence of a submerged object can be calculated using a multi-pole expansion, i.e. expanding the propagator, or Stokeslet, in a Taylor series,

$$p^{(s)}(\mathbf{r}-\mathbf{r}_S) = p^{(s)}(\mathbf{r}) - p^{(D)}(\mathbf{r}) + \cdots; \qquad \mathbf{v}^{(s)}(\mathbf{r}-\mathbf{r}_S) = \mathbf{v}^{(s)}(\mathbf{r}) - \mathbf{v}^{(D)}(\mathbf{r}) + \cdots,$$

where the first correction terms,

$$p^{(D)}(\mathbf{r}) = \mathbf{r}_S \cdot \nabla p^{(s)}(\mathbf{r}); \qquad v_j^{(D)}(\mathbf{r}) = \mathbf{r}_S \cdot \nabla v_j^{(s)}(\mathbf{r}). \qquad (\text{F.62})$$

constitute the Stokes dipole, or *doublet* [3, 4]. The latter is obtained by superimposing two Stokeslets of infinitely large, equal, and opposite strengths, separated by an infinitesimal distance d, such that the separation is normal to the force direction. Accordingly, considering the gradient of the Stokeslet, (F.57) and (F.58), and denoting

$$\alpha = \frac{F}{8\pi\eta}; \qquad \beta = -\mathbf{r}_S, \qquad (\text{F.63})$$

we see that the Stokes doublet can be written as:

$$v_j^{(D)}(\mathbf{r}; \boldsymbol{\alpha}, \boldsymbol{\beta}) = \beta_i V_{ijk}^{(D)}(\mathbf{r})\alpha_k, \qquad (\text{F.64})$$

and

$$p^{(D)}(\mathbf{r}; \boldsymbol{\alpha}, \boldsymbol{\beta}) = \beta_i P_{ik}^{(D)}(\mathbf{r})\alpha_k, \qquad (\text{F.65})$$

where

$$V_{ijk}^{(D)}(\mathbf{r}) = \frac{1}{r^3}\left(\delta_{ik} - 3\frac{r_i r_k}{r^2}\right)r_j + \frac{\delta_{ij}r_k - \delta_{jk}r_i}{r^3}, \tag{F.66}$$

and

$$P_{ik}^{(D)}(\mathbf{r}) = 2\eta\frac{1}{r^3}\left(\delta_{ik} - 3\frac{r_i r_k}{r^2}\right). \tag{F.67}$$

Since the Stokes flow is not irrotational, the velocity gradient tensor resulting from the Stokes dipole has both a symmetric (strain) and an antisymmetric (rotation) component, with [see (8.12)] $\nabla\mathbf{v} = \boldsymbol{\epsilon} : \boldsymbol{\Omega} + \mathbf{S}$, where $\boldsymbol{\Omega} = \frac{1}{2}\boldsymbol{\epsilon} : \nabla\mathbf{v}$ is the angular velocity, while $\mathbf{S}$ is the shear rate tensor. Then, we see that the Stokes doublet can be written as the sum of two terms, i.e. $\mathbf{v}^{(D)} = \mathbf{v}^{(R)} + \mathbf{v}^{(S)}$ and $p^{(D)} = p^{(R)} + p^{(R)}$, where,

$$\mathbf{v}^{(R)}(\mathbf{r}; \boldsymbol{\gamma}) = \frac{\boldsymbol{\gamma} \times \mathbf{r}}{r^3}; \qquad p^{(R)}(\mathbf{r}; \boldsymbol{\gamma}) = 0, \tag{F.68}$$

with $\boldsymbol{\gamma} = \boldsymbol{\beta} \times \boldsymbol{\alpha}$, is the *rotlet*,[6] while,

$$\mathbf{v}^{(S)}(\mathbf{r}; \boldsymbol{\alpha}, \boldsymbol{\beta}) = \frac{1}{r^3}\left[(\boldsymbol{\alpha}\cdot\boldsymbol{\beta}) - 3\frac{(\boldsymbol{\alpha}\cdot\mathbf{r})(\boldsymbol{\beta}\cdot\mathbf{r})}{r^2}\right]\mathbf{r}, \tag{F.69}$$

$$p^{(S)}(\mathbf{r}; \boldsymbol{\alpha}, \boldsymbol{\beta}) = \frac{2\eta}{r^3}\left[\boldsymbol{\alpha}\cdot\boldsymbol{\beta} - 3\frac{(\boldsymbol{\alpha}\cdot\mathbf{r})(\boldsymbol{\beta}\cdot\mathbf{r})}{r^2}\right], \tag{F.70}$$

is the *stresslet*.

The rotlet can be regarded as the flow field generated by a singular point torque at the origin, as it coincides with the velocity field generated by a sphere of radius a rotating with angular velocity $\boldsymbol{\omega} = \boldsymbol{\gamma}/a^3$ and subjected to a torque $\boldsymbol{\Gamma}_0 = 8\pi\eta a^3\boldsymbol{\omega}$. Since the flow field is potential, it exerts zero force on the fluid, just like the potential doublet.

These statements can be proved considering that, for any closed surface S containing the rotlet, we have:

$$\boldsymbol{\Gamma}_0 = -\oint_S \mathbf{r}_0 \times \left(\mathbf{n}\cdot\mathbf{T}^{(R)}\right)d^2\mathbf{r} = -\int_V \mathbf{r}_0 \times \left(\nabla\cdot\mathbf{T}^{(R)}\right)d^3\mathbf{r},$$

where $\mathbf{T}^{(R)} = -p^{(R)}\mathbf{I} + 2\eta\mathbf{S}^{(R)}$ is the rotlet stress tensor.[7] Now, define $\mathbf{f}^{(R)}$ as the singular forcing function that generates the rotlet, i.e.,

$$-\nabla\cdot\mathbf{T}^{(R)} = \nabla p^{(R)} - \eta\nabla^2\mathbf{v}^{(R)} = \mathbf{f}^{(R)} = 4\pi\eta\nabla\times\left[\boldsymbol{\gamma}\delta(\mathbf{r})\right].$$

---

[6] Also called *couplet* by [1].

[7] The minus sign comes from the fact that the unit vector $\mathbf{n}$ here is directed inward.

Consequently,

$$\mathbf{\Gamma}_0 = \int_V \mathbf{r}_0 \times \mathbf{f}^{(R)} \, d^3\mathbf{r} = 8\pi \eta \boldsymbol{\gamma}. \tag{F.71}$$

This confirms that if a torque $\mathbf{\Gamma}_0$ is exerted on a sphere of radius a immersed in an otherwise quiescent fluid, the sphere will rotate with angular velocity $\boldsymbol{\omega} = \mathbf{\Gamma}_0/8\pi \eta a^3$.

The stresslet represents straining motion of the fluid symmetric about the $\alpha\beta$ plane, with the principle axes of strain lying in the $\alpha + \beta$, $\alpha - \beta$ and $\alpha \times \beta$ directions. In virtue of these symmetries, the stresslet exerts zero force and zero torque on the fluid. Sometimes it is more convenient to rewrite the stresslet as:

$$v_j^{(S)}(\mathbf{r}; \boldsymbol{\alpha}, \boldsymbol{\beta}) = \beta_i V_{ijk}^{(S)}(\mathbf{r})\alpha_k; \quad V_{ijk}^{(S)}(\mathbf{r}) = -3V_{ik}^{(d)}(\mathbf{r})r_j, \tag{F.72}$$

$$p^{(S)}(\mathbf{r}; \boldsymbol{\alpha}, \boldsymbol{\beta}) = \beta_i P_{ik}^{(S)}(\mathbf{r})\alpha_k; \quad P_{ik}^{(S)}(\mathbf{r}) = -6\eta V_{ik}^{(d)}(\mathbf{r}), \tag{F.73}$$

where $\mathbf{V}^{(d)}$ is the potential doublet (F.40). Therefore, we obtain,

$$\mathbf{v}^{(S)}(\mathbf{r}; \boldsymbol{\alpha}, \boldsymbol{\beta}) = \frac{1}{2\eta}\mathbf{r}p^{(S)}(\mathbf{r}; \boldsymbol{\alpha}, \boldsymbol{\beta}). \tag{F.74}$$

Consequently, referring to the fundamental solution (F.17) of Stokes flow, we see that the Stokes doublet can be written as

$$\mathbf{v}^{(D)} = \frac{1}{2\eta}\mathbf{r}p^{(S)} + \mathbf{v}^{(R)}, \tag{F.75}$$

where we have considered that $\mathbf{v}^{(R)}$ is a harmonic function, so that $p^{(D)} = p^{(S)}$.

The Stokes quadruple distribution, $\nabla\nabla\mathbf{v}^{(s)}$ is more complicated; however, one component is particular useful, namely the potential doublet $\nabla^2\mathbf{v}^{(s)}$, as we have seen in Eq. (F.39).

F.3 Further Topics

F.3.1 Reciprocal Theorems

Consider the smooth velocity and stress fields, $(\mathbf{v}'; \mathbf{T}')$ and $(\mathbf{v}''; \mathbf{T}'')$, with

$$\mathbf{T} = -p\mathbf{I} + \eta(\nabla\mathbf{v} + \nabla\mathbf{v}^+); \qquad \nabla \cdot \mathbf{T} = \mathbf{0} \quad \text{and} \quad \nabla \cdot \mathbf{v} = 0,$$

corresponding to any two flows of the same fluid in the volume V outside a closed surface S. Clearly we have:

$$\int_V \mathbf{T}' : \nabla\mathbf{v}'' \, d^3\mathbf{r} = \int_V \mathbf{T}'' : \nabla\mathbf{v}' \, d^3\mathbf{r}. \tag{F.76}$$

Now, consider the following general vector identity:

$$\nabla_i \left(T'_{ij} v''_j - T''_{ij} v'_j \right) = v''_j \nabla_i T'_{ij} - v'_j \nabla_i T''_{ij} + T'_{ij} \nabla_i v''_j - T''_{ij} \nabla_i v'_j, \tag{F.77}$$

where the first two terms on the RHS are identically zero. Therefore, taking the volume integral and integrating by parts, we obtain:

$$\oint_S \mathbf{n} \cdot \mathbf{T}' \cdot \mathbf{v}'' \, d^2\mathbf{r} = \oint_S \mathbf{n} \cdot \mathbf{T}'' \cdot \mathbf{v}' \, d^2\mathbf{r}. \tag{F.78}$$

This is the famous reciprocal theorem, obtained in 1892 by Lorentz, who generalized the analogous Maxwell-Betti theorem.

Identical results can be obtained in heat and mass transport. In fact, considering the temperature (or concentration) and heat (or mass) flux fields, $(\theta'; \mathbf{J}')$ and $(\theta''; \mathbf{J}'')$, with $\mathbf{J} = -k\nabla\theta$ and $\nabla \cdot \mathbf{J} = 0$, we obtain:

$$\int_V \mathbf{J}' \cdot \nabla\theta'' \, d^3\mathbf{r} = \int_V \mathbf{J}'' \cdot \nabla\theta' \, d^3\mathbf{r}, \tag{F.79}$$

that is,

$$\oint_S \mathbf{n} \cdot \mathbf{J}'' \theta'' \, d^2\mathbf{r} = \oint_S \mathbf{n} \cdot \mathbf{J}'' \theta' \, d^2\mathbf{r}. \tag{F.80}$$

F.3.2 Faxen's Law

As a first application of the reciprocal theorem, let us apply Eq. (F.80) to calculate the net heat flux $\dot{Q}$ released by a perfectly conducting body (i.e. one with a very large heat conductivity) immersed in an arbitrary undisturbed temperature field, $T^\infty(\mathbf{r})$. In fact, let us denote: $(\theta'; \mathbf{J}') \equiv (T^\infty; \mathbf{J}^\infty)$, while $(\theta''; \mathbf{J}'')$ denotes the temperature field induced by a constant temperature difference T_0 between the body surface and infinity. Accordingly, Eq. (F.80) gives:

$$T_0 \dot{Q} = \oint_S \mathbf{n} \cdot \mathbf{J}'' T^\infty \, d^2\mathbf{r}. \tag{F.81}$$

In particular, when the body is a sphere of radius a and centered in the origin, $\mathbf{n} \cdot \mathbf{J}'' = k T_0 / a$, so that we obtain:

$$\dot{Q} = -\frac{k}{a} \oint_{r=a} T^\infty(\mathbf{r}) \, d^2\mathbf{r}. \tag{F.82}$$

Now, expanding $T^\infty(\mathbf{r})$ in a Taylor expansion in terms of the values of T^∞ and all its gradients at the origin, we can write:

$$T^\infty(\mathbf{r}) = T^\infty(0) + \mathbf{r} \cdot \left(\nabla T^\infty \right)_0 + \frac{1}{2}\mathbf{r}\mathbf{r}:\left(\nabla\nabla T^\infty \right)_0 + \cdots. \tag{F.83}$$

Therefore, considering that $\nabla^2 T^\infty = 0$, as well as all the other higher-order even gradients, we obtain:

$$\dot{Q} = -4\pi a k T^\infty(0).$$ (F.84)

This shows that if we introduce a perfectly conducting sphere with known temperature T in an undisturbed temperature distribution $T^\infty(\mathbf{r})$, then the net heat flux released by the sphere can be calculated directly, without actually solving the heat transfer problem, and it depends only on the undisturbed temperature at the center of the sphere. This law is equivalent to saying that this net heat flux is determined only by a monopole located at the center of the sphere, as one would expect, since all the higher-order multi-poles release no net fluxes.

In the previous analysis we have assumed that the temperature vanishes at infinity or, equivalently, that it is kept equal to zero at the surface of the sphere (remind that the temperature is uniform, as the sphere is perfectly conducting). In general, denoting by T the temperature of the sphere, Eq. (F.84) can be rewritten as:

$$T = T^\infty(0) + \frac{1}{4\pi a k}\dot{Q}.$$ (F.85)

Here we see that, when $\dot{Q} = 0$, i.e. in the absence of any flux release, we find $T = T^\infty(0)$, revealing that the temperature of the sphere is equal to the unperturbed temperature at the center.

A similar result can be obtained applying the reciprocal theorem (F.78) to calculate the hydrodynamic force on a rigid body (i.e. one with a very large viscosity), which is kept fixed in an arbitrary undisturbed flow field, $\mathbf{v}^\infty(\mathbf{r})$. Proceeding like before, we obtain:

$$\mathbf{V}_0 \cdot \mathbf{F} = \oint_S \mathbf{n} \cdot \mathbf{T}''\mathbf{v}^\infty d^2\mathbf{r},$$ (F.86)

where $\mathbf{T}''$ is the stress tensor induced by the uniform translation of the body, with velocity $\mathbf{V}_0$. In particular, when the body is a sphere of radius a, centered in the origin, Eq. (F.37) yields: $\mathbf{n} \cdot \mathbf{T}'' = (3/2a)\eta\mathbf{V}_0$, so that we obtain:

$$\mathbf{F} = \frac{3\eta}{2a} \oint_{r=a} \mathbf{v}^\infty(\mathbf{r}) d^2\mathbf{r}.$$ (F.87)

Now, expanding $\mathbf{v}^\infty(\mathbf{r})$ in a Taylor expansion as in (F.83), applying the identity (F.25), and considering that $\nabla^4\mathbf{v}^\infty = 0$, as well as all the other higher-order even gradients, we finally obtain:

$$\mathbf{F} = 6\pi a\eta\left(1 + \frac{a^2}{6}\nabla^2\right)\mathbf{v}^\infty(0).$$ (F.88)

This result is generally referred to as *Faxen's* law, stating that the force exerted on a rigid sphere by a known unperturbed flow field can be calculated directly, without actually solving the flow field problem, and it depends only on the undisturbed temperature at the center of the sphere and its Laplacian. This law is equivalent to saying that this net force is determined only by a monopole and a potential doublet located at the center of the sphere, as one would expect, since all the other Stokes multi-poles exert zero force.

In the previous analysis we have assumed that the flow field vanishes at infinity or, equivalently, that the velocity of the sphere is zero. In general, denoting by $\mathbf{V}$ the velocity of a sphere centered in the origin, Eq. (F.88) can be rewritten as:

$$\mathbf{V} = \left(1 + \frac{a^2}{6}\nabla^2\right)\mathbf{v}^\infty(\mathbf{0}) + \frac{1}{6\pi a\eta}\mathbf{F}. \tag{F.89}$$

Here we see that, when $\mathbf{F} = \mathbf{0}$, i.e. for a neutrally buoyant sphere, we find $\mathbf{V} = [1 + \frac{a^2}{6}\nabla^2]\mathbf{v}^\infty(\mathbf{0})$. Consequently, if a sphere is immersed in a linear flow field, its velocity coincides with the unperturbed velocity at its center. However, this is not true in general. For example, the velocity of a sphere immersed in a Poiseuille flow is slower than the unperturbed fluid velocity at its center.

Using the same procedure, the angular velocity $\mathbf{\Omega}$ of a sphere centered in the origin can be determined, obtaining,

$$\mathbf{\Omega} = \frac{1}{2}\nabla \times \mathbf{v}^\infty(\mathbf{0}) + \frac{1}{8\pi a^3\eta}\mathbf{\Gamma}, \tag{F.90}$$

where $\mathbf{\Gamma}$ is the torque exerted on the sphere, while $\nabla \times \mathbf{v}^\infty$ is the vorticity (i.e. twice the angular velocity) of the fluid flow the origin, that is at the center of the sphere. Here we see that, when $\mathbf{\Gamma} = \mathbf{0}$, i.e. for a neutrally buoyant sphere, we find $\mathbf{\Omega} = \frac{1}{2}\nabla \times \mathbf{v}^\infty(\mathbf{0})$, that is the sphere rotates with the same angular velocity as that of the unperturbed fluid flow at the center of the sphere. When a torque is exerted on the sphere, conversely, it will increment its angular velocity, with respect to its unperturbed rotation, by an amount $\mathbf{\Gamma}/8\pi a^3\eta$. Equation (F.90) is identical to (F.85), since the vorticity, like the temperature, satisfies the Laplace equation.

F.4 Sedimentation Velocity in Dilute Suspensions

In a dilute suspension of identical rigid spheres, particles fall through the liquid due to gravity. Under the assumption of negligible inertia and Brownian motion effects, the average velocity of the particles in a well-mixed suspension can be expressed, to leading order in the particle number density n_0, as:

$$\overline{\mathbf{V}} = \mathbf{V}_0 + n_0 \int_{r\geq 2}\left[\mathbf{V}(\mathbf{x}_0|\mathbf{x}_0 + \mathbf{r}) - \mathbf{V}_0\right]d^3\mathbf{r}, \tag{F.91}$$

where $\mathbf{V}_0$ is the Stokes terminal velocity of an isolated sphere and $\mathbf{V}(\mathbf{x}_0|\mathbf{x}_0 + \mathbf{r})$ is the velocity of a test sphere located at $\mathbf{x}_0$ in an unbounded fluid when another sphere is located at $\mathbf{x}_0 + \mathbf{r}$.

As is well known, the integral in the expression given above is divergent since $\mathbf{V} - \mathbf{V}_0$ decays as $1/r$ as $r \to \infty$. Therefore, since the average sedimentation velocity is finite, a proper renormalization of the integral is required. Although such a renormalization was given by Batchelor [2], here we present an alternate derivation which, perhaps, might be of interest in a wider context.

The velocity of the test sphere in the presence of another sphere can be determined reminding that the disturbance $\mathbf{v}(\mathbf{x}_0|\mathbf{x}_0 + \mathbf{r})$ of the fluid velocity at $\mathbf{x}_0$ due to the motion of sphere at $(\mathbf{x}_0 + \mathbf{r})$, can be determined by replacing the second sphere with a Stokeslet and a potential doublet at $(\mathbf{x}_0 + \mathbf{r})$, i.e., [cf. Eq. (F.37)]

$$\mathbf{v}(\mathbf{x}_0|\mathbf{x}_0 + \mathbf{r}) = \mathbf{V}_0\left(\frac{3}{4r} + \frac{1}{4r^3}\right) + \mathbf{r}\frac{\mathbf{r} \cdot \mathbf{V}_0}{r^2}\left(\frac{3}{4r} - \frac{3}{4r^3}\right), \tag{F.92}$$

where, without loss of generality, the radius of the particles has been set equal to unity. At this point, applying Faxen's law (see Sect. F.3.2), we obtain:

$$\mathbf{V} = \mathbf{V}_0 + \mathbf{v}(\mathbf{x}_0|\mathbf{x}_0 + \mathbf{r}) + \frac{1}{6}\nabla^2\mathbf{v}(\mathbf{x}_0|\mathbf{x}_0 + \mathbf{r}) + \mathbf{w}(\mathbf{x}_0|\mathbf{x}_0 + \mathbf{r}), \tag{F.93}$$

where $\mathbf{w}(\mathbf{x}_0|\mathbf{x}_0 + \mathbf{r})$ is the difference between the exact value and Faxen's law approximation, which decays exponentially as r^{-4}. In fact, Faxen's law applies only to an unbounded fluid and therefore cannot account for the fact that the second sphere has a non zero size.

Substituting (F.93) into (F.91) yields

$$\overline{\mathbf{V}} = \mathbf{V}_0 + \mathbf{V}_1 + \mathbf{V}_2 + \mathbf{V}_3, \tag{F.94}$$

where

$$\mathbf{V}_1 = n_0\int_{r \geq 2} \mathbf{v}(\mathbf{x}_0|\mathbf{x}_0 + \mathbf{r})\,d^3\mathbf{r}, \tag{F.95}$$

$$\mathbf{V}_2 = n_0\int_{r \geq 2} \frac{1}{6}\nabla^2\mathbf{v}(\mathbf{x}_0|\mathbf{x}_0 + \mathbf{r})\,d^3\mathbf{r}, \tag{F.96}$$

and

$$\mathbf{V}_3 = n_0\int_{r=2}^{\infty} \mathbf{w}(\mathbf{x}_0|\mathbf{x}_0 + r)\,d^3\mathbf{r}. \tag{F.97}$$

Since $\mathbf{v}$ is of order $1/r$ and $\nabla^2\mathbf{v}$ is of order $1/r^3$, the integrals $\mathbf{V}_1$ and $\mathbf{V}_2$ are divergent, while $\mathbf{V}_3$ converges, since $\mathbf{w}$ decays as r^{-4}.

Clearly, in a dilute suspension, $\mathbf{V}_1$ equals the average velocity of a fluid point when a sphere is located in the infinite domain $r \geq 2$, i.e. outside the exclusion region. Hence, the average fluid velocity within the suspension, $\overline{\mathbf{V}}^{(f)}$, is given by

$$\overline{\mathbf{V}}^{(f)} = \mathbf{V}_1 + n_0\int_{1 \leq r \leq 2} \mathbf{v}(\mathbf{x}_0|\mathbf{x}_0 + \mathbf{r})\,d^3\mathbf{r} = \mathbf{V}_1 + \frac{9}{2}\phi\mathbf{V}_0, \tag{F.98}$$

with $\phi \equiv \frac{4\pi}{3}n_0$ being the volume fraction occupied by the spheres in the suspension.

Thus, the divergent integral V_1 represents physically the average velocity of the fluid outside the exclusion regions of the spheres in a very dilute suspension, and is related to the average fluid velocity within the suspension by (F.98). Now, we know that the bulk velocity of the suspension on a macroscopic scale is zero relative to the container, so that,

$$(1 - \phi)\overline{\mathbf{V}}^{(f)} + \phi\overline{\mathbf{V}}^{(p)} = \mathbf{0}, \tag{F.99}$$

where $\overline{\mathbf{V}}^{(p)}$ is the average velocity of the particles. Up to $O(\phi)$, (F.99) gives

$$\overline{\mathbf{V}}^{(f)} = -\phi\mathbf{U}_0 + O(\phi^2). \tag{F.100}$$

Thus, the divergent integral V_1 must be renormalized such that the above constraint is satisfied. Therefore, by (F.98) and (F.100),

$$\mathbf{V}_1 = -\frac{11}{2}\phi\mathbf{U}_0. \tag{F.101}$$

This is identical to Batchelor's equation (5.3) and in fact the whole derivation parallels his.

The integral V_2, can be renormalized in a similar way by noting that, in the fluid phase, this term is related to the dynamic pressure[8] disturbance p due to a single sphere by the Stokes equation $\nabla^2\mathbf{v} = \frac{1}{\eta}\nabla p$. Therefore, aside from the factor $1/6\eta$, V_2 equals the average pressure gradient of the fluid at the point $\mathbf{x}_0$ in the infinite domain $r \geq 2$, outside the exclusion region. Consequently, the average pressure gradient over the whole domain occupied by the fluid phase in the dilute suspension, $\overline{\nabla p}^{(f)}$, is given by

$$\overline{\nabla p}^{(f)} = 6\eta\mathbf{V}_2 + n_0 \int_{1 \leq r \leq 2} \nabla p(\mathbf{x}_0|\mathbf{x}_0 + \mathbf{r}) \, d^3\mathbf{r} = 6\eta\mathbf{I}_2, \tag{F.102}$$

where the integral is evaluated using

$$p(\mathbf{x}_0 + \mathbf{r}|\mathbf{x}_0) = 3\eta\frac{\mathbf{r} \cdot \mathbf{V}_0}{2r^4}. \tag{F.103}$$

Thus, the integral V_2 is closely related to the average pressure gradient in the fluid phase, which naturally leads to the examination of the bulk pressure gradient in the suspension. On a macroscopic scale, however, the suspension as a whole is akin to a static effective fluid with total pressure gradient $\phi(\rho^{(p)} - \rho^{(f)})\mathbf{g}$, in addition to that of the stagnant pure fluid, where $\rho^{(p)}$ and $\rho^{(f)}$ are the mass densities of the particles and of the fluid, respectively, and $\mathbf{g}$ is the gravitational acceleration. Thus, we have the constraint,

$$(1 - \phi)\overline{\nabla p}^{(f)} + \phi\overline{\nabla p}^{(p)} = \phi(\rho^{(p)} - \rho^{(f)})\mathbf{g} = \frac{9}{2}\phi\eta\mathbf{V}_0, \tag{F.104}$$

---

[8]The dynamic pressure is defined as the difference between the total pressure and the static pressure of a pure fluid.

where $\overline{\nabla p}^{(p)}$ is the average pressure gradient inside the spheres,

$$\overline{\nabla p}^{(f)} \equiv \frac{1}{\frac{4}{3}\pi} \int_{r \leq 1} \nabla p(\mathbf{x_0}|\mathbf{x_0} + \mathbf{r}) \, d^3\mathbf{r}. \tag{F.105}$$

The above can be converted into an integral over the volume inside a fixed sphere. Noting that $\nabla p(\mathbf{x_0}|\mathbf{x_0} + \mathbf{r})$ depends only on the position of the sphere relative to that of the sample point and is independent of their absolute positions, we obtain

$$\int_{r \leq 1} \nabla p(\mathbf{x_0}|\mathbf{x_0} + \mathbf{r}) \, d^3\mathbf{r} = \int_{r \leq 1} \nabla p(\mathbf{x_0} - \mathbf{r}|\mathbf{x_0}) \, d^3\mathbf{r} = \int_{r \leq 1} \nabla p(\mathbf{x_0} + \mathbf{r}|\mathbf{x_0}) \, d^3\mathbf{r},$$

where account has been taken of the symmetry of the domain of integration.

To evaluate the last integral, which refers to the integral over the position of the sample point $\mathbf{x_0} + \mathbf{r}$ within the sphere located at $\mathbf{x_0}$, we make use of the divergence theorem and the fact that, on the surface of an isolated sphere translating with velocity $\mathbf{U_0}$, the fluid pressure is given by (F.103). Hence,

$$\int_{r \leq 1} \nabla p(\mathbf{x_0}|\mathbf{x_0} + \mathbf{r}) \, d^3\mathbf{r} = 2\pi \eta \mathbf{V_0}, \tag{F.106}$$

and therefore,

$$\overline{\nabla p}^{(p)} = \frac{3}{2}\eta \mathbf{V_0}. \tag{F.107}$$

From (F.104) and (F.107), we obtain the constraint

$$\overline{\nabla p}^{(f)} = 3\phi \eta \mathbf{V_0} + O(\phi^2), \tag{F.108}$$

so that, on account of (F.102),

$$\mathbf{V_2} = \frac{1}{2}\phi \mathbf{V_0}. \tag{F.109}$$

Although this is identical to Batchelor's equation (5.4), its derivation here is slightly different.

Finally, the last integral (F.97) was evaluated numerically by Batchelor [2], obtaining:

$$\mathbf{V_3} = -1.55\phi \mathbf{V_0}. \tag{F.110}$$

At the end, by substituting (F.101), (F.109) and (F.110) into (F.94), we obtain:

$$\overline{\mathbf{V}} = \mathbf{V_0}\big[1 - 6.55\phi + O(\phi^2)\big]. \tag{F.111}$$

In conclusion, Batchelor showed that, imposing two constraints on the volumetric flux and the pressure gradient, the expression for the average sedimentation velocity of a monodisperse suspension of spheres can be renormalized, thus obtaining the analytical result (F.111), exact up to $O(\phi^2)$-terms.

References

1. Batchelor, G.K.: J. Fluid Mech. **41**, 545 (1970)
2. Batchelor, G.K.: J. Fluid Mech. **52**, 245 (1972)
3. Chwang, A.T., Wu, T.W.: J. Fluid Mech. **63**, 607 (1974)
4. Leal, L.G.: Laminar Flow and Convective Transport Processes, p. 239. Butterworth, Stoneham (1992)

Appendix G
Solutions of the Problems

G.1 Chapter 1

Problem 1.1

$$\left\langle (\delta A)^4 \right\rangle = \sum_{i,j,k,\ell=1}^{N} \langle \delta a_i \delta a_j \delta a_k \delta a_\ell \rangle = \sum_{i,j,k,\ell=1}^{N} (\delta a)^4 (\delta_{ij}\delta_{k\ell} + \delta_{ik}\delta_{j\ell} + \delta_{i\ell}\delta_{jk}). \quad \text{(G.1)}$$

Therefore:

$$\left\langle (\delta A)^4 \right\rangle = 3N^2 (\delta a)^4. \quad \text{(G.2)}$$

Problem 1.2 Multiplying Eq. (1.21) by $-\mathbf{g}^{-1}$, and considering that $\mathbf{x} = -\mathbf{g}^{-1} \cdot \mathbf{X}$, we obtain:

$$\langle x_i \, x_k \rangle = -g_{kj}^{-1} \langle x_i \, X_j \rangle = g_{ik}^{-1}, \quad \text{(G.3)}$$

where we have considered that $\mathbf{g}$ is a symmetric matrix.

In the same way:

$$\langle X_i \, X_k \rangle = -g_{ij} \langle x_j \, X_k \rangle = g_{ik}. \quad \text{(G.4)}$$

Problem 1.3 Choosing S and P as independent variables, consider the following equalities,

$$\Delta T = \left(\frac{\partial T}{\partial S} \right)_P \Delta S + \left(\frac{\partial T}{\partial P} \right)_S \Delta P = \frac{T}{n c_p} \Delta S + \left(\frac{\partial T}{\partial P} \right)_S \Delta P, \quad \text{(G.5)}$$

and

$$\Delta V = \left(\frac{\partial V}{\partial S} \right)_P \Delta S + \left(\frac{\partial V}{\partial P} \right)_S \Delta P = \left(\frac{\partial T}{\partial P} \right)_S \Delta S - V \kappa_S \Delta P, \quad \text{(G.6)}$$

R. Mauri, *Non-Equilibrium Thermodynamics in Multiphase Flows*,
Soft and Biological Matter, DOI 10.1007/978-94-007-5461-4,
© Springer Science+Business Media Dordrecht 2013

where we have applied Maxwell's relation $(\frac{\partial T}{\partial P})_S = (\frac{\partial V}{\partial S})_P$. Finally we obtain:

$$\Delta G = \frac{1}{2}\left[\frac{T}{nc_P}(\Delta S)^2 + V\kappa_S(\Delta P)^2\right],\tag{G.7}$$

from which it is easy to obtain the final result, confirming that $c_P > 0$ and $\kappa_S > 0$.

Problem 1.4 Choosing T and P as independent variables, for $\langle(\Delta T)^2\rangle$ and $\langle(\Delta P)^2\rangle$ we will obtain the results (1.43) and (1.47) that we have already derived. The cross term can be found as follows:

$$\langle\Delta T\Delta P\rangle = \langle(\Delta T)^2\rangle\left(\frac{\partial P}{\partial T}\right)_V + \langle\Delta T\Delta V\rangle\left(\frac{\partial P}{\partial V}\right)_T = \frac{kT^2\alpha_P}{c_V\kappa_T}.\tag{G.8}$$

Here, we have considered that $\langle\Delta T\Delta V\rangle = 0$, while $\langle(\Delta T)^2\rangle = \frac{kT^2}{nc_V}$, and we have substituted the thermodynamic equality:

$$\left(\frac{\partial P}{\partial T}\right)_V = -\left(\frac{\partial P}{\partial V}\right)_T\left(\frac{\partial V}{\partial T}\right)_P = \frac{\alpha_P}{\kappa_T},\tag{G.9}$$

where we have defined the thermal expansion coefficient $\alpha_P = \frac{1}{V}(\partial V/\partial T)_P$.

In the same way, choosing V and P as independent variables, for $\langle(\Delta V)^2\rangle$ and $\langle(\Delta P)^2\rangle$ we will obtain the results (1.43) and (1.47) that we have already derived, while the cross term can be found as follows:

$$\langle\Delta V\Delta P\rangle = \langle(\Delta V)^2\rangle\left(\frac{\partial P}{\partial V}\right)_T + \langle\Delta T\Delta V\rangle\left(\frac{\partial P}{\partial T}\right)_V = -kT.\tag{G.10}$$

Note that this variance is always negative, as to an increase in pressure must correspond a decrease in volume.

In alternative, it is instructive to obtain the first of these results using a *brute force* approach. Consider the following equalities,

$$\Delta S = \left(\frac{\partial S}{\partial T}\right)_P\Delta T + \left(\frac{\partial S}{\partial P}\right)_T\Delta P = \frac{T}{nc_P}\Delta T - V\alpha_P\Delta P,\tag{G.11}$$

and

$$\Delta V = \left(\frac{\partial V}{\partial T}\right)_P\Delta T + \left(\frac{\partial V}{\partial P}\right)_T\Delta P = V\alpha_P\Delta T - V\kappa_T\Delta P,\tag{G.12}$$

where we have applied Maxwell's relation $-(\frac{\partial S}{\partial P})_T = (\frac{\partial V}{\partial T})_P$. Finally we obtain:

$$\Delta G = \frac{1}{2}\left[\frac{nc_P}{T}(\Delta T)^2 - 2V\alpha_P\Delta T\Delta P + V\kappa_T(\Delta P)^2\right].\tag{G.13}$$

Comparing this relation with (1.30), with $x_1 = \Delta T$ and $x_2 = \Delta P$, we see that:

$$g_{11} = \frac{nc_P}{kT^2}; \qquad g_{22} = \frac{V \kappa_T}{kT}; \qquad g_{12} = g_{21} = -\frac{V \alpha_P}{kT}, \qquad \text{(G.14)}$$

or,

$$\mathbf{g} = \frac{V}{kT} \begin{pmatrix} \frac{nc_P}{TV} & -\alpha_P \\ -\alpha_P & \kappa_T \end{pmatrix}. \qquad \text{(G.15)}$$

Therefore, calculating the inverse of the $\mathbf{g}$-matrix, we obtain:

$$\mathbf{g}^{-1} = \frac{kT^2}{n c_V \kappa_T} \begin{pmatrix} \kappa_T & \alpha_P \\ \alpha_P & \frac{nc_P}{TV} \end{pmatrix} \qquad \text{(G.16)}$$

where we have considered the thermodynamic identity,

$$n\kappa_T(c_P - c_V) = V T \alpha_P^2.$$

This relation shows that,

$$\langle (\Delta T)^2 \rangle = \frac{kT^2}{nc_V}; \qquad \langle (\Delta P)^2 \rangle = \frac{kT}{V \kappa_S}; \qquad \langle \Delta T \, \Delta P \rangle = \frac{kT^2 \alpha_P}{c_V \kappa_T}, \qquad \text{(G.17)}$$

where we have considered that $\kappa_T / \kappa_S = c_P / c_V$.

G.2 Chapter 2

Problem 2.1 Denoting by $\mathbf{F}$ and $\mathbf{\Gamma}$ the force and the torque exerted by the fluid on the particle and by $\mathbf{V}$ and $\mathbf{\Omega}$ its velocity and angular velocity, respectively,[1] in creeping flow regime the entropy production rate is

$$T\sigma^{(S)} = \mathbf{F} \cdot \mathbf{V} + \mathbf{\Gamma} \cdot \mathbf{\Omega}. \qquad \text{(G.18)}$$

Therefore, the general phenomenological equations relating forces and fluxes is:

$$\mathbf{F} = -\boldsymbol{\zeta}^{(tt)} \cdot \mathbf{V} - \boldsymbol{\zeta}^{(tr)} \cdot \mathbf{\Omega}, \qquad \text{(G.19)}$$

$$\mathbf{\Gamma} = -\boldsymbol{\zeta}^{(rt)} \cdot \mathbf{V} - \boldsymbol{\zeta}^{(rr)} \cdot \mathbf{\Omega}, \qquad \text{(G.20)}$$

where $\boldsymbol{\zeta}^{(tt)}$ is the translation resistance tensor, $\boldsymbol{\zeta}^{(rr)}$ is the rotation resistance tensor, while $\boldsymbol{\zeta}^{(tr)}$ and $\boldsymbol{\zeta}^{(rt)}$ are coupling resistance tensors. This shows that, in general, an applied force induces in the body both a velocity and an angular velocity, and

---

[1]In general, $\mathbf{V}$ is the difference between the particle velocity and the local fluid velocity, and analogously for $\mathbf{\Omega}$.

likewise for an applied torque (think, for example, of a body with a corkscrew geometry).[2]

Now, in order to apply the Onsager reciprocity relations, let us define the following sixth-order generalized forces velocity vectors as:

$$\mathcal{F}_i = (F_1, F_2, F_3, \Gamma_1, \Gamma_2, \Gamma_3); \qquad V_i = (V_1, V_2, V_3, \Omega_1, \Omega_2, \Omega_3). \tag{G.21}$$

So, the entropy production rate is expressed as:

$$T\sigma^{(S)} = \sum_{i=1}^{6} \mathcal{F}_i V_i, \tag{G.22}$$

with the phenomenological equations and the Onsager reciprocity relations becom-
ing:

$$\mathcal{F}_i = \sum_{i=1}^{6} \zeta_{ij} V_i; \qquad \zeta_{ij} = \zeta_{ji}. \tag{G.23}$$

Considering for sake of simplicity a 2D case. The phenomenological equations become:

$$\begin{pmatrix} F_1 \\ F_1 \\ \Gamma_1 \\ \Gamma_2 \end{pmatrix} = \begin{pmatrix} \zeta_{11}^{(tt)} & \zeta_{12}^{(tt)} & \zeta_{11}^{(tr)} & \zeta_{12}^{(tr)} \\ \zeta_{21}^{(tt)} & \zeta_{22}^{(tt)} & \zeta_{21}^{(tr)} & \zeta_{22}^{(tr)} \\ \zeta_{11}^{(rt)} & \zeta_{12}^{(rt)} & \zeta_{11}^{(rr)} & \zeta_{12}^{(rr)} \\ \zeta_{21}^{(rt)} & \zeta_{22}^{(rt)} & \zeta_{11}^{(rr)} & \zeta_{12}^{(rr)} \end{pmatrix} \begin{pmatrix} V_1 \\ V_2 \\ \Omega_1 \\ \Omega_2 \end{pmatrix}. \tag{G.24}$$

Therefore, the Onsager reciprocity relations can be written as:

$$\zeta_{ij}^{(tt)} = \zeta_{ji}^{(tt)}, \qquad \zeta_{ij}^{(rr)} = \zeta_{ji}^{(rr)}, \qquad \zeta_{ij}^{(tr)} = \zeta_{ji}^{(rt)}, \tag{G.25}$$

that is,

$$\zeta^{(tt)} = \zeta^{(tt)+}, \qquad \zeta^{(rr)} = \zeta^{(rr)+}, \qquad \zeta^{(tr)} = \zeta^{(rt)+}. \tag{G.26}$$

The first two equalities indicate that the translation and rotation resistance tensors
are symmetric; so, for example, the ratio between the force applied along x_1, F_1 and
the resulting velocity along x_2, V_2, is equal to the ratio between F_2 and V_1. The last
equality is very interesting, as it relates the rotation induced by an applied linear
force with the translation induced by an applied torque.

Problem 2.2 The equation of motion is

$$m\ddot{x} + \zeta\dot{x} + m\omega_0^2 x = f, \tag{G.27}$$

---

[2] A beautiful study of these effects, with all the involved symmetry properties, can be found in [6].

or, in frequency domain,

$$\widehat{x}(\omega) = \frac{1}{kT}\widehat{\kappa}(\omega)\widehat{f}(\omega),$$ (G.28)

where

$$\widehat{\kappa}(\omega) = \frac{kT}{m(\omega_0^2 - \omega^2) - i\zeta}.$$ (G.29)

Therefore,

$$(\widehat{ff})(\omega) = (kT)^2\frac{2}{\omega}\mathrm{Im}\left\{\frac{1}{\kappa^*(\omega)}\right\} = 2kT\zeta.$$ (G.30)

So, as expected, the random force is delta-correlated.

G.3 Chapter 3

Problem 3.1 When $t \gg m/\zeta$, the inertial term of the Langevin equation can be neglected and therefore the Langevin equation reduces to:

$$\zeta\dot{x} = f(t),$$ (G.31)

where $f(t)$ satisfies the relations (3.25) and (3.26). Squaring this equation and taking its average we obtain:

$$\langle\dot{x}(t)\dot{x}(t')\rangle = \frac{\langle f(t)f(t')\rangle}{\zeta^2} = 2D\delta(t-t'); \quad D = \frac{kT}{\zeta},$$ (G.32)

where we have substituted Eq. (3.26). Now,

$$x(t) = \int_0^t \dot{x}(t')\,dt',$$ (G.33)

with $x(0) = x_0 = 0$, and therefore

$$\langle x^2\rangle_t^0 = \int_0^t \int_0^t \langle\dot{x}(t')\dot{x}(t'')\rangle_t^0\,dt'\,dt'',$$ (G.34)

obtaining again Stokes-Einstein relation,

$$\langle x^2\rangle_t^0 = 2Dt; \quad D = \frac{kT}{\zeta}.$$ (G.35)

Problem 3.2 The Stokes-Einstein relation can be obtained also in the following way.

$$\langle v_0v(\tau)\rangle = \langle v_0\langle v\rangle_\tau^{v_0}\rangle = \langle v_0^2\rangle e^{-\zeta\tau/m} = \frac{kT}{m}e^{-\zeta\tau/m},$$ (G.36)

where we have applied Eqs. (2.18) and (3.30). Therefore,

$$\langle x^2 \rangle_t^0 = \int_0^t \int_0^t \langle v(t')v(t'')\rangle_t^0 \, dt' \, dt'' = \frac{kT}{m} \int_0^t dt' \int_{-t'}^{t-t'} d\tau e^{-\zeta|\tau|/m}, \qquad \text{(G.37)}$$

that is

$$\langle x^2 \rangle_t^0 = 2\frac{kT}{\zeta}t\left[1 - \frac{m}{\zeta t}(1 - e^{-\zeta t/m})\right]. \qquad \text{(G.38)}$$

For $t \gg m/\zeta$, we find again the Stokes-Einstein relation.

Problem 3.3 Defining $\delta p = p - p_0$, from the general solution of the Langevin equation, Eq. (3.29), for $\tilde{t} = t\zeta/m \ll 1$, we have:

$$\langle \delta p \rangle_t^{p_0} = p_0\left[e^{-\zeta t/m} - 1\right] = -\frac{\zeta}{m}p_0 t + O\left(\tilde{t}^2\right), \qquad \text{(G.39)}$$

and

$$\langle (\delta p)^2 \rangle_t^{p_0} = \int_0^t \int_0^t e^{[-\frac{\zeta}{m}(t-t')-\frac{\zeta}{m}(t-t'')]}\langle f(t')f(t'')\rangle dt'dt'' + O\left(\tilde{t}^2\right), \qquad \text{(G.40)}$$

that is,

$$\langle (\delta p)^2 \rangle_t^{p_0} = \frac{Qm}{\zeta}(1 - e^{-2\zeta t/m}) = 2kT\zeta t + O\left(\tilde{t}^2\right). \qquad \text{(G.41)}$$

Problem 3.4 Consider the Brownian motion of a particle with mass m immersed in a fluid and attracted to the origin through a linear force,

$$m\ddot{z} + \zeta\dot{z} + Az = f(t), \qquad \text{(G.42)}$$

where $f(t)$ is a random noise. The question is whether $\langle f^2 \rangle$ is the same that we have found in the absence of any external force, that is defined through Eqs. (3.25) and (3.26). Equation (3.54) can be written as:

$$\dot{z} = v, \qquad \text{(G.43)}$$

$$m\dot{v} = -\zeta v - Az + f(t), \qquad \text{(G.44)}$$

that is Eq. (3.37) with:

$$\mathbf{x} = (z, mv); \qquad \mathbf{M} = \begin{pmatrix} 0 & -m^{-1} \\ A & \zeta m^{-1} \end{pmatrix}; \qquad \tilde{\mathbf{J}} = (0, f). \qquad \text{(G.45)}$$

At equilibrium we have the Boltzmann distribution,

$$\Pi^{eq} = C\exp\left(-\frac{mv^2}{2kT} - \frac{Ax^2}{2kT}\right), \qquad \text{(G.46)}$$

so that:

$$\mathbf{g} = \frac{1}{kT}\begin{pmatrix} A & 0 \\ 0 & m^{-1} \end{pmatrix}. \tag{G.47}$$

Finally we obtain:

$$\mathbf{Q} = \left(\mathbf{M} \cdot \mathbf{g}^{-1}\right)^{sym} = kT \left[\begin{pmatrix} 0 & -m^{-1} \\ A & \zeta m^{-1} \end{pmatrix}\begin{pmatrix} A^{-1} & 0 \\ 0 & m \end{pmatrix}\right]^{sym}, \tag{G.48}$$

that is

$$\mathbf{Q} = kT\begin{pmatrix} 0 & -1 \\ 1 & \zeta \end{pmatrix}^{sym} = kT\zeta\begin{pmatrix} 0 & 0 \\ 0 & 1 \end{pmatrix}. \tag{G.49}$$

Therefore, we find again the usual relation,

$$\left\langle f(t)f(t+\tau)\right\rangle_{\tau}^{0} = 2kT\zeta\delta(\tau). \tag{G.50}$$

G.4 Chapter 4

Problem 4.1

- M can be determined from the phenomenological relation $\langle \dot{v}\rangle_{t}^{0} = -M\langle v\rangle_{t}^{0}$, with $M = \zeta/m$;
- Q can be determined from the fluctuation-dissipation theorem, $Q = Mg^{-1}$, with $g = m/kT$, so that $Q = kT\zeta/m^{2}$.

Then, the Fokker-Planck equation for $\Pi(v, t|v_0)$ (note that v is the independent random variable) becomes the Kramers equation:

$$\dot{\Pi} + \frac{\partial J_v}{\partial v} = 0; \quad J_v = -\frac{\zeta}{m}v\Pi - \frac{kT\zeta}{m^2}\frac{\partial \Pi}{\partial v}. \tag{G.51}$$

At equilibrium, $J_v = 0$ and $d\Pi/\Pi = -(m/kT)dv$, so that we obtain the Maxwell distribution,

$$\Pi^{eq}(v) = C\exp\left(-\frac{mv^2}{2kT}\right). \tag{G.52}$$

The general solution (4.55) in this case becomes:

$$\Pi[v(t)|v_0] = C\exp\left(-\frac{1}{2}V(\Delta v)^2\right), \tag{G.53}$$

where

$$\Delta v = v - \langle v\rangle_{t}^{0} = v - v_0 e^{-\zeta t/m}; \tag{G.54}$$

$$V^{-1} = \langle(\Delta v)^2\rangle_{t}^{0} = \frac{kT}{m}\left[1 - e^{-2\zeta t/m}\right]. \tag{G.55}$$

Problem 4.2 Considering Eqs. (3.52) and (3.53), from the definitions (4.7) and (4.8) we obtain, as $t \to 0$,

$$\langle \delta v \rangle_t^{v_0} = -\frac{\zeta}{m} v_0 t = -M v_0 t, \tag{G.56}$$

and

$$\left\langle (\delta v)^2 \right\rangle_t^{v_0} = \frac{2kT\zeta}{m^2} t = 2Qt, \tag{G.57}$$

with,

$$M = \frac{\zeta}{m}; \qquad Q = \frac{kT\zeta}{m^2}. \tag{G.58}$$

Problem 4.3 In the presence of a linear external force $F = -Ax$ and for long timescales, $t \gg m/\zeta$, we can repeat the same analysis as in Problem 4.1, as the Langevin equation becomes:

$$\zeta \dot{x} + Ax = f. \tag{G.59}$$

Again, the coefficient M in the Fokker-Planck equation can be easily determined from the phenomenological relation $\langle \dot{x} \rangle_t^0 = -M \langle x \rangle_t^0$, with $M = A/\zeta$, while Q can be determined from the fluctuation-dissipation theorem, $Q = Mg^{-1}$, with $g = A/kT$, so that $Q = kT/\zeta = D$, where D is the diffusivity. So we obtain the Smoluchowski equation:

$$\dot{\Pi} + \frac{\partial J_x}{\partial x} = 0; \quad J_x = -\frac{A}{\zeta} x \Pi - \frac{kT}{\zeta} \frac{\partial \Pi}{\partial x}, \tag{G.60}$$

and we may conclude:

$$\left\langle (\Delta x)^2 \right\rangle_t^0 = V^{-1} = \frac{kT}{A} \left[1 - e^{-2At/\zeta} \right], \tag{G.61}$$

where $\Delta x = x(t) - x_0 e^{-2At/\zeta}$. In particular, when $A = 0$, there is no equilibrium configuration and this procedure seems invalid. However, taking the limit when $A \to 0$, Eq. (G.61) becomes:

$$\left\langle \left[x(t) - x_0 \right]^2 \right\rangle_t^0 = 2\frac{kT}{\zeta} t = 2Dt, \tag{G.62}$$

i.e. the "usual" Stokes-Einstein result.

G.5   Chapter 5

Problem 5.1 Ito's lemma (5.11) with $x = W$, $f = x^4$, $V = 0$ and $B = 1$, reduces to:

$$dW^4 = 6W^2 \, dt + 4W^3 \, dW, \tag{G.63}$$

so that we obtain by integration:

$$\mathcal{I} \int_0^t W^3(s)\,dW(s) = \frac{1}{4}W^4(t) - \frac{3}{2}\int_0^t W^2(s)\,ds. \tag{G.64}$$

Again, the second term on the RHS would be absent by the rules of standard calculus.

Problem 5.2 Ito's lemma (5.11) with $x = W$, $f = x^{n+1}$, $V = 0$ and $B = 1$, reduces to:

$$dW^{n+1} = \frac{n(n+1)}{2}W^{n-1}\,dt + (n+1)W^n\,dW, \tag{G.65}$$

so that we obtain by integration:

$$\mathcal{I} \int_0^t W^n(s)\,dW(s) = \frac{1}{n+1}W^{n+1}(t) - \frac{n}{2}\int_0^t W^{n-1}(s)\,ds. \tag{G.66}$$

Again, the second term on the RHS would be absent by the rules of standard calculus.

Problem 5.3 The stochastic differential equation

$$dS = rS\,dt + \sigma S\,dW \tag{G.67}$$

can be solved applying Ito's lemma (5.11) with $x = S$, $f = \ln S$, $V = rS$ and $B = \sigma S$, obtaining:

$$d\ln S = \left(r - \frac{1}{2}\sigma^2\right)dt + \sigma\,dW. \tag{G.68}$$

Integrating, with $S(0) = S_0$, we obtain:

$$\ln\frac{S(t)}{S_0} = \left(r - \frac{1}{2}\sigma^2\right)t + \sigma W(t), \tag{G.69}$$

that is,

$$S(t) = S_0 e^{(r - \frac{1}{2}\sigma^2)t + \sigma W(t)}. \tag{G.70}$$

This is called the log-normal, or Black-Scholes model, of the market for a particular stock.

Note that the basic assumption of the Black-Scholes model is that when $\sigma = 0$, i.e. in the absence of fluctuations, the system is well behaved, which contradicts all experimental data. This is particularly alarming, as most of the predictions of the financial analysts are based on this model.

G.6 Chapter 6

Problem 6.1 Consider a Brownian particle immersed in an elongational incompressible flow field, $V_1 = \gamma x_2$ and $V_2 = \gamma x_1$. Then, the minimum path obeys the same Eq. (6.41), with $M_i = \gamma$, obtaining again Eq. (6.42). Substituting this result into Eq. (6.34), i.e. $\mathcal{L}_{min} = \zeta(\dot{\mathbf{y}} - \mathbf{V})^2$, we obtain:

$$\mathcal{L}_{min} = \zeta \gamma^2 \left[\left(X_1^2 + X_2^2 \right) \left[\coth^2(\gamma\tau) + 1 \right] - 4X_1 X_2 \coth(\gamma\tau) \right], \qquad (G.71)$$

and finally leading to the Gaussian distribution [9]:

$$\Pi(\mathbf{X}, t|0) = W(t) \exp\left[-\frac{\gamma}{4D} \left[\coth(\gamma t)\left(X_1^2 + X_2^2 \right) - 2X_1 X_2 \right] \right]. \qquad (G.72)$$

The same result can be obtained [5] by solving directly the Fokker-Planck equation. The variances of this distribution read:

$$\langle X_1^2 \rangle = \langle X_2^2 \rangle = \frac{D}{\gamma} \sinh(2\gamma t); \qquad \langle X_1 X_2 \rangle = \frac{2D}{\gamma} \sinh^2(\gamma t). \qquad (G.73)$$

Clearly, up to $O(\gamma t)$-term, we find: $\langle X_1^2 \rangle = \langle X_2^2 \rangle = 2Dt$ and $\langle X_1 X_2 \rangle = 0$.

Problem 6.2 Changing coordinate system along the principle axis of the elongational flow of Problem 6.1, with $X_1 = X_2$ (stretching) and $X_1 = -X_2$ (squeezing), define $Y_1 = (X_1 + X_2)/\sqrt{2}$ and $Y_2 = (X_1 - X_2)7\sqrt{2}$, so that we obtain the straining flow, $V_{Y_1} = Y_1$ and $V_{Y_2} = -Y_2$. Then, the distribution (G.72) becomes [8]:

$$\Pi(Y_1, Y_2, t|0) = W(t) \exp\left[-\frac{\gamma}{4D \sinh(\gamma t)} \left[e^{-\gamma t} Y_1^2 + e^{\gamma t} Y_2^2 \right] \right], \qquad (G.74)$$

with the following variances:

$$\langle Y_1^2 \rangle = \frac{D}{\gamma}\left(e^{2\gamma t} - 1 \right); \qquad \langle Y_2^2 \rangle = \frac{D}{\gamma}\left(1 - e^{-2\gamma t} \right); \qquad \langle Y_1 Y_2 \rangle = 0. \qquad (G.75)$$

Here, too, for short times, at leading order, we find: $\langle Y_1^2 \rangle = \langle Y_2^2 \rangle = 2Dt$ and $\langle Y_1 Y_2 \rangle = 0$.

G.7 Chapter 7

Problem 7.1 As we saw in Eq. (7.21), for non reactive mixtures the law of mass conservation of a chemical species k is:

$$\rho \frac{D\phi^{(k)}}{Dt} = -\nabla \cdot \mathbf{J}_d^{(k)},$$

where $\rho^{(k)}$ and $\rho = \sum \rho^{(k)}$ are the mass density (i.e. mass per unit volume) of component k and the total mass density, respectively, $\phi^{(k)} = \rho^{(k)}/\rho$ is the mass fraction of component k, while $\mathbf{J}_d^{(k)} = \rho\phi^{(k)}(\mathbf{v}^{(k)} - \mathbf{v})$ is its diffusive flux, with $\mathbf{v}^{(k)}$ denoting the mean velocity of component k and $\mathbf{v}$ the barycentric velocity, $\mathbf{v} = \sum \phi^{(k)}\mathbf{v}^{(k)}$. If the chemical species k carries a charge per unit mass, $z^{(k)}$, then the total electric current flux (i.e. the electric charge crossing per unit time a unit surface cross section) $\mathbf{I}$ can be written as the sum of a convective and a diffusive component,

$$\mathbf{I} = \sum_{k=1}^{n} \rho^{(k)}z^{(k)}\mathbf{v}^{(k)} = \rho z\mathbf{v} + \mathbf{i},$$

where,

$$z = \frac{1}{\rho}\sum_{k=1}^{n} \rho^{(k)}z^{(k)} = \sum_{k=1}^{n} \phi^{(k)}z^{(k)}$$

is the total charge per unit mass, while

$$\mathbf{i} = \sum_{k=1}^{n} z^{(k)}\mathbf{J}_d^{(k)}$$

is the diffusive electric flux, due to as electric conduction.

Finally, multiplying the equation of mass conservation of a chemical species by $z^{(k)}$ and summing over k we obtain:

$$\rho\frac{Dz}{Dt} = -\nabla \cdot \mathbf{i},$$

showing that the law of conservation of electric charge follows directly, without any further assumptions, from the law of mass conservation.

Problem 7.2 Consider a non-reactive charged mixture immersed in an electric and magnetic fields, $\mathbf{E}$ and $\mathbf{B}$, respectively.[3] The momentum balance equation is again given by Eqs. (7.25)–(7.26), where

$$\mathbf{F}^{(k)} = z^{(z)}\left(\mathbf{E} + \mathbf{v}^{(k)} \times \mathbf{B}\right)$$

is the Lorentz force acting on component k per unit mass.[4]

---

[3]Here we neglect all polarization phenomena, so that the electric and magnetic fields coincide with their respective displacement vectors, $\mathbf{D}$ and $\mathbf{H}$.

[4]The speed of light, that sometimes appears in the Maxwell equations, and therefore in the Lorentz force as well, here is incorporated within the magnetic field.

Now, the entropy production expression (7.72) will include an additional term, i.e. the product of this force by the diffusive flux $\mathbf{J}_d^{(k)}$, i.e.,

$$T\sigma^{(S)} = \sum_{k=1}^{n} \mathbf{J}_d^{(k)} \cdot \mathbf{F}^{(k)}.$$

This expression can be simplified considering that by definition

$$\mathbf{v}^{(k)} = \mathbf{v} + \frac{1}{\rho\phi^{(k)}}\mathbf{J}_d^{(k)}.$$

Therefore, since $\mathbf{J}_d^{(k)} \cdot \mathbf{J}_d^{(k)} \times \mathbf{B} = 0$, the additional term becomes:

$$T\sigma^{(S)} = \sum_{k=1}^{n} z^{(z)}\mathbf{J}_d^{(k)} \cdot (\mathbf{E} + \mathbf{v} \times \mathbf{B}).$$

Note that, expressing the electric field in terms of an electric potential ϕ, defined as $\mathbf{E} = -\nabla\phi$, its contribution can be included in Eq. (7.72) directly, considering the electrochemical potential $\tilde{\mu}^{(k)}$ as the sum of the chemical potential, $\mu^{(k)}$, and the electrostatic contribution, $\psi^{(k)} = z^{(k)}\phi$.

G.8 Chapter 8

Problem 8.1 Considering the symmetry of $\check{P}_{ij}^{(s)}$ and $\check{S}_{k\ell}$, we see that

$$\eta_{ijk\ell} = \eta_{jik\ell} = \eta_{ij\ell k}.$$

In addition, since $\check{P}_{ij}^{(s)}$ and $\check{S}_{k\ell}$ are traceless by definition, we have:

$$\delta_{ij}\,\eta_{ijk\ell} = \eta_{jik\ell}\,\delta_{k\ell} = 0.$$

Finally, since from (8.7) and (8.14) we see that the entropy production rate includes a term $T\sigma_S = \check{P}_{ij}^{(s)}\check{S}_{ij}$, the Onsager reciprocity relations yield:

$$\eta_{ijk\ell} = \eta_{\ell kji}.$$

Problem 8.2 Considering that the very dilute case consists of the diffusion of a single Brownian particle that moves along a trajectory $\tilde{\mathbf{r}}(t)$, we have:

$$\tilde{\mathbf{J}}^{(1)} = m_1 n^{(1)}\tilde{\mathbf{v}}, \qquad\qquad\qquad\qquad (G.76)$$

with $n^{(1)}(\mathbf{r}, t) = \delta[\mathbf{r} - \tilde{\mathbf{r}}(t)]$, while $\tilde{\mathbf{v}}(t)$ is the velocity of the Brownian particle.[5] Therefore, we may conclude:

$$\langle \tilde{v}_i(t_1)\tilde{v}_k(t_2)\rangle = 2D\delta_{ik}\delta(t_1 - t_2),\tag{G.77}$$

which coincides with Eq. (G.32).

G.9   Chapter 9

Problem 9.1 At the end of a tedious, but elementary power expansion in terms of the reduced variables, we see that the Van der Waals equation reads, neglecting higher order terms,[6]

$$\tilde{p} = 4\tilde{t} - 5\tilde{t}\tilde{v} - \frac{3}{2}\tilde{v}^3.\tag{G.78}$$

Note that we cannot have any term proportional to $\tilde{v}$ or $\tilde{v}^2$, in agreement with the conditions $(\partial P/\partial \tilde{v})_{T_C} = (\partial^2 P/\partial \tilde{v}^2)_{T_C} = 0$, while the coefficient of the $\tilde{v}^3$-term must be negative, as $(\partial^3 P/\partial \tilde{v}^3)_{T_C} < 0$. When $\tilde{t} > 0$, all states of the system are stable, that is there is no phase separation and the system remains homogeneous. That means that, when $\tilde{t} > 0$, it must be $(\partial P/\partial \tilde{v})_{T_C} < 0$, and therefore the coefficient of the $\tilde{t}\tilde{v}$-term must be negative. Finally, note that the $\tilde{t}\tilde{v}^2$ and $\tilde{t}^2\tilde{v}$-terms have been neglected because they are much smaller than $\tilde{t}\tilde{v}$, while the $\tilde{t}\tilde{v}$-term must be kept, despite being much smaller than $\tilde{t}$, for reasons that will be made clear below.

At the critical point, where $\tilde{t} = 0$, from (G.78) we obtain: $\tilde{p} \propto \tilde{v}^\delta$, where $\delta = 3$ is a critical exponent.

Consider an isotherms below the critical point, with $\tilde{t} = -a^2$, where the system is unstable and separates into two coexisting phases. The spinodal volumes can be determined imposing that $(\partial \tilde{p}/\partial \tilde{v})_{\tilde{T}} = 0$, obtaining: $\tilde{v}_s^\alpha = -\tilde{v}_s^\beta = \sqrt{-\frac{4}{3}\tilde{t}}$, i.e. $\beta = 1/2$.[7]

---

[5]Here, $\tilde{\mathbf{J}}^{(1)}$ is a diffusive flux, since the solvent is quiescent and therefore the average velocity of the mixture is zero.

[6]This expression can be generalized as $\tilde{p} = b_c\tilde{t} - 2a_c\tilde{t}\tilde{v} - 4B_c\tilde{v}^3$, which is the basis of Landau's mean field theory [7, Chaps. 146, 148].

[7]This critical property can also be determined from the free energy f_{Th}. In fact, integrating $(df_{Th})_T = -Pdv$ and substituting (G.78), we see that in the vicinity of the critical point the free energy has the following expression:

$$f_{Th}(\tilde{v}, \tilde{T}) = P_C v_C\left[h(\tilde{T}) + (1 + 4\tilde{T})\tilde{v} + 3\tilde{T}\tilde{v}^2 + \frac{3}{8}\tilde{v}^4\right],\tag{G.79}$$

where $h(\tilde{t}) = (1 + \tilde{t})\ln[3/(2v_C)] - 9/8$. From this expression we can determine $\tilde{v}_e$, finding again $\beta = 1/2$.

From the definition of the isothermal compressibility coefficient, we obtain:

$$\kappa_T^{-1} = -v\left(\frac{\partial P}{\partial v}\right)_T = -P_C(1+\tilde{v})\left(\frac{\partial \tilde{p}}{\partial \tilde{v}}\right)_{\tilde{T}} \cong 2a_c P_C\tilde{t}, \qquad (G.80)$$

showing that $\gamma = 1$.

Finally, considering that the molar internal energy u_{Th} remains finite at the critical point, we find $\alpha = 0$.[8]

Problem 9.2 This problem continues the previous one. From,

$$c_p - c_v \propto \left(\frac{\partial \tilde{p}}{\partial \tilde{t}}\right)_{\tilde{v}}^2 \bigg/ \left(\frac{\partial \tilde{p}}{\partial \tilde{v}}\right)_{\tilde{t}}, \qquad (G.82)$$

we see that, since $(\partial \tilde{p}/\partial \tilde{t})_{\tilde{t}=0,\tilde{v}=0} = b_c$ and $(\partial \tilde{p}/\partial \tilde{v})_{\tilde{t}=0,\tilde{v}=0} = 0$, the specific heat c_p diverges. In fact, we find:

$$c_p \propto \frac{1}{(\partial \tilde{p}/\partial \tilde{t})_{\tilde{t}}} = \frac{1}{-2a_c\tilde{t} - 12B_c\tilde{v}^2} \propto \frac{1}{\tilde{t}}, \qquad (G.83)$$

where we have considered that on the equilibrium curve, $\tilde{v} \propto \sqrt{\tilde{t}}$.

Problem 9.3 Substituting $u = 2\phi - 1$ and $\psi = \Psi - 2$ into (9.104) and (9.117) and imposing that $u \ll 1$, we easily obtain at leading order:

$$\frac{\partial u}{\partial t} = -2\psi\nabla^2 u - \nabla^4 u,$$

where the spatial and temporal variables have been made non dimensional in terms of a and $2a^2/D$. Assuming a periodic perturbation $u = u_0 \exp(i\mathbf{k}\cdot\mathbf{r} + \sigma t)$ we find:

$$\sigma = k^2(2\psi - k^2),$$

where $k = |\mathbf{k}|$. Therefore, we conclude that the homogeneous state $\phi = \phi_0$ is unstable when $k < \sqrt{2\psi}$. Among the infinite unstable modes, the one that grows fastest will eventually prevail. That corresponds to the wave vector that maximizes the exponential growth σ, which is given by $k_{max} = \sqrt{\psi}$ [1, 2].

---

[8]From $u_{Th} = f_{Th} + Ts = f_{Th} - T(\partial f_{Th}/\partial T)_v$, we obtain from Eq. (G.79)

$$c_v^{sat} = \frac{P_C v_C}{T_C}\left[5 + f_{Th}(0) - \frac{df_{Th}}{d\tilde{t}}(0)\right]. \qquad (G.81)$$

Therefore, we see that c_v^{sat} remains finite at the critical point.

G.10 Chapter 10

Problem 10.1 Considering that $J_i = \nabla_k(r_i J_k)$, together with the continuity of T and J_i, integrating by parts we obtain:

$$\langle J_i \rangle = \frac{1}{V} \int_V \nabla_k(r_i J_k) \, d^3\mathbf{r} = \frac{1}{V} \oint_{r=R} n_k r_i J_k \, d^2\mathbf{r}, \qquad (G.84)$$

and

$$\langle \nabla_i T \rangle = \frac{1}{V} \int_V \nabla_i T \, d^3\mathbf{r} = \frac{1}{V} \oint_{r=R} n_i T \, d^2\mathbf{r}, \qquad (G.85)$$

where the volume V has been assumed to be a sphere of radius $R \gg a$, and $n_k = R_k/R$ is the unit vector perpendicular to the sphere of radius R. At this point, substituting into these expressions the outer solution (10.5), we find,

$$\langle J_i \rangle = -k_0 \frac{3}{4\pi R^4} \oint_R R_k R_i \left[(1 + K\phi)\delta_{k\ell} - 3K\phi \frac{R_i R_\ell}{R^2} \right] G_\ell \, d^2\mathbf{R},$$

and

$$\langle \nabla_i T \rangle = \frac{3}{4\pi R^4} \oint_R R_i R_k (1 + K\phi) \, d^2\mathbf{R},$$

where $\phi = a^3/R^3$ is the volume fraction of the inclusions. Considering that $\oint_R R_i R_k d^2\mathbf{R} = \frac{4}{3}\pi R^4$, we obtain:

$$\langle J_i \rangle = -k_0 G_i (1 - 2K\phi), \qquad (G.86)$$

and

$$\langle \nabla_i T \rangle = G_i (1 + K\phi). \qquad (G.87)$$

Therefore, considering the definition (10.8) of the effective conductivity, i.e. $\langle J_i \rangle = -\kappa^* \langle \nabla_i T \rangle$, we obtain:

$$\frac{\kappa^*}{\kappa_0} = \frac{1 - 2K\phi}{1 + K\phi} = 1 - 3K\phi + O(\phi^2). \qquad (G.88)$$

Problem 10.2 Proceeding as in Problem 10.1, considering the continuity of **S** and **T** at the interfaces, together with the equality $T_{ij} = \nabla_k(r_i T_{kj})$ [cf. Eq. (10.25)], we obtain:

$$\langle S_{ij} \rangle = \frac{1}{2V} \int_V (\nabla_i v_j + \nabla_j v_i) \, d^3\mathbf{r} = \frac{1}{2V} \oint_R (n_i v_j + n_j v_i) \, d^2\mathbf{r},$$

and

$$\langle T_{ij} \rangle = \frac{1}{V} \int_V T_{ij} \, d^3\mathbf{r} = \frac{1}{V} \oint_R n_k r_i T_{kj} \, d^2\mathbf{r}.$$

The shear flow past a sphere is given by Eq. (F.47). At large distances, i.e. when $R \gg a$, at leading order it reduces to:

$$v_i - G_{ik}r_k = -\frac{5}{2}a^3 G_{jk}\frac{r_i r_j r_k}{r^5},$$

$$T'_{ij} = -\frac{5}{2}\eta_0 a^3 G_{k\ell}\left[\frac{r_i r_k \delta_{j\ell} + r_j r_k \delta_{i\ell} + r_i r_\ell \delta_{jk} + r_j r_\ell \delta_{ik}}{r^5} - 10\frac{r_i r_j r_k r_\ell}{r^7}\right],$$

where $T'_{ij} = T_{ij} + p\delta_{ij} - 2\eta_0 G_{ij}$ is the perturbation of the shear stress due to the presence of the particle.

Therefore,

$$\langle S_{ij}\rangle - G_{ij} = -\frac{3}{4\pi R^4}\oint_R \frac{5}{2}\phi G_{k\ell}\frac{r_i r_j r_k r_\ell}{r^2}\,d^2\mathbf{r},$$

and considering that

$$\oint_R r_i r_j r_k r_\ell\,d^2\mathbf{r} = \frac{4}{15}\pi R^6(\delta_{ij}\delta_{k\ell} + \delta_{ik}\delta_{j\ell} + \delta_{i\ell}\delta_{jk}),$$

with $G_{ii} = 0$, we obtain:

$$\langle S_{ij}\rangle = G_{ij}(1 - \phi).$$

This shows that the mean shear rate decreases, due to the presence of rigid particles, as one should expect.

In the same way, we find:

$$\langle T'_{ij}\rangle = -\frac{3}{4\pi R^4}\oint_R \frac{5}{2}\eta_0 a^3 G_{k\ell}\left(\frac{r_i r_k \delta_{j\ell} + r_i r_\ell \delta_{jk}}{r^3} - 8\frac{r_i r_j r_k r_\ell}{r^5}\right)d^2\mathbf{r},$$

i.e.,

$$\langle T'_{ij}\rangle = -\frac{3}{4\pi R^4}\frac{5}{2}\eta_0 \phi G_{k\ell}\oint_R\left(r_i r_k \delta_{j\ell} + r_i r_\ell \delta_{jk} - 8\frac{r_i r_j r_k r_\ell}{r^2}\right)d^2\mathbf{r},$$

and therefore,

$$\langle T'_{ij}\rangle = -\frac{15}{8\pi R^4}\eta_0 \phi G_{k\ell}\left[\frac{4}{3}\pi R^4(\delta_{ik}\delta_{j\ell} + \delta_{i\ell}\delta_{jk}) - 8 \times \frac{4}{15}\pi R^4(\delta_{ik}\delta_{j\ell} + \delta_{i\ell}\delta_{jk})\right],$$

that is,

$$\langle T'_{ij}\rangle = -\frac{15}{8\pi R^4}\eta_0 \phi G_{ij}\left(\frac{8}{3} - \frac{64}{15}\right)\pi R^4 = 3\phi\eta_0 G_{ij}.$$

Consequently, we conclude:

$$\langle T_{ij} + p\delta_{ij}\rangle = 2\eta_0 G_{ij}\left(1 + \frac{3}{2}\phi\right).$$

So, finally, defining the effective viscosity η^* as

$$\langle T_{ij} + p\delta_{ij}\rangle = 2\eta^*\langle S_{ij}\rangle,$$

we obtain the celebrated Einstein's result:[9]

$$\eta^* = \eta_0\left(1 + \frac{5}{2}\phi\right),$$

where $O(\phi^2)$ terms have been neglected.

Problem 10.3 Substituting the equations of motion (10.76) into the RHS of Eq. (10.101), we obtain,

$$\dot{\eta}^{(2)}_{ijkl} = \int x_i \, \dot{\Pi}_{jk} \, x_l \, d^3\mathbf{r} = \int x_i \left(\pi_{k,j} - \eta\nabla^2 \Pi_{jk}\right)x_l \, d^3\mathbf{r}, \qquad (G.89)$$

where $\pi_{i,j} = \partial\pi_i/\partial x_j$. Now, considering (10.85)–(10.87), we find:

$$\dot{\eta}^{(2)}_{ijkl} = \int x_i \left(-\eta\nabla^2 P^*_{jk}\right)x_l \, d^3\mathbf{r}. \qquad (G.90)$$

Finally, integrating by parts, applying the normalization condition (10.90) and considering the exponential decay of $\mathbf{P}^*$, we rederive the dynamical definition of viscosity (10.101).

G.11 Chapter 11

Problem 11.1 Here the Brownian particle can sample only transversal position $-(Y-a) < y < (Y-a)$. Accordingly, as the mean particle velocity V_p is calculated over all the accessible values of $V(y)$ we have:

$$V_p = \frac{1}{2(Y-a)} \int_{-(Y-a)}^{Y-a} \frac{3}{2}V\left(1 - \frac{y^2}{Y^2}\right)dy,$$

i.e.,

$$V_p = \frac{3V}{4(1-\epsilon)}\left[(1-\epsilon) - \frac{1}{3}(1-\epsilon)^3\right],$$

so that at leading order we obtain: $V_p = V(1+\epsilon)$. This shows that the suspended particle has a mean velocity that is larger than that of the fluid.

---

[9]In his first calculation of the effective viscosity [3], Einstein neglected to take into account the variation of the mean shear rate, so that he obtained a factor 3/2, which only later [4] he corrected into 5/2. The fact that even Him made a mistake has always been a source of great consolation to all of us.

Similar calculations show that the Taylor diffusivity decreases as ϵ increases, since by "cutting out" the wall region, where the velocity gradient is larger, we keep only the central part of the velocity profile, which is rather flat and does not influence diffusion too much (in the limit case of an ideal plug flow, Taylor diffusion is identically zero).

Problem 11.2 (a) Eulerian approach.

Repeat the analysis of Sect. 11.2.2, assuming that the Brownian particles are convected by a linear Couette flow, $V(y) = 2Vy/Y$, with $0 \leq y \leq Y$.[10] In this case, as $\tilde{v} = 2\tilde{y} - 1$, we find $d\tilde{B}/dy = \tilde{y}^2 - \tilde{y}$ and $\alpha = 1/30$. Therefore, we finally obtain:

$$D^* = D + \frac{1}{30}\frac{V^2Y^2}{D}. \tag{G.91}$$

(b) Lagrangian approach.

In this case, when $V(y) = 2Vy/Y$, with $0 \leq y \leq Y$, we find,

$$v_n = \frac{4\sqrt{2}V}{\pi^2 n^2} \quad \text{for } n = 2m + 1; \qquad v_n = 0 \quad \text{for } n = 2m. \tag{G.92}$$

Finally, considering that $\sum_{m=0}^{\infty}(2m + 1)^{-4} = \pi^4/960$, substituting this result into (11.91) we obtain:

$$D^* = D + \frac{1}{30}\frac{V^2Y^2}{D}, \tag{G.93}$$

which coincides with (G.91).

Problem 11.3 Creeping flow of a Newtonian fluid of viscosity η past a random array of fixed particles is governed by the Stokes equations,

$$\nabla p = \eta\nabla^2\mathbf{v}; \qquad \nabla \cdot \mathbf{v} = 0,$$

subjected to the no-slip boundary condition at the particle surface s_p,

$$\mathbf{v} = \mathbf{0} \quad \text{at } s_p.$$

Consider the following non dimensional quantities,

$$\tilde{\mathbf{r}} = \frac{\mathbf{r}}{L}; \qquad \tilde{p} = \frac{pL}{\eta V}; \qquad \tilde{\mathbf{v}} = \frac{\mathbf{v}}{V},$$

where L is a characteristic macroscopic length, V a characteristic velocity, while the tilde indicates non dimensionality. At this point, assume that all quantities depend separately on macroscopic and microscopic length, $\mathbf{x}$ and $\mathbf{y}$, respectively, defined in

---

[10]Obviously, in this case the no-flux boundary condition is applied at $y = 0$ and $y = Y$.

(11.101), where $\epsilon = \ell/L$, with ℓ denoting the micro, or pore, scale. Now impose the following scaling [10]:

$$\widetilde{p} = p_0(\mathbf{x}, \mathbf{y}) + \epsilon p_1(\mathbf{x}, \mathbf{y}) + \cdots ,$$

and

$$\widetilde{\mathbf{v}} = \epsilon^2 \mathbf{v}_0(\mathbf{x}, \mathbf{y}) + \epsilon^3 \mathbf{v}_1(\mathbf{x}, \mathbf{y}) + \cdots ,$$

together with the expansion (11.103):

$$\widetilde{\nabla} = \nabla_x + \frac{1}{\epsilon} \nabla_y.$$

At leading order we find:

$$p_0 = p_0(\mathbf{x}).$$

At the next order we obtain:

$$\nabla_y p_1 - \nabla_y^2 \mathbf{v}_0 = -\nabla_x p_0; \qquad \nabla_y \cdot \mathbf{v}_0 = 0; \quad \text{with B.C.} \quad \mathbf{v}_0 = \mathbf{0} \quad \text{at } s_p.$$

Define:

$$\mathbf{v}_0(\mathbf{x}, \mathbf{y}) = -\mathbf{W}(\mathbf{y}) \cdot \nabla_x p_0,$$

and,

$$p_1(\mathbf{x}, \mathbf{y}) = -\mathbf{P}(\mathbf{y}) \cdot \nabla_x p_0.$$

Here, $\mathbf{W}$ and $\mathbf{P}$ satisfy the following cell problem:

$$\nabla_y \mathbf{P} - \nabla_y^2 \mathbf{W} = \mathbf{I}; \qquad \nabla_y \cdot \mathbf{W} = \mathbf{0}; \quad \text{with B.C.} \quad \mathbf{W} = \mathbf{0} \quad \text{at } s_p.$$

Averaging we obtain:

$$\langle \mathbf{v}_0 \rangle = -\widetilde{\mathbf{k}} \cdot \nabla_x p_0,$$

where $\widetilde{\mathbf{k}}$ is the non-dimensional permeability,

$$\widetilde{\mathbf{k}} = \langle \mathbf{W} \rangle.$$

Clearly, this shows that permeability is the mean velocity induced by a normalized pressure gradient.

Going back to dimensional variables, we find the Darcy law,

$$-\eta \langle \nabla p \rangle = \mathbf{k} \cdot \langle \mathbf{V} \rangle,$$

where $\mathbf{k} = \ell^2 \widetilde{\mathbf{k}} = \ell^2 \langle \mathbf{W} \rangle$, thus confirming that permeability has the typical dimension of the square of the pore size.

References

1. Cahn, J., Hilliard, J.: J. Chem. Phys. **28**, 258 (1958)
2. Cahn, J., Hilliard, J.: J. Chem. Phys. **31**, 688 (1959)
3. Einstein, A.: Ann. Phys. **19**, 289 (1906)
4. Einstein, A.: Ann. Phys. **34**, 591 (1911)
5. Foister, R.T., Van de Ven, T.G.M.: J. Fluid Mech. **96**, 105 (1980)
6. Happel, J., Brenner, H.: Low Reynolds Number Hydrodynamics. Springer, Berlin (1963), Chap. 5
7. Landau, L.D., Lifshitz, E.M.: Statistical Physics, Part I. Pergamon, Elmsford (1980)
8. Liron, N., Rubinstein, J.: SIAM J. Appl. Math. **44**, 493 (1984)
9. Mauri, R., Haber, S.: SIAM J. Appl. Math. **46**, 49 (1986)
10. Sanchez-Palencia, E.: Non-Homogeneous Media and Vibration Theory. Springer, Berlin (1980)

Background Reading

- Einstein, A.: Investigations on the Theory of the Brownian Movement. Dover, New York (1956) (First Edition: 1926)
- Gardiner, C.W.: Handbook of Stochastic Methods, 2nd edn. Springer, Berlin (1985)
- de Groot, S.R., Mazur, P.: Non-Equilibrium Thermodynamics. Dover, New York (1984) (First Edition: 1961)
- Kreuzer, H.J.: Nonequilibrium Thermodynamics and Its Statistical Foundations. Clarendon Press, Oxford (1981)
- Landau, L.D., Lifshitz, E.M.: Statistical Physics, Part I. Pergamon, Elmsford (1980), Chap. XII (First Edition: 1978)
- Prosperetti, A.: Advanced Mathematics for Applications. Cambridge University Press, Cambridge (2011)
- Reif, F.: Fundamentals of Statistical and Thermal Physics. McGraw Hill, New York (1965)
- van der Waals, J.D.: The thermodynamic theory of capillarity under the hypothesis of a continuous variation of density. Z. Phys. Chem. Stochiom. Verw. **13**, 657 (1894); reprinted in J. Stat. Phys. **20**, 200 (1979)

R. Mauri, *Non-Equilibrium Thermodynamics in Multiphase Flows*,
Soft and Biological Matter, DOI 10.1007/978-94-007-5461-4,
© Springer Science+Business Media Dordrecht 2013

Index

A

Accessible states, 187
Action, 121, 212

B

Balance equations, 73–85, 219–222
 for angular momentum, 82, 83
 for chemical species, 76, 77
 for entropy, 84, 85
 for internal energy, 81
 for kinetic energy, 78
 for mass, 220
 for mechanical energy, 79
 for momentum, 74, 75, 77, 78, 221
 for potential energy, 79
 for total energy, 80–82, 221
 general, 73, 74
Binomial distribution, 25, 181–183, 188
 mean value, 182
 normalization, 182
 variance, 183
Bohm's quantum potential, 58
Boltzmann's entropy, 190, 193
Brownian motion, 25–28

C

Cahn boundary condition, 121
Casimir reciprocity relations, 19
Chemical potential
 generalized, 118, 130
 non local, 118
 thermodynamic, 118, 129, 143
Conservation
 of angular momentum, 208
 of energy, 207
 of momentum, 208
Constitutive relations, 87–102

binary mixtures, 92–95
multicomponent mixtures, 98–100
non-isotropic media, 101, 102
single-phase fluid, 92
Correlation function, 16, 20, 197, 198
Critical point, 111–113
Curie's principle, 89, 103

D

Diffuse interface approach, 108
Diffusivity
 colloidal suspensions, 141–144
 dynamical definition, 145
 effective, 161, 166, 167, 170, 174
 gradient, 19, 27, 143
 in binary mixtures, 94, 98
 in multicomponent mixtures, 99
 self, 19, 27, 141
Dispersion relations, 21, 22
Drag coefficient, 28
 of a suspension, 140, 242
Drift velocity, 174
Dufour coefficient, 94

E

Effective constitutive relations, 133–144
Entropy production, 6, 19, 85, 87
Equilibrium curve, 114

F

Faxen law, 231, 237–240
Fluctuation-dissipation theorem, 18–24, 28,
 30, 32, 41, 42
 "violation" of, 43, 141
Fluctuations
 Einstein's formula, 4
 equilibrium, 4–10
 in binary solutions, 10

R. Mauri, *Non-Equilibrium Thermodynamics in Multiphase Flows*,
Soft and Biological Matter, DOI 10.1007/978-94-007-5461-4,
© Springer Science+Business Media Dordrecht 2013